Guide to the

MARINE ISOPOD CRUSTACEANS

of the Caribbean

Guide to the

MARINE ISOPOD

CRUSTACEANS

of the Caribbean

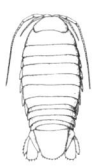

Brian Kensley and
Marilyn Schotte

SMITHSONIAN INSTITUTION PRESS

WASHINGTON, D.C., AND LONDON

© 1989 by the Smithsonian Institution
All rights reserved

Designer: Linda McKnight
Editor: Nancy Dutro

Library of Congress Cataloging-in-Publication Data
Kensley, Brian Frederick.
   Guide to the marine isopod crustaceans of the
   Caribbean / Brian Kensley
and Marilyn Schotte.
    p. cm.
   Bibliography: p.
   Includes index.
   ISBN 0-87474-724-4 (alk. paper)
   1. Isopoda—Caribbean Sea—
Classification. 2. Crustacea—Caribbean Sea—
Classification. I. Schotte, Marilyn. II. Title.
QL444.M34K434 1989
595.3′7209153′35—dc19   88-38647
                                             CIP

British Library Cataloging-in-Publication Data
available

Manufactured in the United States of America
10 9   8   7   6   5   4   3   2   1
98 97 96 95 94 93 92 91 90 89

∞ The paper used in this publication meets the
minimum requirements of the American National
Standard for Performance of Paper for Printed
Library Materials Z39.48-1984

Contents

- 1   Introduction
  - 1   HISTORIC BACKGROUND
  - 3   GEOGRAPHIC AREA COVERED IN THIS GUIDE
  - 4   ARRANGEMENT OF THE GUIDE AND HOW TO USE IT
  - 5   ACKNOWLEDGMENTS
- 7   Glossary of Technical Terms
- 13   Marine Isopods of the Caribbean
  - 13   ORDER ISOPODA
  - 15   SUBORDER ANTHURIDEA
  - 73   SUBORDER ASELLOTA
  - 107   SUBORDER EPICARIDEA
  - 114   SUBORDER FLABELLIFERA
  - 236   SUBORDER GNATHIIDEA
  - 243   SUBORDER MICROCERBERIDEA
  - 246   SUBORDER ONISCIDEA
  - 251   SUBORDER VALVIFERA
- 261   Zoogeography
  - 261   FAUNAL PROVINCES
  - 262   ANALYSIS OF THE ISOPOD FAUNA
  - 266   THE BAHAMAS
  - 269   BERMUDA
  - 269   CAVE ISOPODS
- 275   Appendix
- 277   Literature Cited
- 293   Index

Introduction

The title of this work will no doubt raise several questions in many readers' minds: why the Caribbean? why not the Caribbean *and* the Gulf of Mexico? why only the marine isopods? just what is the "Caribbean area"? We hope that the answers to some of these (and other) questions will become apparent.

There are several works that already deal with the isopods of the Caribbean, as part of a wider treatment of North American isopods (e.g., Richardson, 1905; Schultz, 1969). Why then this "Isopods of the Caribbean"? As partial answer, the following: many new records of isopods from the Caribbean region (in its broadest sense) have appeared in scattered publications in the last few decades. The time has come to pull these together in a single work. The number of marine laboratories in the area has increased, with more and more students exploring especially the shallow marine environment. A single work on a relatively speciose and abundant group of invertebrates would be useful to such investigators, as they build up a comprehensive view of the biology of the region. Concepts of the taxonomy of several isopod groups have changed radically over the last few years; again, there is obvious utility in having these changes summarized in a single source. New species and records are continuously being found. Having a single baseline work decreases the time needed for investigating and establishing the validity of such records.

HISTORIC BACKGROUND

The history of isopod taxonomic research in the Caribbean really starts with a worldwide monographic work on the Cirolanidae by Hansen in 1890. Included here were about 12 species from the Danish West Indies, now the U.S. Virgin Islands. Since then a few major works on Caribbean isopods have appeared, such as Moore's report on the isopods of Puerto Rico (1901) and Menzies and Glynn's report on the same area (1968). Some areas have received considerable attention, such as the aforementioned Puerto Rico and, more recently, Belize. A list of 116 species of isopods from Cuba (including Oniscidea) has been published (Ortiz, Lalana, and Gomez, 1987). At the other extreme, there are no records from a number of localities, especially the

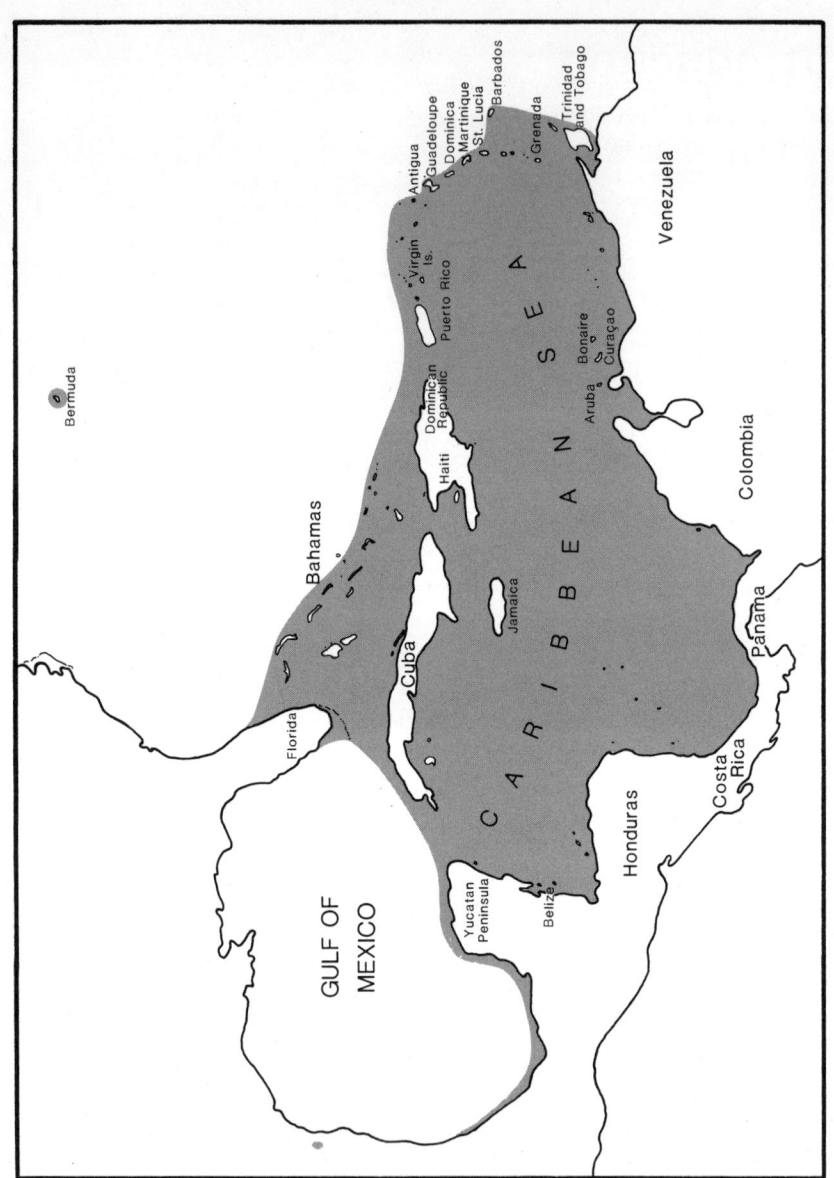

Figure 1. Map of area covered by this guide.

islands of the southeastern chain of the Lesser Antilles. In total there are about 40 publications, varying from descriptions of single species to longer works, that deal with isopods from the Caribbean. These publications will be encountered in the following guide, under the specific taxa.

GEOGRAPHIC AREA COVERED IN THIS GUIDE

The accompanying map (Figure 1) shows the area for which records are included in this guide.

While it may seem logical to include the Gulf of Mexico, and while there are several isopod species common to both areas, this has not been done. There are relatively few isopod records from the Gulf; undoubtedly a great deal of taxonomic work awaits the careful collector in this area. Also, from a zoogeographic point of view, separation of the Gulf may be justified.

Bermuda, on the other hand, situated in the northwestern Atlantic several hundred miles off the coast of the United States, is included. This island, although remote from the Caribbean, is swept by waters that earlier have passed through the Caribbean. Zoogeographically, the shallow-water Bermudan and Caribbean faunas have much in common.

While perhaps not strictly in the Caribbean Sea, the Bahamas and the Florida Keys are included here, their shallow-water marine faunas being overwhelmingly Caribbean in nature.

Turning to depths limits, within the area under discussion, species from the intertidal to 200 meters have been dealt with in some detail. This arbitrary cutoff depth was selected because most Caribbean isopod species inhabit relatively shallow depths. About 30 species have been recorded from below 200 meters in the Caribbean, many of these known only from the type material. A list of species of this very poorly known deeper fauna is included here. Without doubt, many species in the deeper waters of the Caribbean await discovery.

A fascinating group of isopods, while not strictly shallow-water marine forms, is included. These are the true cave forms, found mainly in the suborders Anthuridea and Flabellifera. Given the history of the Caribbean from the Quaternary to the present, it is not surprising that caves are common throughout the region. These caves may be well inland and contain only freshwater, but are more commonly anchialine, that is, having some (frequently subterranean) link to the sea. Less common, and of lesser interest from an isopod taxonomist's point of view, are the fully marine caves in direct communication with the sea or, indeed, under the surface of the sea itself.

ARRANGEMENT OF THE GUIDE AND HOW TO USE IT

A short introduction to the Crustacea Isopoda is provided, followed by a glossary of descriptive terms and morphological features used throughout the guide (see Figure 2).

Keys and diagnoses to the suborders and all lower taxa follow. For ease of usage, except in the keys, all taxa are presented in alphabetical order, regardless of their phylogenetic relationships.

Diagnoses are provided for all suborders, families, genera, and species. The only exceptions to this are in the suborders Epicaridea and Oniscidea. Within each suborder, a key to the families occurring in the Caribbean is provided. Similarly, within each family and genus, keys are provided to the relevant genera and species, respectively.

In whatever context, where an author and date appear, a reference to these is provided in the Literature Cited section. In some cases, reference is made to useful publications such as revisions of families or genera.

As this is not a textbook on the Isopoda, biological information is generally kept to a minimum. In the case of individual species, however, what little ecological information is available, is provided. For general texts on biology, internal anatomy, physiology, and reproduction, the reader is referred to works such as Kaestner (1967), Waterman (1960), Bliss (1982–1985), and Schram (1986).

Within each species discussion, a diagnosis is given, along with maximum (total) middorsal lengths for males and females, where known. The diagnoses are not exhaustive, but provide only the information needed to distinguish the species. Diagnoses thus vary in length from the statement of a single feature to a paragraph concerning several features, depending on the understanding and complexity of the taxonomy of the group. In the longer diagnoses, morphological features are dealt with in order from anterior to posterior on the animal's body. Records are given, rather than geographical distribution, as our knowledge of many species is woefully incomplete. These records are given in a roughly north-to-south order; records outside of the Caribbean region, as here defined, are given on a separate line. A few species not yet recorded from the Caribbean are included, in the strong likelihood that they will eventually be found here. The records include depth distribution information in meters, where known. Records were taken from published papers; in addition, the collections of the United States National Museum of Natural History, Smithsonian Institution, were scoured, and many unpublished records from this source are also included. In the "Remarks" section, ecological information such as substrate preferences is given. Hosts of parasitic species are given. Formal synonymies are not provided, but nomenclatural comments are included in the few cases where a species may be known under a more commonly used name. Usually, a figure of the entire animal of each species is given. Diagnostic features are usually illustrated. Unless otherwise stated, all illustrations are original and by the authors, and were made from actual specimens.

Common and scientific names of fishes that are hosts to parasitic isopods are taken from the American Fisheries Society special publication no. 12 (Robins et al., 1980).

Finally, a word of warning. Difficulties may be experienced in using the keys, for which there may be any of several reasons: characters seen in the animal may not clearly conform to those in the key (in which case refer to the figures, as well as to good recent descriptions or diagnoses); your material may be a new record for the region; or you may have an undescribed species (in which case refer to more comprehensive treatments of the group).

ACKNOWLEDGMENTS

Much of the material covered in this work comes from the many collectors who have deposited specimens from several regions of the Caribbean in the collections of the National Museum of Natural History, Smithsonian Institution. Rather than risk the unwitting omission of a name, we thank all of these individuals collectively. Without their efforts, our knowledge of the Caribbean fauna would be the poorer.

Material was borrowed from several institutions. We thank the following scientists for their assistance in this connection: Jan Stock and Dirk Platvoet of the Instituut voor Taxonomische Zoologie, University of Amsterdam; Jean Just and Torben Wolff of the Zoological Museum, University of Copenhagen; Jacques Forest of the Muséum National d'Histoire Naturelle, Paris; Richard Heard of the Gulf Coast Research Laboratory, Ocean Springs, Mississippi; Willard Hartman of the Peabody Museum of Natural History, Yale University; Herbert W. Levi of the Museum of Comparative Zoology, Harvard University; Harold S. Feinberg of the American Museum of Natural History, New York; John E. Miller of the Harbor Branch Foundation, Florida; and Paula M. Mikkelsen of the Indian River Coastal Museum at Fort Pierce, Florida.

Bruce Collette of the National Marine Fisheries Laboratory at the Smithsonian Institution assisted with fish names used in this work, for which we are grateful.

We thank the staff of the Scanning Electron Microscope Laboratory of the Smithsonian Institution, and especially Susann Braden, who produced the electron micrographs used here.

We are grateful to the Smithsonian Research Opportunities fund, administered by David Challinor, then Assistant Secretary for Research, and to the Smithsonian's Caribbean Coral Reef Ecosystems program administered by Klaus Rützler of the Department of Invertebrate Zoology, for the funding of several fieldtrips to the Caribbean. The second author acknowledges a financial award from the Smithsonian's Women's Committee for a grant to facilitate fieldwork.

Several individuals have provided encouragement, advice, suggestions, critical comments, and missed references, all of which have vastly improved this work. In this regard we are especially grateful to Thomas E. Bowman, C. W. Hart, Jr., Horton H. Hobbs, Jr., and Molly K. Ryan, all of the Department of Invertebrate Zoology, National Museum of Natural History, Smithsonian Institution; Dan Adkison, Bass Harbor, Maine (who provided considerable assistance with the Epicaridea); Richard Heard of the Gulf Coast Research Laboratory, Mississippi; and Paul Delaney of the Los Angeles County Museum of Natural History.

This is Contribution Number 248 of the Caribbean Coral Reef Ecosystems (CCRE) program, Smithsonian Institution, supported in part by the Exxon Corporation.

Glossary of Technical Terms

AESTHETASC. Thin-walled sensory seta usually found on flagellum of antennule
AMBULATORY (as applied to pereopods). Used for walking.
ANCHIALINE. An aqueous habitat near the sea; referring to saltwater or brackish pools fluctuating with the tides, but with no surface connection to the sea.
ANGULATE. Having an angle or an angular shape.
ANTENNA. Paired appendage of the third cephalon segment; sometimes referred to as antenna 2.
ANTENNULE. Paired appendage of the second cephalon segment; sometimes referred to as antenna 1.
APICAL. Relating to the apex or tip.
APPENDAGE. An articulated structure used for feeding, locomotion, sensory reception, e.g., mouthparts, antennae, pereopods, pleopods, uropods.
ARTICLE. A single section of an appendage, with an articulation at one or both ends.
BASIS. Article of appendage adjoining coxa proximally, and carrying endopod distally, i.e., article 2 of pereopod.
BIARTICULATE. Composed of two articles.
BIDENTATE. Having two teeth.
BIFID. Divided into two lobes or parts by a cleft.
BILOBED. Composed of two lobes.
BIRAMOUS. Composed of two rami or branches.
BIUNGUICULATE. Having two claws, as in a bifid dactylus.
CARINA. A keel, or an acute ridge.
CARINATE. Having one or more carinae or acute ridges.
CARPOCHELATE. Having a chela or pincerlike structure formed by the seventh (dactylus) and fifth (carpus) articles of an appendage.
CARPUS. Article 5 of pereopod.
CEPHALON. Anterior region of body or head; more correctly the cephalothorax in isopods, as the first pereonal segment is usually fused with the head.
CHELA. Distal pincerlike part of appendage, often formed by a mobile and an immobile finger.
CHELATE. Having a chela; modified to form a pincer.
CLAVATE. Club shaped; having one end thickened.
CLYPEUS. Platelike structure of cephalon, anterior to upper lip or labrum, sometimes fused with frontal lamina.
CONGLOBATE. Able to roll up into a ball, as in some sphaeromatid and oniscidean isopods.
CONSPECIFIC. Belonging to the same species.
CONTIGUOUS. Touching.

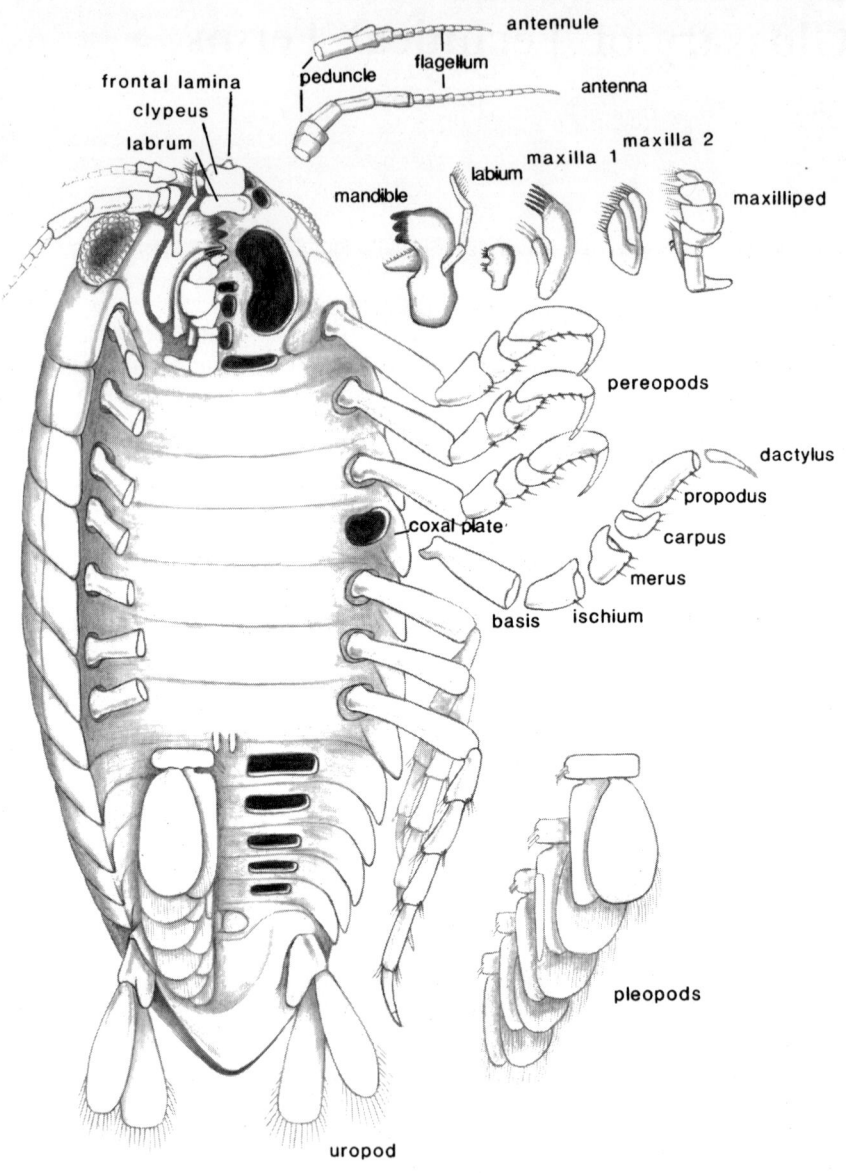

Figure 2. Schematic representation of an isopod illustrating morphological terms.

COPULATORY STYLET. Structure situated on endopod of pleopod 2 in males, used for transfer of spermatophore in some species; also referred to as appendix masculina.
CORDATE. Heart shaped in outline.
COXA. Basal article of an appendage, attached to sternite, sometimes expanded into a lateral coxal plate.
CRENULATE. Having a scalloped edge with rounded teeth, usually used to refer to the margin of a structure.
DACTYLUS. Terminal (7th) article of a pereopod or thoracic appendage.
DENTATE. Edged with teeth.
DENTICLE. A small tooth.
DENTICULATE. Having fine teeth.
DIGITIFORM. Fingerlike.
DISTAL. Situated away from the base or point of origin or attachment.
ECDYSIS. Molting of the integument.
EMARGINATE. Having the margin concave.
ENDITE. Medially directed lobe of coxa or basis of an appendage, especially the maxilliped.
ENDOPOD. Inner ramus of a biramous appendage.
ENTIRE. Complete; usually referring to the margin of a structure that is smooth.
EPIMERON. Lateral part of a somite.
EPIPOD. Lateral extension of a protopodite.
EXCAVATE. Hollowed out.
EXOPOD. Outer ramus of a paired appendage.
FALCATE. Sickle shaped; curved and tapering to a point.
FLAGELLUM. Distal part of antenna or antennule, usually multiarticulate, occasionally reduced to one or a few articles.
FRONTAL LAMINA. Platelike structure of the cephalon immediately anterior to, and sometimes fused with, clypeus.
GENICULATE. Bent at an abrupt angle, as in the body of many arcturid isopods.
GRANULATE. Having the appearance of bearing beadlike or grainlike protuberances; usually applied to the description of a surface.
HIRSUTE. Bearing hairs (elongate hairs in the case of most isopods).
HYPOGEAN. Underground.
IMMERSED. Sunken into, as with one structure into another.
INCISOR. Cutting process of the mandible, usually dentate, sometimes modified for piercing.
INDURATE. Hardened, usually by calcium carbonate or sclerotized protein.
INTEGUMENT. Outer covering, e.g., the exoskeleton.
INTERSTITIAL. Relating to interstices; living in the interstices of sand grains, gravel, or rubble.
ISCHIUM. Article 3 of pereopod.
LABIUM. Lower lip; usually consisting of a pair of lobes posterior to the mouth.
LABRUM. Unpaired projection anterior to mouth, attached to the clypeus; upper lip.
LACINIA MOBILIS. Small, usually toothed, process articulating at base of incisor in left or both mandibles.
LAMELLAR. In the shape or structure of a thin plate or lamella.

LAMINA DENTATA. Serrate platelike structure in the mandible of anthurideans, formed by the fusion of spines of the spine-row.
LANCEOLATE. Lance shaped; narrow and tapering to a point.
LINGUIFORM. Tongue shaped.
MANCA. Young of some peracaridean crustaceans (including isopods), lacking last thoracic appendage at time of release from broodpouch.
MANDIBLE. First pair of mouthparts, functioning as jaws, often sclerotized.
MARSUPIUM. Structure in which eggs are retained by female; the broodpouch.
MAXILLA (1 and 2). Two sets of paired mouthpart appendages immediately posterior to mandible.
MAXILLIPED. First paired appendage of the thorax; usually incorporated into the mouthparts.
MEDIAN. At, near, or directed toward the middle or midline.
MERUS. Article 4 of pereopod.
METAMORPHOSED. Transformed; changed in appearance, structure, or function.
MESIAL. Near or toward the middle or midline.
MOLAR. Grinding, and sometimes piercing, structure of the mandible.
MULTIARTICULATE. Composed of many articles.
NATATORY. Adapted for swimming.
OBSOLETE. Becoming vestigial, and losing original function.
OMMATIDIA. Individual visual components of the compound eye.
OOSTEGITE. Medially directed lamellar structure arising from coxa of pereopod in the female, forming part of the broodpouch or marsupium.
OPERCULIFORM. In the form of a cover or lid.
OVATE. Egg shaped or oval.
PALM. Cutting edge of the propodus, often defined proximally by a spine, in a subchelate appendage.
PALP. Articulated ramus consisting of one to three articles in mandible, of up to five articles in the maxilliped.
PECTINATE. Having teeth like a comb.
PEDUNCLE. Stalk or proximal part of an appendage, as in antennae.
PENIAL RAMI. Paired submedian process on sternite 7 of male.
PEREON. Middle or thoracic region of the body, consisting of seven segments or pereonites, first fused with cephalon in isopods.
PEREONITE. Segment of the pereon.
PEREOPOD. Paired appendage of the pereon, consisting of seven articles when unmodified.
PILOSE. Covered with short hairs or setae.
PLEON. Posterior or abdominal region of the body, primitively consisting of six segments or pleonites, and bearing paired pleopod and uropod appendages.
PLEONITE. Segment of the pleon.
PLEOPOD. Paired appendage of the pleon, five pairs being present in the primitive condition.
PLEOTELSON. Structure resulting from the fusion of the telson and one or more pleonal segments.
PLICATE. Pleated or folded.
PRANIZA. Juvenile, immature stage of gnathiideans.
PREHENSILE. Adapted for holding or clinging.

PRODUCED. Extended or lengthened.
PROPODUS. Article 6 of pereopod.
PROTANDROUS. In hermaphroditic forms, becoming a functional male producing spermatozoa before becoming a functional female producing eggs.
PROTOGYNOUS. In hermaphroditic forms, becoming a functional female producing eggs before becoming a functional male producing spermatozoa.
PROTOPODITE. Proximal part of an appendage, consisting of the coxa and basis.
PROXIMAL. Situated near the point of attachment.
PYLOPOD. First pereopod of the Gnathiidea, modified to form part of the mouthparts.
RAMUS. Branch of an appendage.
RENIFORM. Kidney shaped.
RETICULATE. Resembling or forming a network.
RETINACULAE. Small hooks on an appendage, used to link the left and right members of a pair of appendages.
ROSTRUM. Anterior middorsal projection of cephalon.
SAGITTATE. Arrow shaped.
SCLEROTIZED. Hardened, usually with chitin.
SERRATE. Edged with toothlike projections as in a saw.
SETIFEROUS. Bearing setae.
SETOSE. Bearing setae.
SINUATE. Having a wavy margin.
SINUOUS. Having curves.
SOMITE. Body segment, usually having a pair of appendages.
SPATULATE. Shaped like a spatula.
SPICATE. Shaped like a spike.
SPINE-ROW. Row of spines situated between the incisor and molar processes of the mandible.
SPINOSE. Bearing spines.
STATOCYST. Small saclike sensory organ, often containing granules, used to indicate to the animal its orientation.
STYGOBIONT. Cave organism.
STYLIFORM. Having a long, slender, stilettolike shape.
SUB-. A prefix indicating "almost" or "just less than," e.g., submarginal—almost on the margin.
SUBCHELATE. Having a subchela, forming a pincerlike structure, especially by the dactylus folding back on the propodus.
SUTURE. A line indicating an area of articulation, or of incomplete fusion.
SYMPOD. Proximal part of an appendage, often formed by the fusion of the coxa and basis.
TELSON. Terminal part of the body, usually bearing the anus.
THORAX. Tagma or body region between the cephalon and the abdomen.
TRACHEATE. Bearing tubular respiratory trachea (more correctly pseudotrachea) on pleopods, as in Oniscidea.
TRICUSPID. Bearing three cusps or points.
TRIDENTATE. Having three teeth.
TRIFID. Divided into three parts or lobes.
TRILOBED. Divided into three lobes.

TRISINUATE. Having three curves.
TRIUNGUICULATE. Bearing three claws, as in a trifid dactylus.
TRUNCATE. Having the appearance of having been abruptly cut off.
TUBERCULATE. Bearing knoblike or wartlike prominences or tubercles.
UNIARTICULATE. Composed of one article.
UNIRAMOUS. Having one ramus or branch.
UNIUNGUICULATE. Having a single claw, as in a dactylus.
UROPOD. Paired pleonal appendage of the last pleonite, usually situated at the base of the telson.

Marine Isopods of the Caribbean

Phylum Arthropoda
Superclass Crustacea Pennant, 1777
Class Malacostraca Latreille, 1806
Subclass Eumalacostraca Grobben, 1892
Superorder Peracarida Calman, 1904
Order Isopoda Latreille, 1817

DIAGNOSIS Body usually dorsoventrally depressed, occasionally subcylindrical, rarely bilaterally compressed. Carapace lacking. Antennules and antennae uniramous (scale on antenna in some asellotes may represent rudimentary second ramus). Eyes sessile (although situated on nonmobile stalks in some asellotes). Mouthparts consisting of one pair of mandibles, two pairs of maxillae, one pair of maxillipeds; latter appendages of first thoracic segment fused with cephalon. Mandible usually with palp consisting of one to three articles; incisor, lacinia mobilis, and molar usually present; lacinia mobilis often differing on left and right sides, sometimes absent from right mandible; molar variable. Maxilliped usually consisting of palp of no more than five articles, lamellar endite often with coupling hooks, lamellar epipod. Pereonites usually separate, although pereonite 1 sometimes fused with cephalon. Coxae of pereopods variously fused with, and forming expanded lateral processes of, pereonites. Pereopod 1 forming additional mouthpart (pylopod) only in Gnathiidea. Pereopods generally similar, ambulatory; pereopods 1–3 secondarily variously modified and becoming subchelate or prehensile; pereopods 4–7 occasionally modified, becoming natatory or prehensile. Pereopod 7 occasionally not developed (neotenous condition). Broodpouch or marsupium formed by varying number of oostegites attached ventrally and medially to coxae of pereopods; eggs held in anterior or posterior pockets or internal pouches in gnathiids and some sphaeromatids. Pleon consisting of six pleonites, free or variously fused, plus telson; if one or more pleonites fused with telson, resulting structure referred to as pleotelson.

Key to suborders of Isopoda

1. Parasitic on crustaceans; body of ♀ nearly always asymmetrical
   .... Epicaridea
   Free-living or parasitic on fishes; body of ♀ bilaterally symmetrical, or
   if parasitic, ♀ somewhat distorted 2

2. Body more or less bilaterally compressed .... Phreatocoidea*
   Body more or less dorsoventrally depressed or subcylindrical 3

3. With six pereonites and five pairs of pereopods .... Gnathiidea
   With seven pereonites and six or seven pairs of pereopods 4

4. Body usually more than six times longer than wide, subcylindrical,
      uropods never operculiform 5
   Body usually less than six times longer than wide, usually
      dorsoventrally depressed; if subcylindrical, uropods operculiform .. 6

5. Uropodal exopod often folding dorsally over pleotelson; rarely
      interstitial forms .... Anthuridea
   Uropods terminal, exopod lacking; minute interstitial forms
      .... Microcerberidea

6. Antennules minute; terrestrial forms, with pleopods tracheate
      .... Oniscidea
   Antennules rarely minute; aquatic forms, pleopods never tracheate .. 7

7. Uropods ventral, operculiform, covering pleopods .... Valvifera
   Uropods never operculiform over pleopods 8

8. Uropods lateral or ventrolateral, forming tailfan with pleotelson;
      pleopods 1 and 2 rarely operculiform .... Flabellifera
   Uropods terminal or subterminal; pleopods 1 and 2 variously
      operculiform .... Asellota

   * The suborder Phreatocoidea contains freshwater forms, and has a
   Gondwanian distribution, primarily in the Southern Hemisphere.

Pleopods (on pleonites 1–5) biramous, lamellar, primarily for respiration; anterior pleopods occasionally operculiform. Pleopod 2 in male (and occasionally also pleopod 1 in Oniscidea and Asellota) with endopod bearing copulatory stylet. One pair of uropods on pleonite 6. Young leave broodpouch as manca, i.e., resembling adult but lacking pereopod 7; in Epicaridea, manca stage represented by epicaridium stage; latter transforms into microniscium and then cryptoniscium stage, before becoming adult.

Suborder Anthuridea Leach, 1814

DIAGNOSIS Body generally elongate and subcylindrical. Eyes absent in some genera. Antennular peduncle of three articles; antennal peduncle of five articles. Mandible with palp of one to three articles, or absent; body of mandible either styliform and lacking molar and lacinia mobilis, or with molar variously specialized or reduced, lacinia mobilis absent, and spine-row modified to form platelike lamina dentata. Maxilla 1 with inner ramus reduced, outer ramus slender. Maxilla 2 rudimentary. Maxilliped variable, with palp of one to five articles, endite present, modified, reduced, or absent. Pereonite 1 free. Pereopod 1, or pereopods 1–3 subchelate; pereopods 4–7 generally ambulatory. Pleonites 1–5 free or fused, pleonite 6 partly or completely fused with telson. Pleopods 1–5 similar, or pleopod 1 variously modified to form operculum. Uropodal exopod often folded dorsally over pleotelson. Pleotelson with pair of statocysts, with single statocyst, or lacking statocysts.

REMARKS Protogyny has been demonstrated in several species of Anthuridea. The order of development in these cases is: egg, manca (both in the broodpouch), immature subadult, ovigerous female, premale, male, with varying numbers of molts between each stage. At least one molt takes place between ovigerous female and premale, the latter being distinguished by the loss of the oostegites and by the elongation of, and acquisition of more flagellar articles in, the antennule. One or two molts take place between premale and sexually mature male, the latter being characterized by the possession of elongate antennular flagella bearing dense whorls of aesthetascs, a more setose and/or spinose pereopod 1, and sometimes by an elongation of the pleon and uropods. In some genera, the males have somewhat atrophied mouthparts, suggesting that they do not feed at this stage. As a result of this seemingly widespread protogyny, sex ratios are strongly biased toward females, and in several species males are not yet known.

The number of families in the suborder Anthuridea has not been settled. At present, three families are recognized. Doubtless, further families will be defined and the genera reshuffled.

Key to families of Anthuridea

1. Mouthparts adapted for piercing and sucking, together forming conelike structure .... **Paranthuridae**
   Mouthparts adapted for cutting, lamina dentata and molar usually present 2

2. Pereopod 1 subchelate, with propodus expanded; pleonites generally fused; if free, much shorter than wide .... **Anthuridae**
   Pereopods 1–3 subchelate, subsimilar; pleonites free, often as long as wide .... **Hyssuridae**

Family Anthuridae Leach, 1814

DIAGNOSIS Mouthparts adapted for cutting. Pereopod 1 usually markedly different from remaining pereopods, subchelate with propodus more or less inflated. Exopod of pleopod 1 operculiform, covering remaining pleopods. Pleonites 1–5 fused, with fusion marked ventrolaterally by short slits, occasionally with dorsal grooves marking lines of fusion, or free; if free, length of each pleonite much less than width. Pleotelson with pair of statocysts, or single medial statocyst, or lacking statocysts.

Key to genera of Anthuridae

1. Pleopod 1, both rami contributing to operculum 2
   Pleopod 1, only exopod operculiform 6

2. Antennal peduncle bearing serrate process .... *Licranthura*
   Antennal peduncle lacking any serrate process 3

3. Pleopod 1 in ♀, rami to some degree fused .... *Eisothistos*
   Pleopod 1 in ♀, rami free 4

4. Pereopods 1 and 2 subchelate, of similar size, propodi not noticeably inflated 5
   Pereopod 1 much larger and propodus more expanded, than pereopod 2 .... *Minyanthura*

5. Integument noticeably pitted; mandible on one side lacking molar, on other side with spicate molar .... *Apanthuroides*
   Integument not pitted; mandibles similar on both sides   *Chalixanthura*

6. Telson with single statocyst .... *Anthomuda*
   Telson with two statocysts, or lacking statocysts 7

7. Pereopods 4–7, carpi roughly rectangular 8
   Pereopods 4–7, carpi triangular, with anterior margin considerably shorter than posterior margin 9

8. Maxillipedal palp of two articles .... *Haliophasma*
   Maxillipedal palp of three articles .... *Malacanthura*

9. Maxillipedal palp of one article; mandibular palp of one article
   .... *Pendanthura*
   Maxillipedal palp of more than one article; mandibular palp of more than one article 10

10. Maxillipedal palp of two articles 11
    Maxillipedal palp of three articles 12

11. Mandibular palp of two articles .... *Cortezura*
    Mandibular palp of three articles .... *Cyathura*

12. Cephalon with midventral process at base of mouthparts .. *Skuphonura*
    Cephalon lacking midventral process 13

13. Pleon lacking dorsal grooves or lines indicating boundaries of fused pleonites; species-specific persistent dorsal pigment pattern usually present .... *Mesanthura*
    Pleon with dorsal grooves or lines indicating boundaries of fused pleonites; persistent pigment pattern lacking 14

14. Pleon with complete dorsal lines separating pleonites
    1–5 .... *Apanthura*
    Pleon with incomplete dorsal line between pleonites 4 and 5
    .... *Amakusanthura*

Amakusanthura Nunomura, 1977

DIAGNOSIS Integument sometimes with pigment. Eyes present. Antennular flagellum of three articles. Antennal flagellum of two to four articles. Mandibular palp of three articles, terminal articles bearing distal spines. Maxillipedal palp of three articles; endite small or absent. Pereopod 1, propodal palm with step or tubercle. Pereopods 4–7, carpus triangular. Pleonites short, pleonites 1–4 fused, boundaries demarked by complete dorsal folds, pleonites 4 and 5 separated only by lateral fold, not demarked dorsally; pleonite 6 dorsally demarked from telson. Pleopod 1 exopod operculiform. Uropodal exopod often notched or excavate distally. Pleotelson with two basal statocysts.

Key to species of *Amakusanthura*

1. Telson thickened, with raised area at midlength, widening and sloping posteroventrally; uropodal exopod distally shallowly excavate .... *magnifica*
   Telson dorsally flat, not thickened; uropodal exopod distinctly notched or barely excavate distally 2

2. Integument pigmented; uropodal endopod length 1.5 times basal width .... *signata*
   Integument not pigmented 3

3. Uropodal exopod distinctly notched 4
   Uropodal exopod, outer margin weakly excavate; mandibular palp article 3 bearing three spines .... *lathridia*

4. Mandibular palp article 3 bearing three spines .... *significa*
   Mandibular palp article 3 bearing two spines .... *geminsula*

Amakusanthura geminsula (Kensley, 1982)
  Figure 3A–E

DIAGNOSIS ♀: 8.1 mm. Shallow middorsal pit on pereonites 4–6, appearing chalky white in life. Antennular flagellum of three articles. Antennal flagellum of two articles. Mandibular palp article 3 bearing two spines. Maxilliped with short slender endite; terminal palp article set obliquely at outer distal angle of penultimate article. Pereopod 1, propodal palm with step; carpus distally a rounded lobe. Uropodal exopod with notch. Pleotelson tapering in posterior half to subtruncate apex. ♂: 4.8 mm. Eyes larger than ♀. Antennular flagellum of 8–9 articles. Pereopod 1, carpus with distal lobe

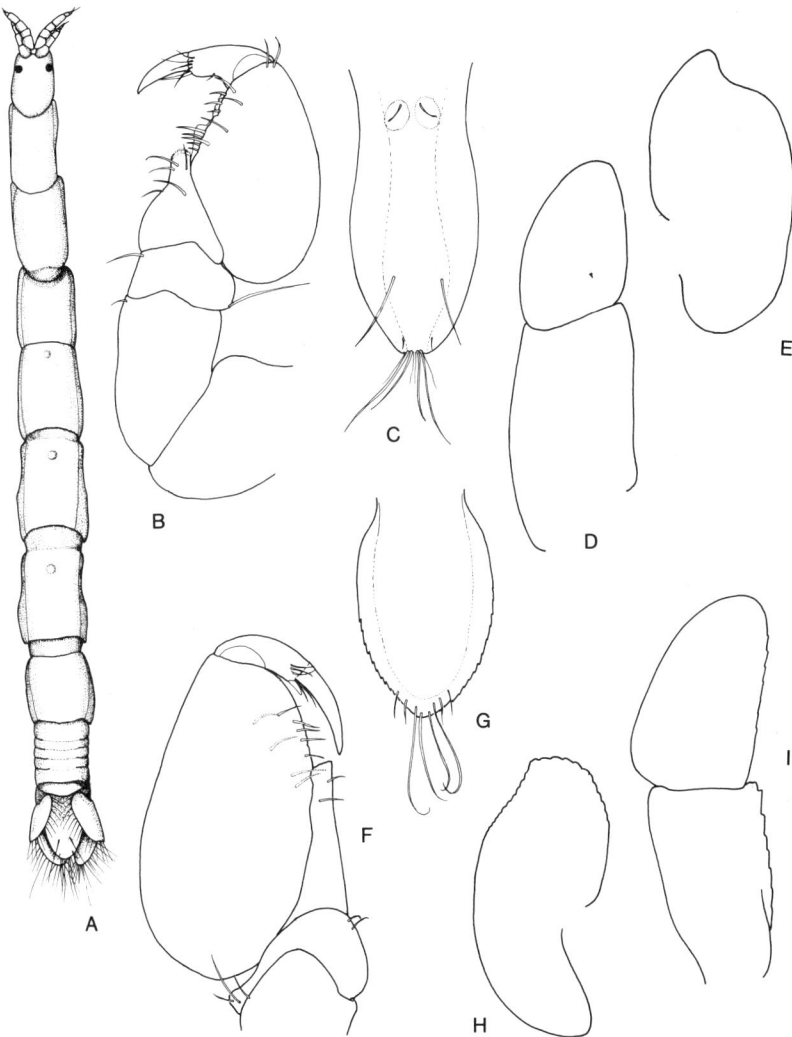

Figure 3. *Amakusanthura geminsula*: A, ♀; B, pereopod 1; C, telson; D, uropodal sympod and endopod; E, uropodal exopod. *Amakusanthura lathridia*: F, pereopod 1; G, telson; H, uropodal exopod; I, uropodal sympod and endopod.

narrowly rounded; propodal palm with rounded tubercle; irregular band of setae on mesial surface.

RECORDS Carrie Bow Cay and Twin Cays, Belize, intertidal to 1.5 m; Jamaica.

Amakusanthura lathridia (Wägele, 1982)
Figure 3F–I

DIAGNOSIS ♀ 2.7 mm. Antennular flagellum of three articles. Antennal flagellum of three articles. Mandibular palp article 3 bearing three spines. Maxilliped lacking endite. Pereopod 1, propodal palm with distal step; carpus with triangular posterodistal lobe. Uropodal exopod apically acute, outer margin slightly excavate, sinuate; endopod length slightly less than twice basal width. Pleotelson elliptical, widest at about midlength, margins weakly serrate.

RECORDS Cuba, interstitial beach sand just above water line.

Amakusanthura magnifica (Menzies and Frankenberg, 1966)
Figures 4, 5

DIAGNOSIS ♀: 13.9 mm. Maxilliped with short endite, terminal article of palp short, set obliquely at outer distal angle of penultimate article. Pereopod 1, propodal palm with strong tubercle; carpus with low distal sclerotized lobe. Uropodal exopod ovate, with shallow distal notch; endopod twice longer than wide. Pleotelson with posterior margin evenly rounded to subtruncate; dorsally with raised area at midlength, broadening and sloping

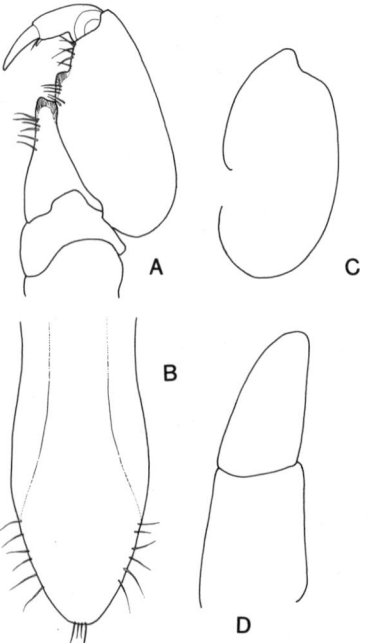

Figure 4. *Amakusanthura magnifica:* A, pereopod 1; B, telson; C, uropodal sympod and endopod (setae omitted); D, uropodal exopod (setae omitted)

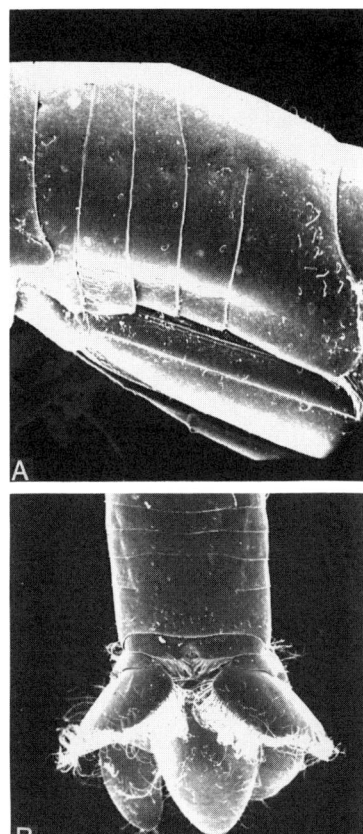

Figure 5. *Amakusanthura magnifica:* A, pleonites in lateral view; B, pleon in dorsal view.

away posteriorly. ♂: 10.3 mm. Antennular flagellum of 24 articles. Pereopod 1, carpus with sclerotized distal rounded lobe; propodal palm with strong sclerotized tubercle longer than in ♀; numerous setae on mesial surface. Pleopod 2, copulatory stylet of endopod not reaching beyond ramus.

RECORDS   Off Georgia, 17–137 m; off Florida, 7–11 m; Cuba; Gulf of Mexico.

Amakusanthura signata (Menzies and Glynn, 1968)
   Figure 6A–E

DIAGNOSIS   ♀: 4.9 mm. Integument with strong patches of pigment on cephalon; pigment sparse on pereonites; two bars on pleonite 6. Antennular

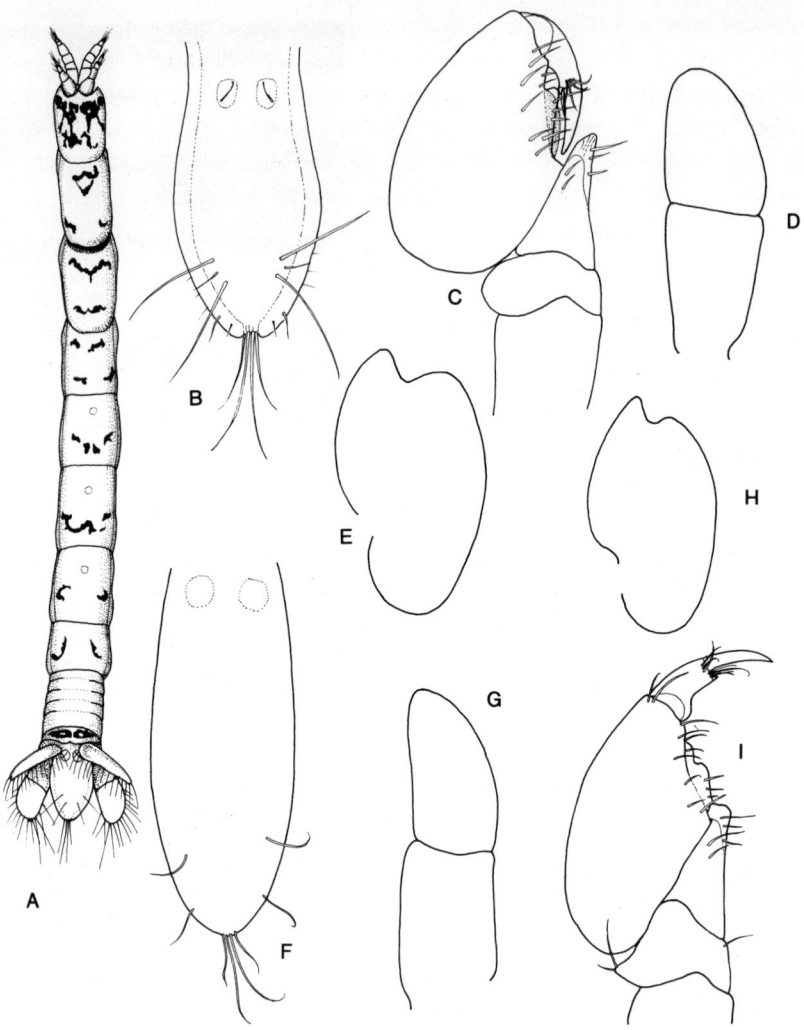

Figure 6. *Amakusanthura signata:* A, ♀; B, telson; C, pereopod 1; D, uropodal sympod and endopod; E, uropodal exopod. *Amakusanthura significa:* F, telson; G, uropodal sympod and endopod; H, uropodal exopod; I, pereopod 1.

flagellum of three articles. Antennal flagellum of three articles. Mandibular palp article 3 bearing three spines. Maxilliped with short endite not reaching base of palp article 2. Pereopod 1, carpus with strong subacute sclerotized lobe distally; propodal palm with transparent convex flange and step.

Uropodal exopod with strong distal notch; endopod length about 1.5 times basal width. Pleotelson widest in posterior half, tapering posteriorly to shallowly notched apex. ♂: 4.5 mm. Eyes larger than in ♀. Antennular flagellum of nine articles. Pereopod 1, carpus with strong distal lobe, propodal palm with step, convex transparent flange, and numerous setae on mesial surface. Pleopod 2, copulatory stylet of endopod reaching well beyond ramus.

RECORDS   Cuba; Puerto Rico, intertidal to 1.5 m; Carrie Bow Cay, Belize, intertidal to 24 m.

Amakusanthura significa (Paul and Menzies, 1971)
    Figure 6F–I

DIAGNOSIS   ♀ 5.0 mm. Antennular flagellum of three articles. Antennal flagellum of four or five articles. Mandibular palp article 3 bearing three spines. Maxilliped lacking endite. Pereopod 1, propodal palm with step; carpus distally rounded. Uropodal exopod ovate, with distal notch; endopod length slightly more than twice basal width. Telson elliptical, posterior margin narrowly rounded.

RECORDS   Off Venezuela, 95 m.

Anthomuda Schultz, 1979

DIAGNOSIS   Eyes present. Mandibular palp of three articles. Maxillipedal palp of four articles; endite reaching well beyond base of palp article 2. Pereopods 1 and 2 similar, subchelate. Pleonites 1–6 short, free; pleonite 6 dorsally demarked. Telson with single semiovate hollow representing statocyst.

Anthomuda stenotelson Schultz, 1979
    Figure 7A

DIAGNOSIS   ♀ 8.8 mm. Antennular flagellum of three or four articles. Antennal flagellum of six articles. Pereopods 1 and 2 similar, propodi somewhat inflated; pereopod 1, propodal palm straight, unarmed; pereopod 2 propodal palm with three sensory spines. Pereopods 4–7 with rectangular carpi. Pleonite 6 free, posterior margin broadly bilobed. Uropodal exopod elongate-elliptical, distally narrowly rounded; endopod length 2.5 times basal width, distally broadly rounded. Telson narrowly lanceolate, posteriorly narrowly rounded, with single open hollow statocyst.

RECORDS   Off Bermuda, 90 m.

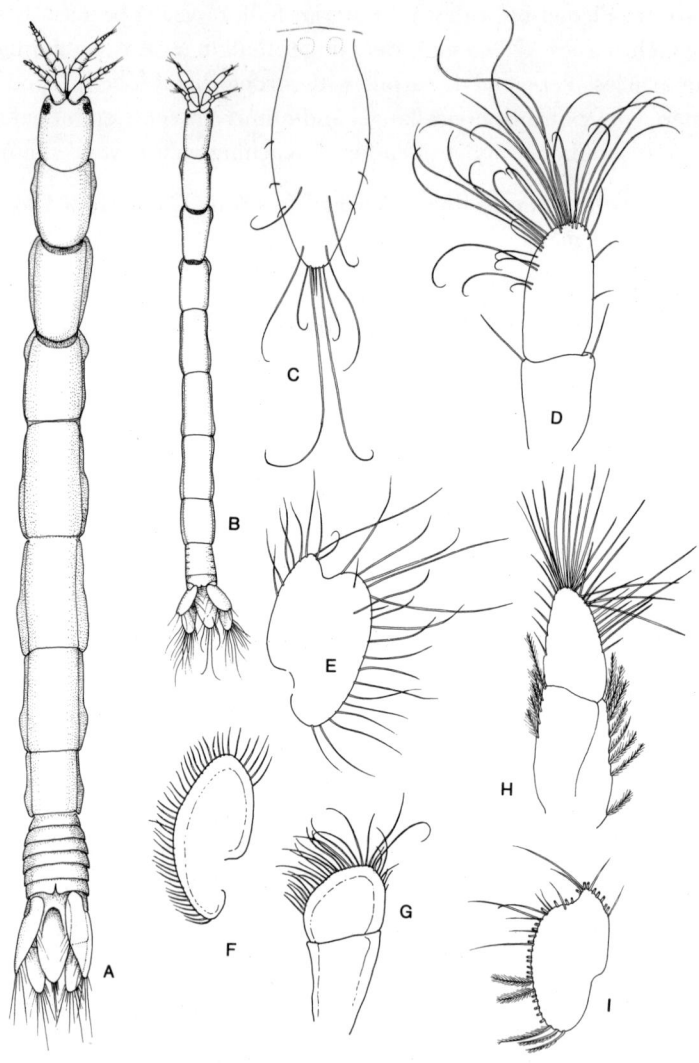

Figure 7. *Anthomuda stenotelson:* A, ♀. *Apanthura cracenta:* B, ♀; C, telson; D, uropodal endopod and sympod; E, uropodal exopod. *Apanthura crucis:* F, uropodal exopod; G, uropodal endopod and sympod. *Apanthura harringtoniensis* (from Wägele, 1981): H, uropodal endopod and sympod; I, uropodal exopod.

Apanthura Stebbing, 1900

DIAGNOSIS   Integument sometimes pigmented. Eyes present. ♀: Antennular flagellum of three articles. Antennal flagellum of two to four articles. Mandibular palp of three articles, terminal article bearing distal spines. Maxillipedal palp of three articles; endite small, or lacking. Pereopod 1, propodal palm usually with step or tubercle; propodus inflated. Pereopods 4–7, carpus triangular. Pleonites 1–5 fused; pleonite 6 dorsally demarked. Pleopod 1, exopod operculiform. Uropodal exopod ovate, sometimes distally notched or excavate. Pleotelson with two basal statocysts. ♂ antennular flagellum of about 10 articles.

REMARKS   Differentiation of species, especially if long preserved and the pigmentation is lost, depends on subtle features of the mandible, maxilliped, pereopods, and pleotelson.

Key to species of *Apanthura*

1. Uropodal exopod distally notched or excavate 2
    Uropodal exopod distally entire; uropodal endopod length subequal to width .... *crucis*
2. Uropodal exopod distally notched; uropodal endopod length 2.4 times basal width .... *cracenta*
    Uropodal exopod faintly excavate or sinuate; uropodal endopod length about twice basal width .... *harringtoniensis*

Apanthura cracenta Kensley, 1984
  Figure 7B–E

DIAGNOSIS   ♀: 4.6 mm. Antennular flagellum of three articles. Antennal flagellum of two articles. Mandibular palp article 3 bearing four spines. Maxilliped with short rounded endite; terminal palp article set at oblique angle on, and less than half length of penultimate article. Pereopod 1, carpus triangular with acute sclerotized tip overlapping base of palm; propodal palm with rounded tubercle near midlength. Uropodal exopod deeply notched; endopod ovate, length $2^{1}/_{3}$ basal length. Pleotelson lanceolate. ♂: 3.8 mm. Antennular flagellum of six articles. Antennal flagellum of two articles. Pleopod 2, copulatory stylet of endopod reaching by half its length beyond ramus.

RECORDS   Turks and Caicos Islands, 1 m; Carrie Bow Cay, Belize, intertidal on reef crest, 2 m.

Apanthura crucis (Barnard, 1925)
Figure 7F,G

DIAGNOSIS   ♀: 5.9 mm. Antennular flagellum of four articles. Antennal flagellum of four articles. Mandibular palp article 3 with eight spines. Maxilliped lacking endite; terminal palp article more than half length of penultimate article. Pereopod 1, propodal palm with low rounded tubercle at midlength. Uropodal exopod ovate, outer margin evenly convex; endopod subcircular. Pleotelson anteriorly narrow, widest at midlength, posterior margin broadly rounded. ♂: 6.2 mm. Antennular flagellum of 12 or 13 articles. Antennal flagellum of four articles. Pereopod 1, propodal palm with low rounded tubercle; dense band of setae on mesial surface. Copulatory stylet on endopod of pleopod 2 reaching a third of its length beyond ramus.

RECORDS   Turks and Caicos Islands, 1 m; St. Croix, U.S. Virgin Islands, 8 m.

Apanthura harringtoniensis Wägele, 1981
Figure 7H,I

DIAGNOSIS   ♀ 6.0 mm. Antennular flagellum of four articles. Antennal flagellum of four articles. Mandibular palp article 3 bearing two spines. Maxillipedal endite short, not reaching base of palp article 2. Pereopod 1, propodal palm stepped; carpus distally bluntly triangular. Uropodal exopod apically acute, outer margin distally sinuate; endopod length about twice basal width. Pleotelson with posterolateral margins faintly denticulate; apex subtruncate and bearing several setae.

RECORDS   Harrington Sound, Bermuda.

Apanthuroides Menzies and Glynn, 1968

DIAGNOSIS   Eyes present. Mandibular palp of three articles; body of mandible anteriorly produced, molar absent on left side, present as narrow spikelike process on right. Maxillipedal palp of three articles; endite present, reaching palp article 2. Pereopod 1, propodus barely expanded, palm unarmed. Pereonite 7 less than half length of pereonite 6. Pereopods 4–7, carpi with free anterior margin shorter than posterior margin. Pleonites 1–5 fused, 6 fused with telson. Pleopod 1, both rami forming operculum.

Apanthuroides millae Menzies and Glynn, 1968
Figure 8

DIAGNOSIS ♀: 4.5 mm. Integument with diffuse brown pigmentation and numerous shallow pits. Antennular flagellum of four articles. Antennal flagellum of seven articles. Pleotelson with strong middorsal ridge. ♂: 5.0 mm. Integumental pigment stronger than in ♀, in irregular brown patches. Eyes larger than in ♀. Antennular flagellum of six or seven articles. Antennal flagellum of seven articles. Pereopod 1, propodal palm armed with five fringed spines.

RECORDS Carrie Bow Cay, intertidal to 30 m; Puerto Rico, intertidal.

REMARKS The highly modified mandible probably indicates some specialized form of feeding, but this has yet to be discovered.

Chalixanthura Kensley, 1984

DIAGNOSIS Eyes present, enormously enlarged, especially ventrally in ♂, resulting in mouthpart reduction. Mandibular palp of three articles; molar small, lamina dentata, and incisor present. Mouthparts reduced in ♂. Pereopods 1–3 similar, propodi barely inflated. Pereopods 4–7, carpi triangular. Pleopod 1, exopod broader than endopod, operculiform, or both rami forming operculum. Pleopods 2–5, exopods biarticulate. Pleonites short, free, longer in ♂ than in ♀. Uropodal rami, margins moderately to strongly incised or serrate. Telsonic margin serrate posteriorly; statocysts lacking.

Key to species of *Chalixanthura*

1. Body pigmented; uropodal exopod deeply incised .... *scopulosa*
   Body not pigmented; uropodal exopod weakly serrate .... *lewisi*

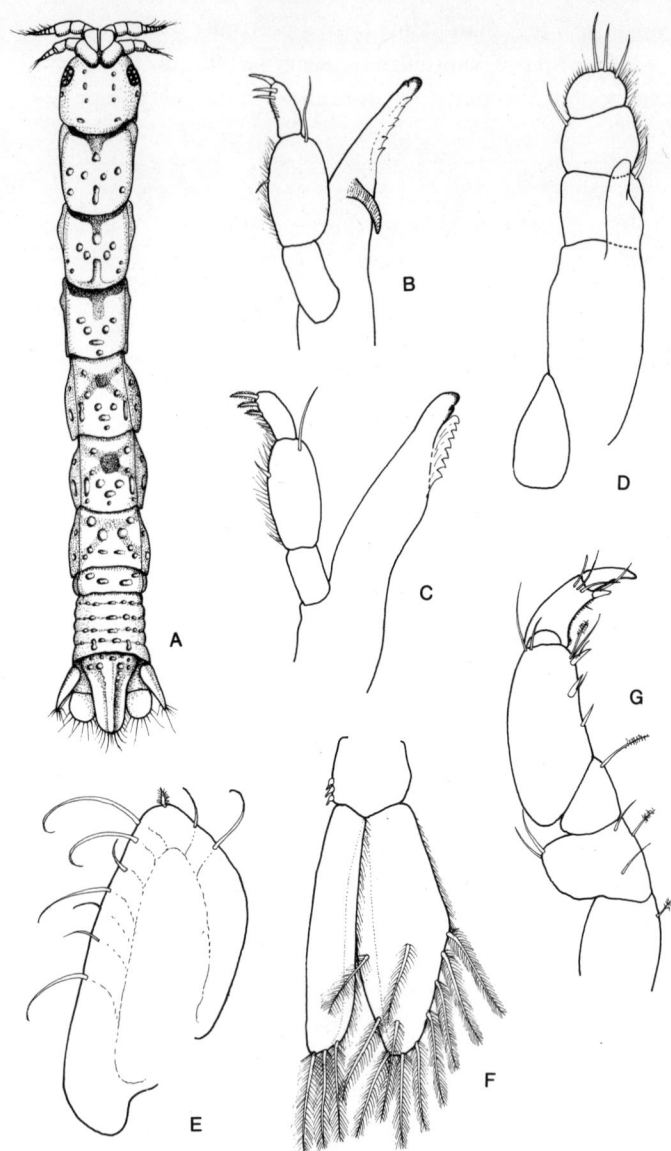

Figure 8. *Apanthuroides millae: A*, ♀; *B*, right mandible; *C*, left mandible; *D*, maxilliped; *E*, uropodal exopod; *F*, pleopod 1; *G*, pereopod 1.

Chalixanthura lewisi Kensley and Snelgrove, 1987
Figure 9A–D

DIAGNOSIS ♀: 3.1 mm. Integument lacking pigment. Antennular flagellum of three articles. Antennal flagellum of seven articles. Maxillipedal endite large, apically acute. Pereopod 1 propodus slightly expanded, palm with few (3) simple setae. Pereopod 7 propodus with two elongate anterodistal fringed spines. Uropodal and telsonic margins serrate. ♂: 2.2 mm. Antennular flagellum of 11 articles. Antennal flagellum of six articles. Maxilliped reduced, lacking endite. Pereopod 1, propodal palm with two spines plus seven spines on mesial surface. Pereopod 7 as in ♀.

RECORDS Barbados, in *Madracis mirabilis* coral, 9–15 m.

Chalixanthura scopulosa Kensley, 1984
Figure 9E–J

DIAGNOSIS ♀: 2.5 mm. Antennular flagellum of three articles. Antennal flagellum of seven articles. Maxillipedal endite short, not reaching base of palp article 4. Pereopod 1, propodal palm unarmed. Pereopod 7, propodus with single elongate anterodistal fringed spine. Uropodal exopod margin deeply incised; endopod ovate, margins serrate. Telson elongate-ovate, posterior margin serrate. ♂: 2.6 mm. Eyes considerably larger and with more ommatidia than in ♀. Antennular flagellum of seven articles. Antennal flagellum of four articles. Pereopod 1 propodal palm with three sensory spines. Uropodal exopod deeply incised.

RECORDS Carrie Bow Cay, Belize, 0.1 m.

Cortezura Schultz, 1977

DIAGNOSIS Eyes small, weakly pigmented. Antennular flagellum of two short articles. Antennal flagellum of single article. Mandibular palp of two articles. Maxillipedal palp of two articles, terminal article small; short endite present. Pereopod 1, propodus inflated. Pereopods 4–7 with carpus having anterior margin shorter than posterior. Pleonites 1–5 fused; pleonite 6 dorsally demarked. Pleopod 1, exopod operculiform. Pleotelson with two basal statocysts.

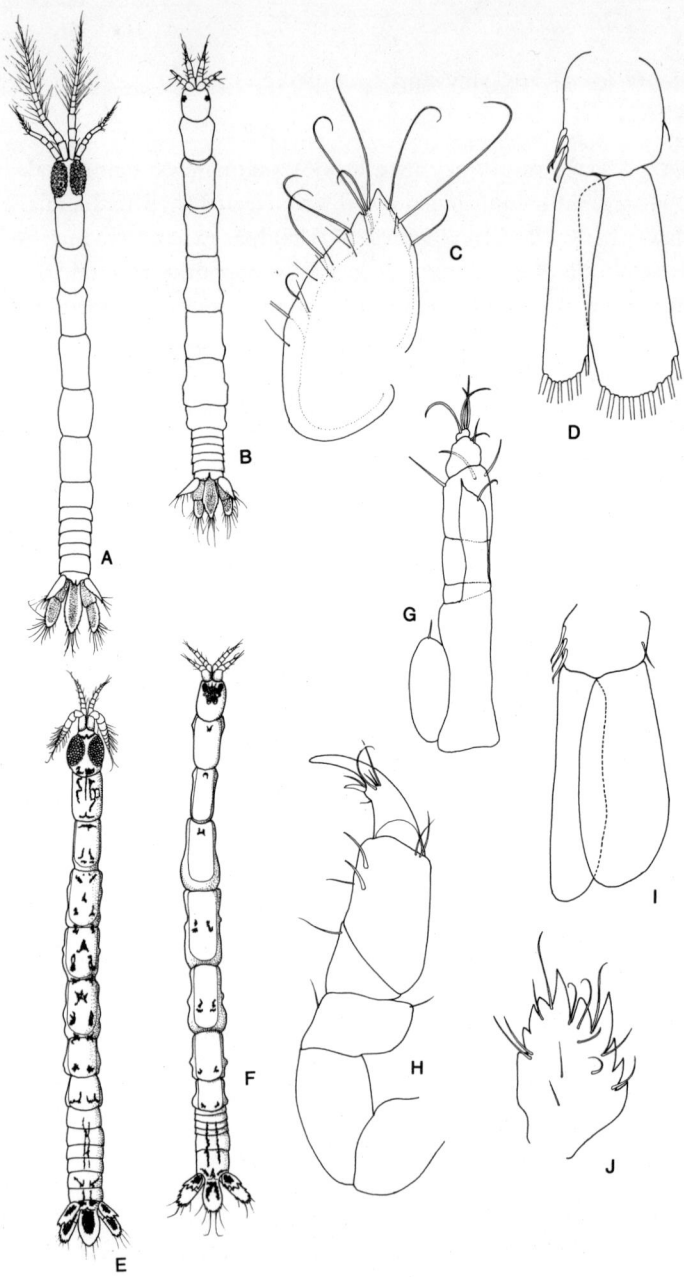

Figure 9. *Chalixanthura lewisi*: *A*, ♂; *B*, ♀; *C*, uropodal exopod; *D*, pleopod 1. *Chalixanthura scopulosa*: *E*, ♂; *F*, ♀; *G*, maxilliped; *H*, pereopod 1; *I*, pleopod 1; *J*, uropodal exopod.

Cortezura confixa (Kensley, 1978)
Figure 10

DIAGNOSIS ♀ 13.4 mm. Antennular peduncle articles 1 and 2 each with clump of ventrally directed setae. Antennal peduncle article 2 and antennular peduncle article 1 locked together. Mandibular palp of two articles, distal article twice length of proximal. Pereopod 1, propodus expanded, especially posteriorly; palm concave, with strong tubercle and irregular band of setae on mesial surface. Uropodal exopod ovate, apically rounded; endopod length 1.5 times greatest width. Telson ovate, posteriorly faintly narrowed, proximally thickened, with faint ridges diverging posteriorly.

RECORDS Cubagua Island, Venezuela, 4–10 m.

REMARKS This species was described under the generic name *Venezanthura* Kensley (1978). The only other known species of *Cortezura* is the type of the genus, *C. penascoensis* Schultz, 1977, from California.

Cyathura Norman and Stebbing, 1886

DIAGNOSIS Eyes present or absent. Antennular flagellum of one to three articles in ♀. Antennal flagellum of one to three articles. Mandibular palp of three articles. Maxillipedal palp of two articles; endite reduced or absent. Pereopod 1, propodus inflated. Pereopods 4–7, carpi triangular. Pleopod 1 exopod operculiform. Pleonites 1–5 short, fused; pleonite 6 fused with telson, sometimes dorsally demarked. Pleotelson with two basal statocysts.

Key to subgenera of *Cyathura*

1. Pleonite 6 not dorsally demarked from telson; articulation of uropodal exopod very short; exopod not adpressed to telson dorsally
   .... *Stygocyathura*
   Pleonite 6 dorsally demarked from telson; articulation of uropodal exopod relatively elongate; exopod adpressed dorsally to telson
   .... *Cyathura*

Cyathura (Cyathura) Norman and Stebbing, 1886

DIAGNOSIS Integument usually strongly pilose or setose; eye and body pig-

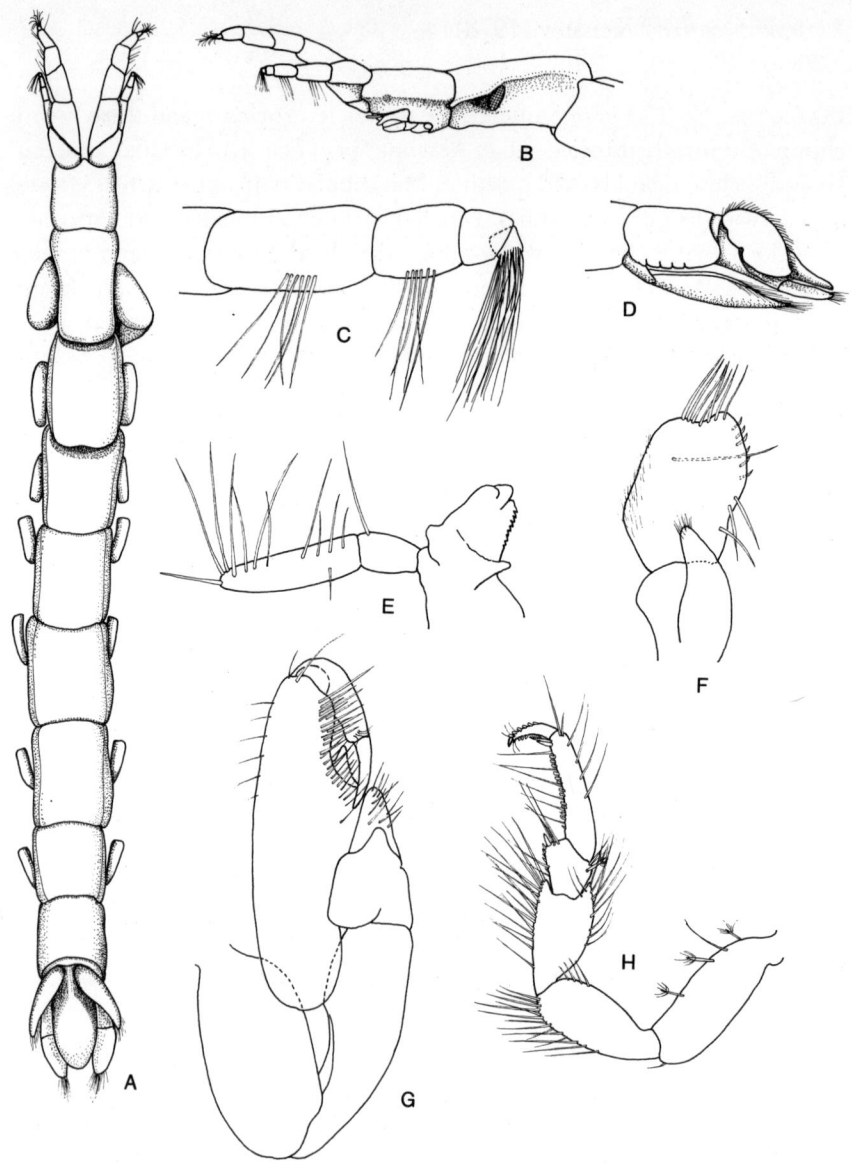

Figure 10. *Cortezura confixa:* A, ♀; B, cephalon and pereonite 1, lateral view (pereopod 1 removed); C, antennule; D, pleon, lateral view; E, mandible; F, maxilliped; G, pereopod 1; H, pereopod 7.

mentation usually present. Pereopod 1, propodal palm armed with tubercle. Pleopod 1, protopod with retinaculae. Uropodal exopod articulation relative elongate along lateral margin of sympod; exopod well developed, ovate. Pleonite 6 dorsally demarked from telson. Marine or estuarine forms.

Cyathura (Cyathura) cubana Negoescu, 1979
Figure 11A,B

DIAGNOSIS Ovigerous ♀: 7.0 mm. Antennular flagellum of two articles. Antennal flagellum of one article. Pereopod 1, propodal palm with rounded lobe in proximal half. Maxillipedal palp with distal article 0.34 times length of proximal article; small rounded endite present. Dorsal pigmentation consisting of irregular brown mottling. ♂: 5.5 mm. Antennular flagellum of four articles. Antennal flagellum of three articles. Pereopod 1, propodal palm with rounded lobe in proximal half. Copulatory stylet elongate-cylindrical, apically narrowed and flexed.

RECORDS Cuba, in mangroves, 2.5–7.0 m; Salt Creek, Belize, in mangroves, 1.5 m.

Cyathura (Stygocyathura) Botosaneanu and Stock, 1982

DIAGNOSIS Eye and body pigmentation absent. Body sparsely pilose or setose. Tendency toward elongation of some appendages, especially propodus of pereopods 2–7. Pereopod 1, propodal palm lacking strong tubercle. Pleopod 1, protopod lacking retinaculae. Pleonite 6 fused with telson, not dorsally demarked. Uropodal exopod with very short articulation on sympod, not adpressed dorsally to telson. Cave or hypogean forms.

REMARKS The ten species of *Stygocyathura* from the area covered in this work are morphologically very similar, with specific differences, although real, being very subtle. A dichotomous key would be cumbersome and require considerable dissection of mouthparts. The copulatory stylet of the male provides a valuable specific feature but males are not always available. Instead of a key, we have provided a list of species with their total lengths and localities (Table 1). Given the very restricted distribution of these cave species, material from localities not listed here should be treated as potentially undescribed, and the material compared with descriptions, especially those of Botosaneanu and Stock (1982).

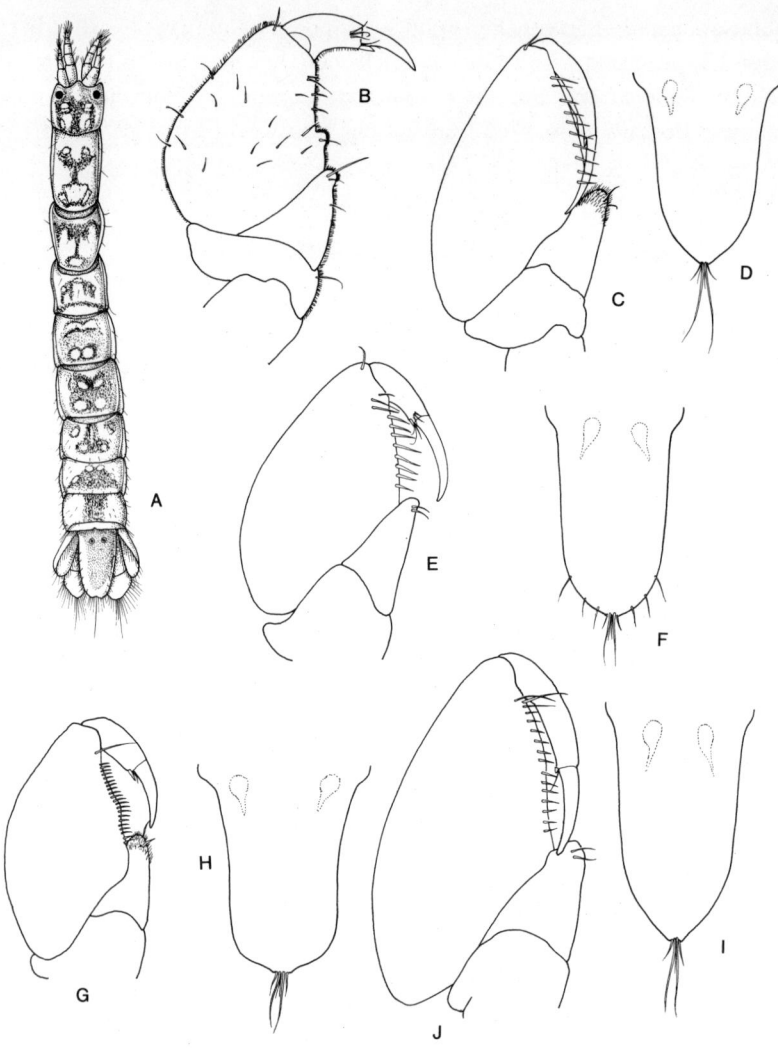

Figure 11. *Cyathura (Cyathura) cubana: A,* ♀; *B,* pereopod 1, ♀. *Cyathura (Stygocyathura) cuborientalis: C,* pereopod 1, ♀; *D,* telson. *Cyathura (Stygocyathura) curassavica: E,* pereopod 1, ♀; *F,* telson. *Cyathura (Stygocyathura) hummelincki: G,* pereopod 1, ♀; *H,* telson. *Cyathura (Stygocyathura) motasi: I,* telson; *J,* pereopod 1, ♀.

Cyathura (Stygocyathura) cuborientalis Botosaneanu and Stock, 1982
  Figure 11C,D

DIAGNOSIS ♀ 6.8 mm. Pereopod 1, propodal palm straight, bearing about

TABLE 1. CARIBBEAN SPECIES OF *Cyathura (Stygocyathura)*, THEIR TOTAL LENGTHS (MM) AND LOCALITIES

C. (S.) cuborientalis	♀ 6.8	Cuba
C. (S.) curassavica	♂ 7.0, ♀ 9.2	Curaçao
C. (S.) hummelincki	♂ 4.75, ♀ 8.5	Aruba
C. (S.) motasi	♂ 6.8, ♀ 10.0	Haiti
C. (S.) orghidani	♀ 8.0	Cuba
C. (S.) parapotamica	ovig. ♀ 3.6, ♀ 4.1	Jamaica
C. (S.) salpiscinalis	♂ 5.6, ♀ 7.3	Haiti
C. (S.) sbordonii	♂, ♀ 9.0	Vera Cruz, Mexico
C. (S.) specus	♂ 18.0, ♀ 19.8	Cuba
C. (S.) univam	♀ 10.0	Venezuela

10 pectinate marginal spines; low triangular ridge present. Pleotelson evenly tapering to notched apex; angle of apex about 90°.

RECORDS   Oriente Province, Cuba, interstitial in river alluvia.

Cyathura (Stygocyathura) curassavica Stork, 1940
Figures 11E,F; 12

DIAGNOSIS   ♂ 7.0 mm, ♀ 9.2 mm. Pereopod 1, propodal palm gently convex, bearing about 10 pectinate spines. Pleotelsonic margins in anterior two-thirds subparallel, tapering gently to finely notched apex; angle of apex less than 90°.

RECORDS   Curaçao, from pits and wells.

Cyathura (Stygocyathura) hummelincki Botosaneanu and Stock, 1982
Figure 11G,H

DIAGNOSIS   ♂ 4.75 mm, ♀ 8.5 mm. Pereopod 1, propodal palm gently sinuate, bearing 11–23 marginal pectinate spines. Pleotelsonic margins faintly concave in midregion, posterior margin evenly convex, apex with slight notch.

RECORDS   Aruba, in pits, wells, and temporary water sources.

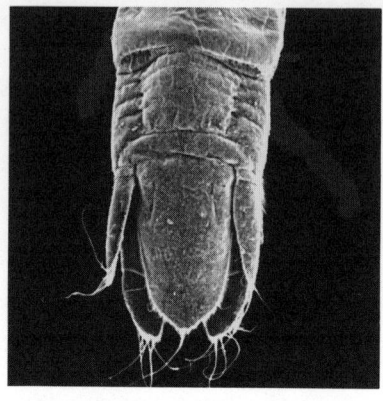

Figure 12. *Cyathura (Stygocyathura) curassavica:* pleon.

Cyathura (Stygocyathura) motasi Botosaneanu and Stock, 1982
Figure 11I,J

DIAGNOSIS ♂ 6.8 mm, ♀ 10.0 mm. Pereopod 1, propodal palm gently sinuate, bearing 11–18 marginal pectinate setae. Pleotelsonic margins tapering evenly to slightly notched apex; angle of apex less than 90°.

RECORDS Haiti, from wells.

Cyathura (Stygocyathura) orghidani Negoescu Vlădescu, 1983
Figure 13A,B

DIAGNOSIS ♀ 8.0 mm. Pereopod 1, propodal palm slightly convex, bearing about 11 marginal pectinate setae. Pleotelson, angle of apex obtuse, with small notch.

RECORDS Pinar del Río Province, Cuba, from freshwater lake in cave.

Cyathura (Stygocyathura) parapotamica Botosaneanu and Stock, 1982
Figure 13D,E

DIAGNOSIS ♀ 4.1 mm (ovig. ♀ 3.6 mm). Pereopod 1, propodal palm sinuate, bearing eight marginal pectinate setae. Pleotelson gently tapering to notched apex; angle of apex about 90°.

RECORDS Jamaica, from river alluvia.

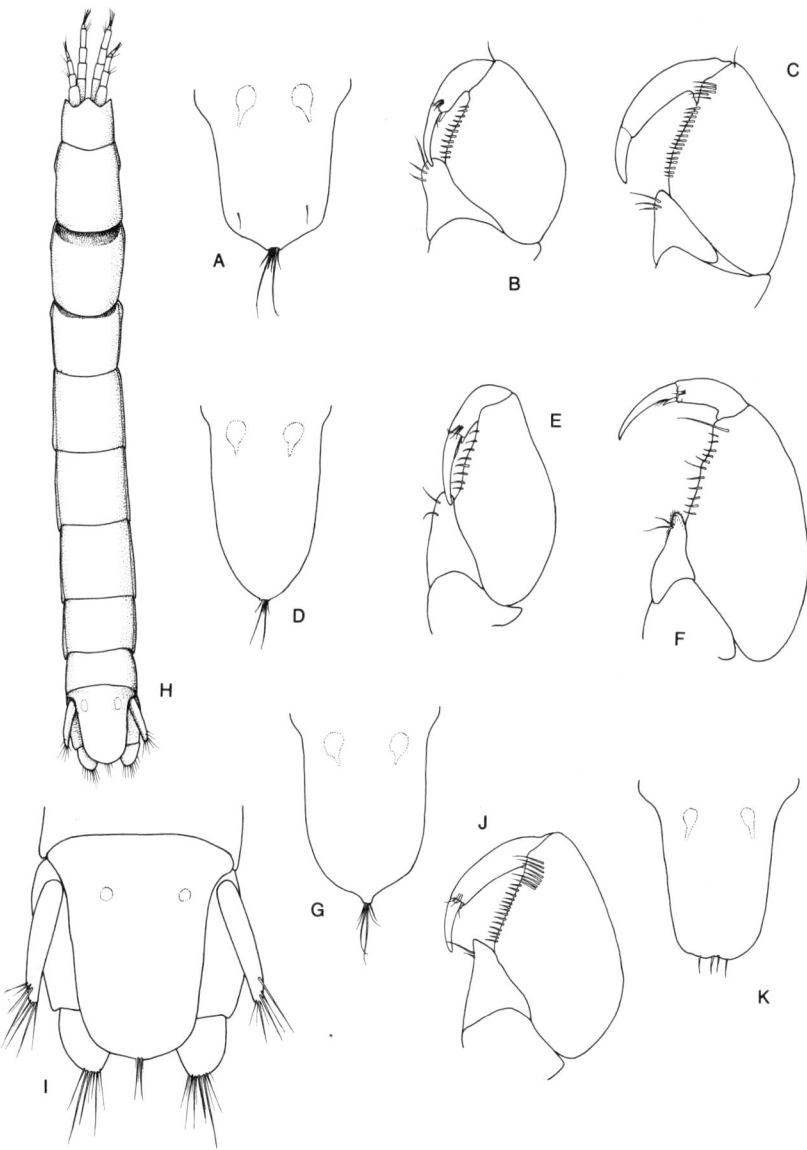

Figure 13. *Cyathura (Stygocyathura) orghidani* (from Negoescu, 1983): *A*, telson; *B*, pereopod 1, ♀. *Cyathura (Stygocyathura) sbordonii: C*, pereopod 1, ♀ (from Argano, 1971). *Cyathura (Stygocyathura) parapotamica: D*, telson; *E*, pereopod 1, ♀. *Cyathura (Stygocyathura) salpicinalis: F*, pereopod 1, ♀; *G*, telson. *Cyathura (Stygocyathura) specus: H*, ♀; *I*, telson and uropods; *J*, pereopod 1, ♀. *Cyathura (Stygocyathura) univam: K*, telson.

Cyathura (Stygocyathura) salpiscinalis Botosaneanu and Stock, 1982
Figure 13F,G

DIAGNOSIS ♂ 5.6 mm, ♀ 7.3 mm. Pereopod 1, propodal palm sinuate, bearing up to 15 marginal pectinate setae, and with distinct triangular ridge. Pleotelson gently tapering, with slight apical eminence.

RECORDS Haiti, from alluvia of lake.

Cyathura (Stygocyathura) sbordonii Argano, 1971
Figure 13C

DIAGNOSIS ♂ and ♀ 9.0 mm. Pereopod 1, propodal palm convex, bearing up to 16 marginal pectinate setae. Pleotelson with angle of apex obtuse.

RECORDS Vera Cruz, Mexico, from freshwater in cave.

Cyathura (Stygocyathura) specus Bowman, 1965
Figure 13H–J

DIAGNOSIS ♂ 18.0 mm, ♀ 19.8 mm. Pereopod 1, propodal palm almost straight, bearing up to 15 marginal pectinate setae. Pleotelson with angle of apex obtuse.

RECORDS Las Villas Province, Cuba, from freshwater lake in cave.

Cyathura (Stygocyathura) univam Botosaneanu, 1983
Figure 13K

DIAGNOSIS ♀ 10.0 mm. Pereopod 1, propodal palm gently convex, bearing 22 marginal pectinate setae. Pleotelson posteriorly broadly rounded, apex emarginate.

RECORDS Peninsula de Morocoy, Venezuela, from phreatic water in cave.

Eisothistos Haswell, 1884

DIAGNOSIS Mouthparts forwardly produced, tailfan spiny, indurate. Eyes present, larger and with more ommatodia in ♂ than in ♀. Pereonites sometimes elongate. Mandible with strong incisor, reduced lamina dentata, palp and molar lacking. Maxilliped lacking endite, palp slender, of three to five articles. Pereopods 1–3 not subchelate, propodi relatively elongate, minimally expanded. Pleopod 1 ♀, rami fused, together forming operculum; ♂ rami separate. Pleonites free, short, longer in ♂ than in ♀. Telson lacking statocysts.

REMARKS   Wägele (1979) first recorded species of *Eisothistos* preying on serpulid polychaete worms in their tubes.

The genus contains about 12 species in the Pacific, Indian Ocean, Caribbean, Antarctic, and Mediterranean.

Key to species of *Eisothistos*

1. ♀ telson with middorsal spines; ♂ pereopod 1 propodal palm with 19 or 20 spines .... *teri*
   ♀ telson lacking middorsal spines; ♂ pereopod 1 propodal palm with 11 spines .... *petrensis*

Eisothistos petrensis Kensley, 1984
    Figure 14A–E

DIAGNOSIS   ♀: 4.0 mm. Antennular flagellum of six articles. Antennal flagellum of six articles. Telson posteriorly faintly bilobed, margin strongly serrate; faint anterior middorsal ridge; middorsally unarmed. Pereopod 1, propodal palm unarmed. Pleopod 1 rami fused for $1/12$ of length. ♂: 2.0 mm. Antennular flagellum of eight articles. Antennal flagellum of six articles. Pereopod 1 propodal palm with 11 fringed spines. Telson narrower than in ♀, with middorsal ridge running almost entire length.

RECORDS   Carrie Bow Cay, Belize, 0.1–36 m; Looe Key, Florida, 5–6 m; Turks and Caicos Islands, 1.0 m; St.Thomas, U.S. Virgin Islands, 7–10 m.

Eisothistos teri Kensley and Snelgrove, 1987
    Figure 14F–H

DIAGNOSIS   ♀: 3.2 mm. Basal antennular peduncle article bearing triangular apically rounded laminate process; flagellum of seven articles. Antennal flagellum of six articles. Mandible with biserrate lamina dentata. Pereopod 1 propodal palm unarmed. Pleopod 1 rami fused for $3/4$ of length. Uropodal exopod with one or two strong slightly recurved spines on dorsal surface.

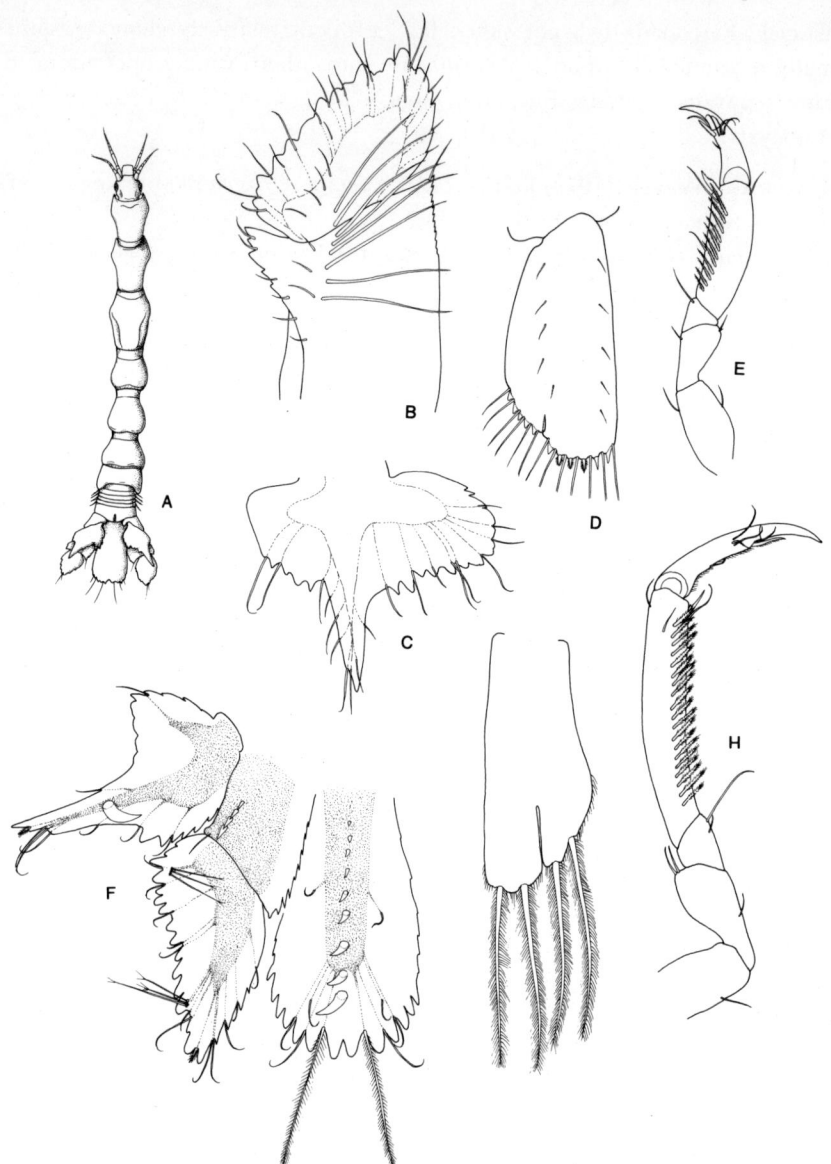

Figure 14. *Eisothistos petrensis:* A, ♀; B, uropodal sympod and endopod; C, uropodal exopod; D, pleopod 1; E, pereopod 1, ♂. *Eisothistos teri:* F, telson and uropod; G, pleopod 1; H, pereopod 1, ♂.

Telson with eight or nine slightly recurved middorsal teeth becoming longer posteriorly. ♂: 2.0 mm. Antennular flagellum of eight articles. Antennal flagellum of six articles. Pereopod 1 propodal palm with 19 or 20 fringed spines. Uropodal exopod lacking dorsal teeth. Telson lacking middorsal teeth; posterior margin incised into 12 acute or narrowly rounded teeth.

RECORDS   Barbados, in *Madracis mirabilis* coral, 9–15 m.

Haliophasma Haswell, 1881

DIAGNOSIS   Eyes present. Integument often indurate, with scattered pitting. Antennular flagellum usually of two articles. Antennal flagellum of 4–7 articles. Mandibular palp of three articles. Maxillipedal palp of two articles, article 2 smaller than article 1. Pereopod 1, propodus expanded. Pereopods 4–7, carpi roughly rectangular. Pleopod 1 exopod operculiform. Pleonites 1–5 short, fused; pleonite 6 usually demarked from telson. Latter with two basal statocysts. ♂ often with more elongate form than ♀. Antennular flagellum multiarticulate. Eyes larger.

Key to species of *Haliophasma*

1. Telson posteriorly narrowly rounded; dactylus of pereopod 1 dentate
   .... *valeriae*
   Telson posteriorly broadly rounded; dactylus of pereopod 1 entire
   .... *curri*

Haliophasma curri Paul and Menzies, 1971
   Figure 15A–C

DIAGNOSIS   ♀ 7.0 mm. Antennular flagellum of three articles. Antennal flagellum of five articles. Mandibular palp article 3 with five spines. Maxillipedal palp, terminal article small, set obliquely at distolateral angle of article 1. Pereopod 1, carpus triangular, posterodistal margin crenulate, distally rounded; propodal palm crenulate, with low rounded proximal lobe. Pleonite 6 dorsally demarked, narrow, with middorsal point in posterior margin. Uropodal exopod elongate, outer margin sinuate, dentate; endopod ovate, distally narrowed, length twice greatest width, outer margin dentate. Telson parallel sided, posterior margin broadly rounded.

RECORDS   Off Venezuela, 95 m; Culebra Island, Bay of Panama, intertidal.

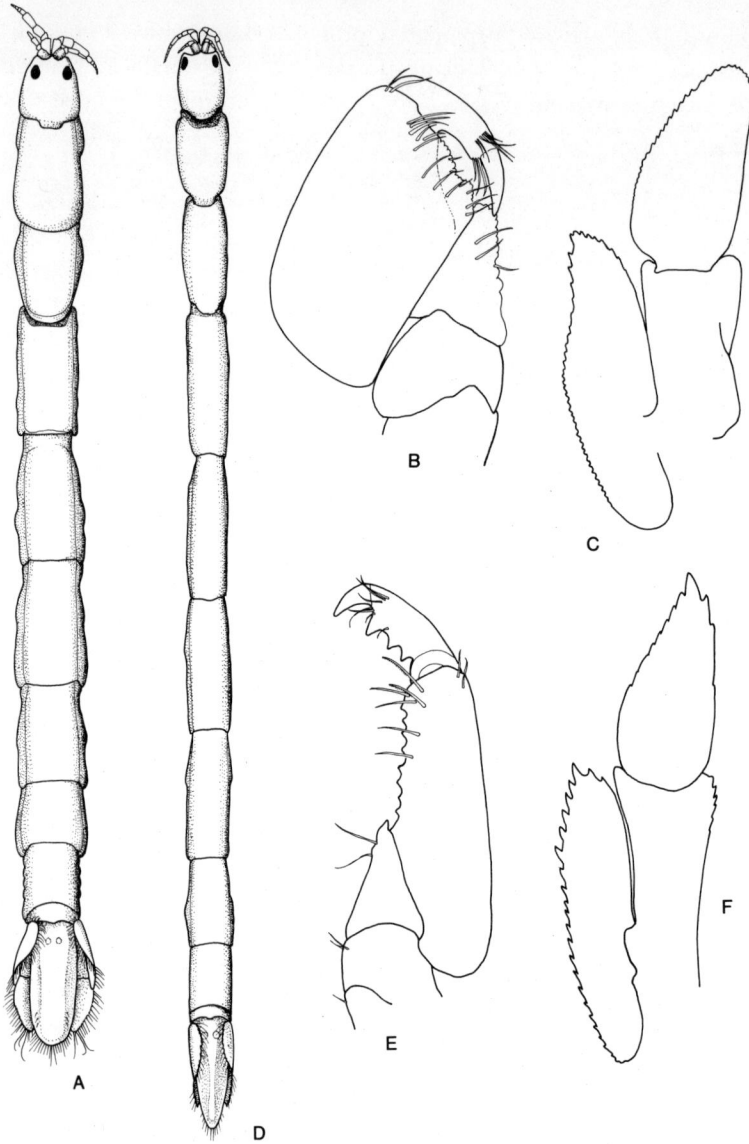

Figure 15. *Haliophasma curri:* A, ♀; B, pereopod 1; C, uropod. *Haliophasma valeriae:* D, ♀; E, pereopod 1; F, uropod.

Haliophasma valeriae Paul and Menzies, 1971
Figure 15D–F

DIAGNOSIS ♀ 6.5 mm. Body long and slender. Antennular flagellum of three articles. Antennal flagellum of six articles. Mandibular palp article 3 with four spines. Pereopod 1, carpus triangular, tipped with acute tooth; propodus elongate, palm with about seven teeth, five fringed spines on mesial surface; unguis of dactylus strongly flexed; margin of dactylus with three strong triangular teeth. Uropodal exopod elongate, apically acute, margins serrate; endopod length little more than twice greatest width, outer margin serrate, apex acute. Telson elongate-elliptical, apically narrowly rounded; strong middorsal longitudinal rounded ridge running almost entire length.

RECORDS Off Venezuela, 95 m.

REMARKS Wägele (1981) made this species the type of his new genus *Nemanthura*, based primarily on the elongate form of the body and appendages. *Haliophasma irmae* Paul and Menzies, 1971, from the same locality as the above species, is probably the same species.

Licranthura Kensley and Schotte, 1987

DIAGNOSIS Serrate process on antennal peduncle article 3. Mandibular palp of three articles; molar lacking. Maxillipedal endite short. Pereopod 1 larger than pereopods 2 and 3. Pereopods 4–7, carpi triangular. Pleonites short, free. Pleopod 1, both rami forming operculum. Pleotelson lacking statocysts.

Licranthura amyle Kensley and Schotte, 1987
Figure 16

DIAGNOSIS ♀ 3.8 mm. Eyes small, pigmented. Antennular flagellum of three articles. Antennal peduncle article 3 with lamellar expanded process, serrate on mesial margin; flagellum of six articles. Maxillipedal palp of five articles; very short endite. Pereopod 1, propodal palm unarmed. Uropodal and telsonic margins serrate.

RECORDS Carrie Bow Cay, Belize, 0–25 m, in coral rubble.

Malacanthura Barnard, 1925

DIAGNOSIS Eyes present. Mandibular palp of three articles. Maxillipedal palp of three articles, terminal article usually broadly ovate. Pereopod 1, propodus expanded. Pereopods 4–7 with carpi roughly rectangular. Pleopod

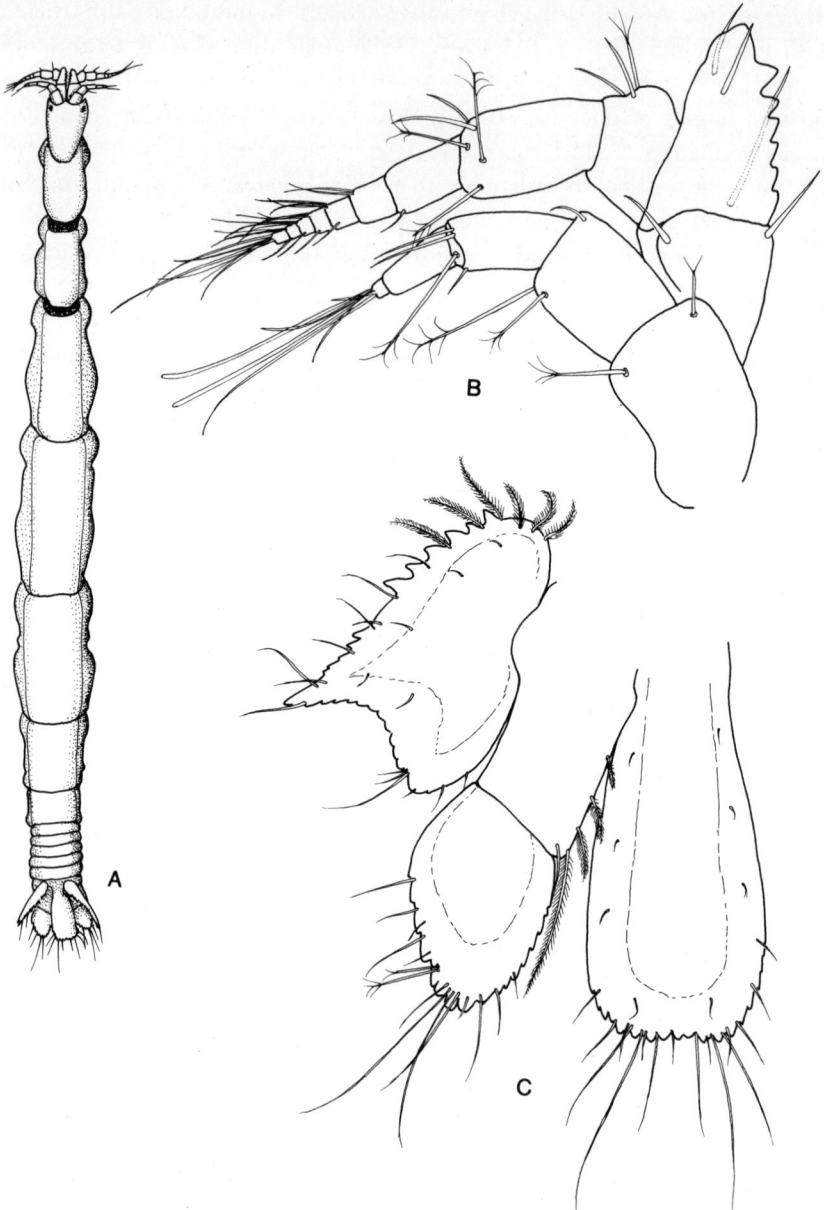

Figure 16. *Licranthura amyle:* A, ♀; B, antennule and antenna; C, telson and uropod.

1, exopod operculiform. Pleonites 1–5 short, fused; pleonite 6 dorsally demarked. Pleotelson with two basal statocysts.

Malacanthura caribbica Paul and Menzies, 1971
Figure 17

DIAGNOSIS ♀ 27.1 mm. Integument moderately indurate. Antennular flagellum of seven articles. Antennal flagellum of four articles. Mandibular palp, article 3 with comb of 11 spines. Maxillipedal palp, terminal article broadly ovate, penultimate article with row of seven spines on mesial margin. Pereopod 1 propodus expanded, palm straight, with few spines on mesial margin. Uropodal exopod barely reaching base of endopod, narrow, apically acute, outer margin sinuate, serrate; endopod set obliquely on sympod, margin serrate, apically acute. Telson lanceolate, apically narrowly rounded, with strong longitudinal middorsal carina.

RECORDS Off Venezuela, 95 m; off Colombia, 42–44 m.

REMARKS *Malacanthura cumanensis* Paul and Menzies, 1971, described from the same locality as *M. caribbica*, was shown to be the latter species (Kensley, 1980).

Mesanthura Barnard, 1914

DIAGNOSIS ♀: Dorsal integument with (usually) species-specific pigment pattern; pigment persistent in alcohol. Mandibular palp of three articles, terminal article with row of spines, number of which specific for species. Maxilliped with endite either very reduced or absent; palp of three articles, with terminal article usually about half length of penultimate article, suture transverse. Pereopod 1, propodus expanded, palm often with step. Pereopods 2 and 3, propodi not expanded. Pereopods 4–7, carpi roughly triangular, with anterior margin shorter than posterior margin. Pleonites 1–5 fused, pleonite 6 dorsally demarked. Pleopod 1, exopod operculiform. Telson with two basal statocysts. ♂: Eyes larger than in ♀. Antennular flagellum of seldom more than 10 articles bearing numerous aesthetascs. Mouthparts, especially body of mandible, reduced. Pereopod 1, propodus bearing dense band of spines on mesial surface near palm. Pigment pattern more diffuse than in ♀, extending onto ventral surface.

REMARKS *Mesanthura* is a relatively large genus of about 30 species, recorded from most tropical and temperate seas, in shallow habitats. The males of few species have been recorded; by themselves, males are difficult to identify as the dorsal pigment pattern characteristic of the female breaks down and spreads onto the ventrum.

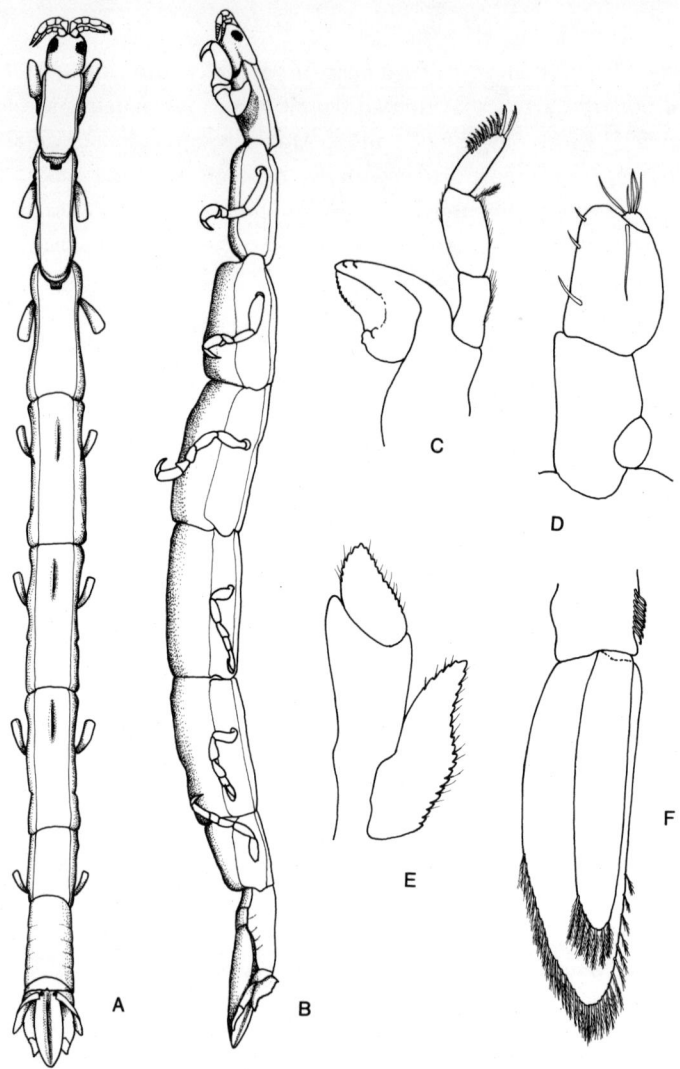

Figure 17. *Malacanthura caribbica*: A, ♀; B, ♀, lateral view; C, mandible; D, maxilliped; E, uropod; F, pleopod 1.

Key to species of *Mesanthura* (♀ only)

1. Pigment in tiny evenly scattered chromatophores over body; cephalon with solid patch of pigment; mandibular palp article 3 with seven spines .... *punctillata*
   Pigment not evenly scattered over body 2

2. Pereonites 4 and 5 with patch of pigment, sometimes with unpigmented area in middle of patch 3
   Pereonites 4 and 5 lacking fairly solid patch of pigment 6

3. Pereonites 4 and 5 lacking unpigmented area in pigment patch; five transverse lines on pigment on pleon; mandibular palp article 3 with six spines .... *paucidens*
   Pereonites 4 and 5 with unpigmented area in pigment patch; pleon lacking transverse pigment lines 4

4. Pigment in obvious double longitudinal bands on pereon and pleon; mandibular palp article 3 with nine spines .... *bivittata*
   Pigment not in obvious double bands 5

5. Unpigmented area in middle of pigment patch of pereonites 1–3; mandibular palp article 3 with 10 spines .... *pulchra*
   No unpigmented area in middle of pigment of pereonites 1–3; mandibular palp article 3 with eight spines .... *looensis*

6. Pigment of pereon in fine reticulate lines; mandibular palp article 3 with six spines .... *reticulata*
   Pigment of pereon not in fine reticulate lines; mandibular palp article 3 with four spines 7

7. Pigment in more or less complete rings on pereonites 1–6 ..... *hopkinsi*
   Pigment in strong transverse posterior bars on pereonites 4–7 .. *fasciata*

Mesanthura bivittata Kensley, 1987a
  Figures 18A, 20A–D

DIAGNOSIS ♂ 5.2 mm, ovigerous ♀ 7.8 mm. Pigment in obvious double longitudinal bands on pereon and pleon. Mandibular palp article 3 with nine spines. Maxilliped lacking endite. Pereopod 1, propodal palm with rounded lobe.

RECORDS Twin Cays, Belize, under red mangroves, 1–2 m.

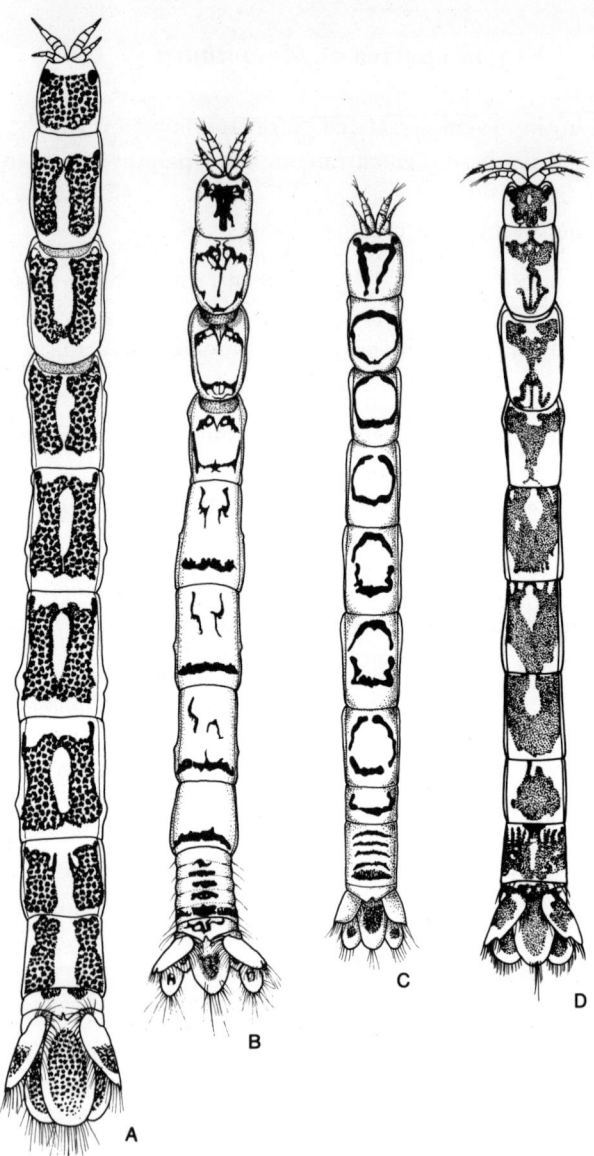

Figure 18. A, *Mesanthura bivittata;* B, *Mesanthura fasciata;* C, *Mesanthura hopkinsi;* D, *Mesanthura looensis.*

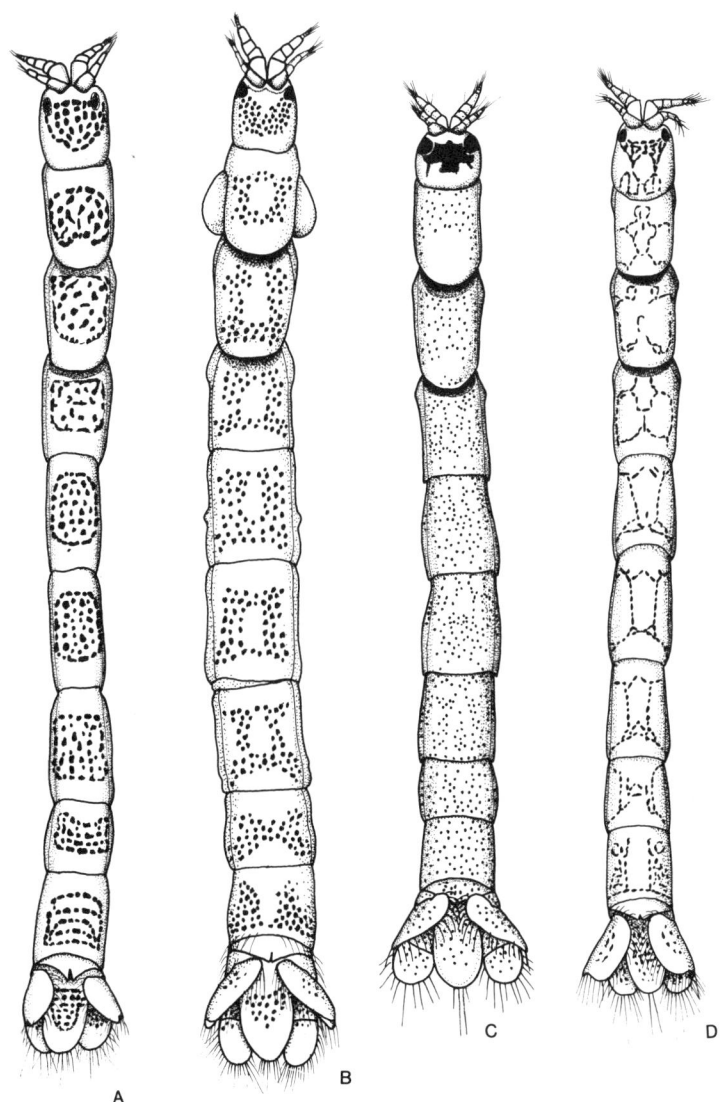

Figure 19. A, *Mesanthura paucidens;* B, *Mesanthura pulchra;* C, *Mesanthura punctillata;* D, *Mesanthura reticulata.*

Mesanthura fasciata Kensley, 1982
Figures 18B, 20E–H

DIAGNOSIS ♀ 4.5 mm. Pigment in triangular patch on head, open irregular rings on pereonites 1–3; pereonites 4–7 with strong transverse posterior bar;

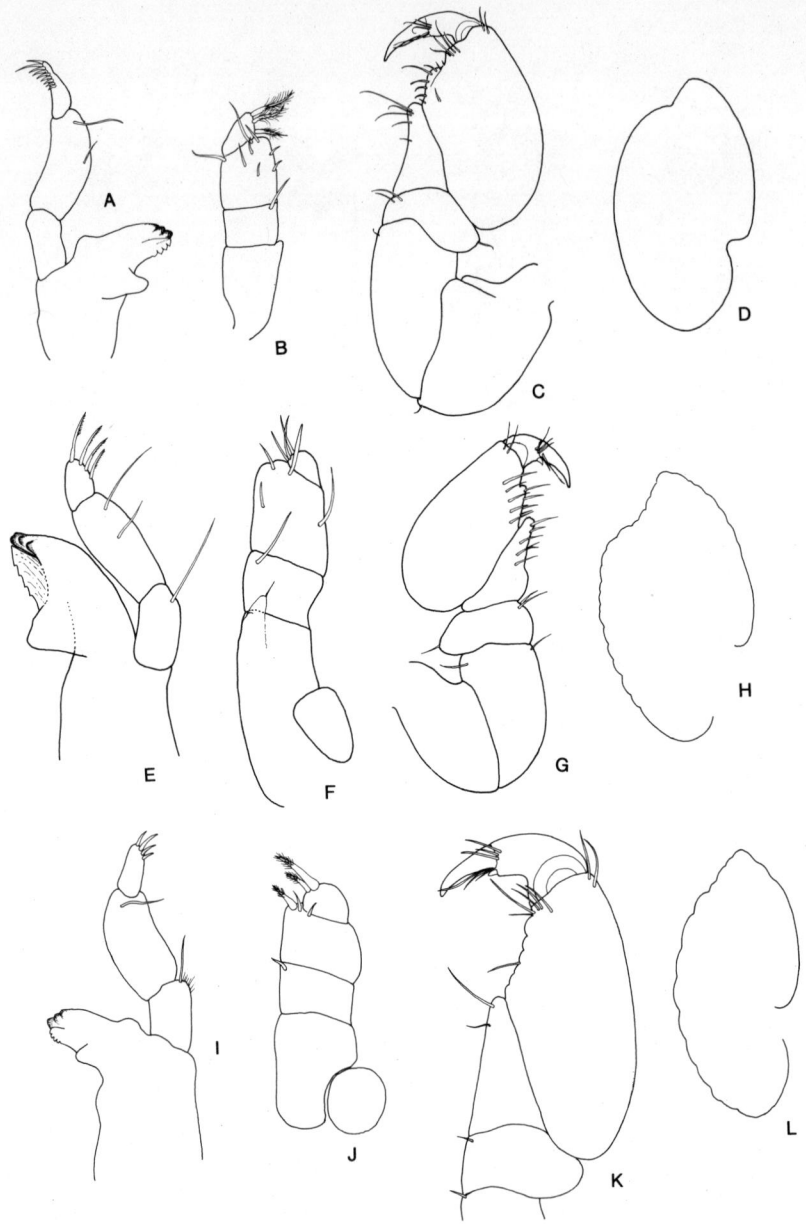

Figure 20. *Mesanthura bivittata:* A, mandible; B, maxilliped; C, pereopod 1; D, uropodal exopod. *Mesanthura fasciata:* E, mandible; F, maxilliped; G, pereopod 1; H, uropodal exopod. *Mesanthura hopkinsi:* I, mandible; J, maxilliped; K, pereopod 1; L, uropodal exopod.

five bars on pleon. Mandibular palp article 3 with four spines. Maxilliped with very reduced endite. Pereopod 1, propodal palm with step.

RECORDS   Looe Key, Florida, 5–6 m; Cozumel, Mexico; Jamaica; Carrie Bow Cay, Belize, 0.2–6 m.

Mesanthura hopkinsi Hooker, 1985
   Figures 18C, 20I–L

DIAGNOSIS   ♀ 2.4 mm. Pigment in triangle on cephalon; in irregular rings on pereonites 1–6; four transverse bars on pleon. Mandibular palp article 3 with four spines. Maxilliped lacking endite. Pereopod 1, palm of propodus lacking step.

RECORDS   Looe Key, Florida, 0.5 m; Florida Middlegrounds, Gulf of Mexico, 55 m.

Mesanthura looensis Kensley and Schotte, 1987
   Figures 18D, 21A–D

DIAGNOSIS   ♀ 10.0 mm. Pigment in solid patches on cephalon, pereonites, pleon, uropods and telson; pereonites 4–6 with open central area in patch. Mandibular palp article 3 with eight spines. Maxilliped lacking endite. Pereopod 1, propodal palm with step.

RECORDS   Looe Key, Florida, 1 m.

Mesanthura paucidens Menzies and Glynn, 1968
   Figures 19A, 21E–I

DIAGNOSIS   ♀: 6.6 mm. Pigment in roughly rectangular to ovate patches on cephalon and pereonites; in five transverse lines connected laterally on pleon. Mandibular palp article 3 with six spines. Maxilliped with short narrow endite. Pereopod 1 propodal palm with step. ♂: 6.4 mm. Antennular flagellum of seven articles. Pigment more diffuse than in ♀ but retaining five pleonal bars.

RECORDS   Looe Key, Florida, 5–6 m; Carrie Bow Cay, Belize, 15.2 m; Puerto Rico, intertidal; Jamaica.

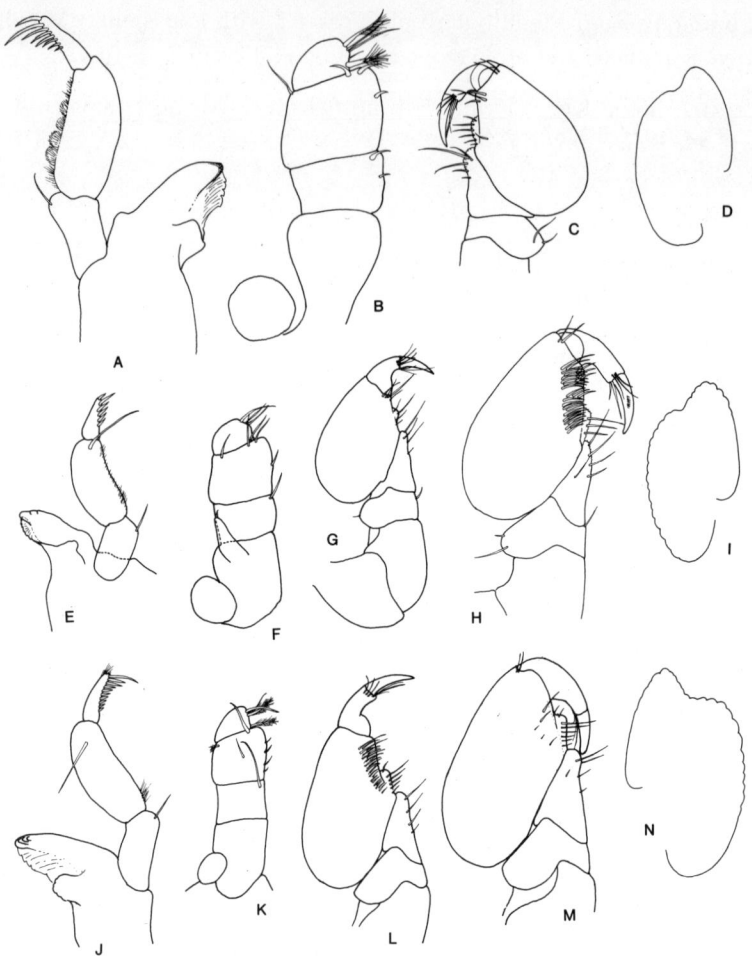

Figure 21. *Mesanthura looensis:* A, mandible; B, maxilliped; C, pereopod 1; D, uropodal exopod. *Mesanthura paucidens:* E, mandible; F, maxilliped; G, pereopod 1, ♀; H, pereopod 1, ♂; I, uropodal exopod. *Mesanthura pulchra:* J, mandible; K, maxilliped; L, pereopod 1, ♂; M, pereopod 1, ♀; N, uropodal exopod.

Mesanthura pulchra Barnard, 1925
Figures 19B, 21J–N

DIAGNOSIS ♀: 9.3 mm. Pigment pattern in roughly rectangular patches on cephalon and pereonites 1–6, with open oval middorsal area in patch. Mandibular palp article 3 with 10 spines. Maxilliped lacking endite. Pereopod 1, propodal palm with step. ♂: 5.4 mm. Antennular flagellum of eight articles.

RECORDS   Egmont Key, Florida, 18.3-36.6 m; Looe Key, Florida, 0.5-12 m; Dry Tortugas; Turks and Caicos Islands, 1 m; Puerto Rico, intertidal to 1.5 m; Carrie Bow Cay, Belize, intertidal to 1.5 m; St.Thomas and St.John's, U.S. Virgin Islands; Cozumel, Mexico.

REMARKS   Menzies and Glynn (1968) recorded this species as *M. decorata* from Puerto Rico, while Menzies and Kruczynski (1983) recorded it as *M. floridensis*.

Mesanthura punctillata Kensley, 1982
    Figures 19C, 22A-F

DIAGNOSIS   ♀: 6.4 mm. Pigment in solid patch between eyes, rest of pigment on pereon and pleon with chromatophores scattered, diffuse, not in any regular pattern. Mandibular palp article 3 with seven spines. Maxilliped lacking endite; terminal palp article semicircular. Pereopod 1 propodal palm with step. ♂: 4.5 mm. Antennular flagellum of 10 articles.

RECORDS   Turks and Caicos Islands, 1 m; Carrie Bow Cay, 0.2-20 m.

Mesanthura reticulata Kensley, 1982
    Figures 19D, 22G-J

DIAGNOSIS   ♀ 6.1 mm. Dorsal pigment pattern a network of chromatophores arranged in fine lines. Mandibular palp article 3 with six spines. Maxilliped lacking endite. Pereopod 1 propodal palm with step.

RECORDS   Carrie Bow Cay, 10-24 m.

Minyanthura Kensley, 1982

DIAGNOSIS   Mandible lacking molar and palp. Maxillipedal palp of five articles; large apically acute endite present. Pereopods 4-7, carpi rectangular. Pleopod 1, both rami forming operculum. Pleonites 1-5 fused; pleonite 6 fused with telson, not dorsally demarked. Telson with two basal statocysts.

Minyanthura corallicola Kensley, 1982
    Figure 23

DIAGNOSIS   ♀: 1.7 mm. Antennular flagellum of one article. Antennal

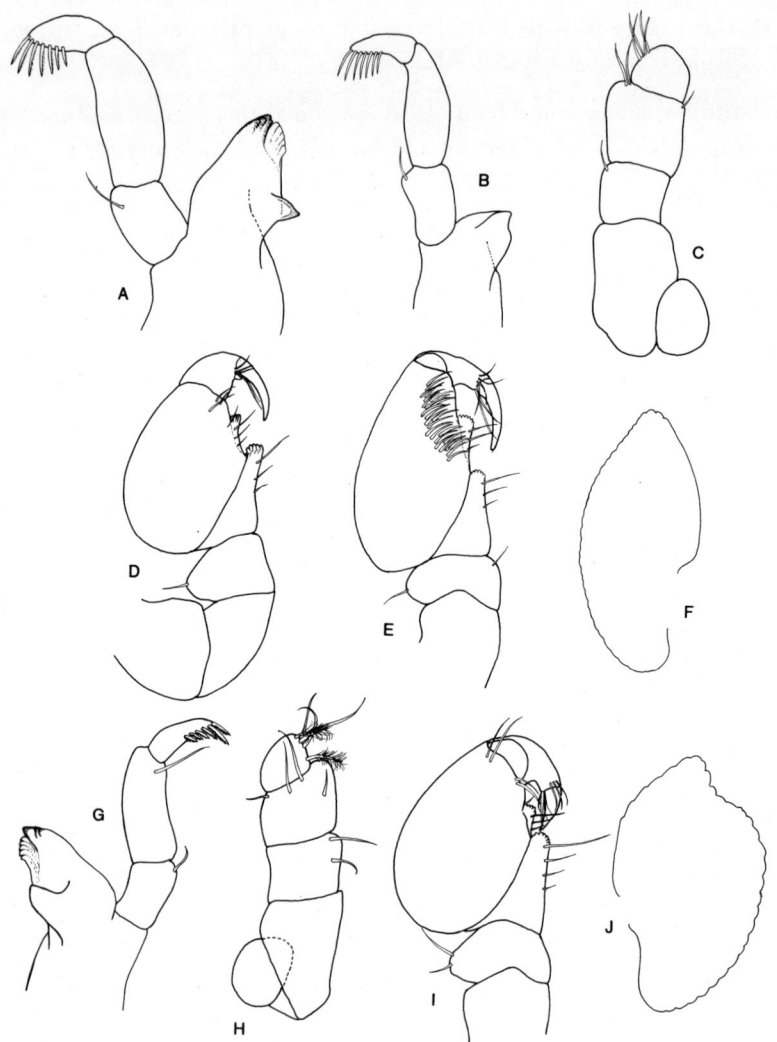

Figure 22. *Mesanthura punctillata:* A, mandible, ♀; B, mandible ♂; C, maxilliped; D, pereopod 1, ♀; E, pereopod 1, ♂; F, uropodal exopod. *Mesanthura reticulata:* G, mandible; H, maxilliped; I, pereopod 1; J, uropodal exopod.

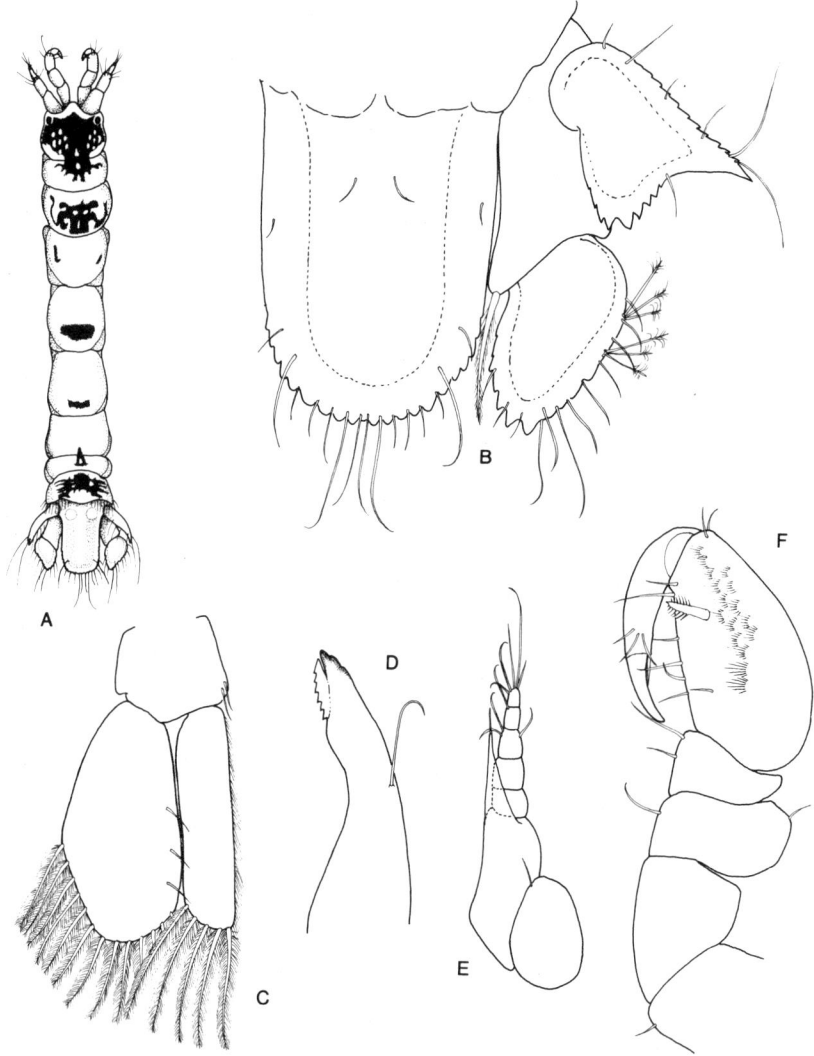

Figure 23. *Minyanthura corallicola:* A, ♀; B, telson and uropod; C, pleopod 1; D, mandible; E, maxilliped; F, pereopod 1.

flagellum of four articles. Pereopod 1, propodus somewhat inflated. Uropodal rami and posterior telsonic margin serrate. ♂: 1.3 mm. Eyes larger than in ♀. Antennular flagellum of three articles. Antennal flagellum of two articles.

RECORDS   Carrie Bow Cay, Belize, 6–24 m; Barbados, 9–15 m; Jamaica.

Pendanthura Menzies and Glynn, 1968

DIAGNOSIS ♀: Integument with some red-brown dorsal pigmentation. Eyes small, pigmented. Antennular flagellum of two articles. Antennal flagellum of one article. Mandibular palp of single reduced article; incisor, molar, and lamina dentata present. Maxillipedal palp of single broad article; small triangular endite present. Pereopod 1, propodus expanded. Pereopods 4–7, carpi short, triangular, lacking free anterior margin. Pleonites 1–5 very short, fused, 6 fused with telson. Pleopod 1, exopod operculiform. Telson basally broad, with two statocysts at about midlength. ♂: Antennular peduncle of four articles. Antennal flagellum of one article. Pereopod 1, propodus with dense clump of spines on mesial surface. Eyes only slightly larger than in ♀.

REMARKS The genus comprises three species, two from the Caribbean and one from the Pacific, all of which have been taken from shallow coral reefs.

Key to species of *Pendanthura*

1. Dorsal pigmentation over entire body; pereopod 1, propodal palm with rounded lobe .... *tanaiformis*
   Dorsal pigmentation on cephalon, pereonite 2, and pleon; pereopod 1, propodal palm lacking rounded lobe .... *hendleri*

Pendanthura hendleri Kensley, 1984
Figure 24A–E

DIAGNOSIS ♀: 3.3 mm. Dorsal pigmentation limited to cephalon, pereonite 2, and very short pleon. Reduced mandibular palp with two setae. Pereopod 1, propodus expanded, palm gently convex, lacking rounded lobe, with four spines on mesial surface near palmar margin. ♂: 2.8 mm. Pigmentation as in ♀. Pereopod 1, propodal palm gently convex, lacking rounded lobe.

RECORDS Carrie Bow Cay, Belize, 9–23 m; Twin Cays, Belize, 0–2 m; Panama, 30 m.

Pendanthura tanaiformis Menzies and Glynn, 1968
Figure 24F–H

DIAGNOSIS Ovigerous ♀: 2.9 mm. Dorsal pigmentation a dense red-brown

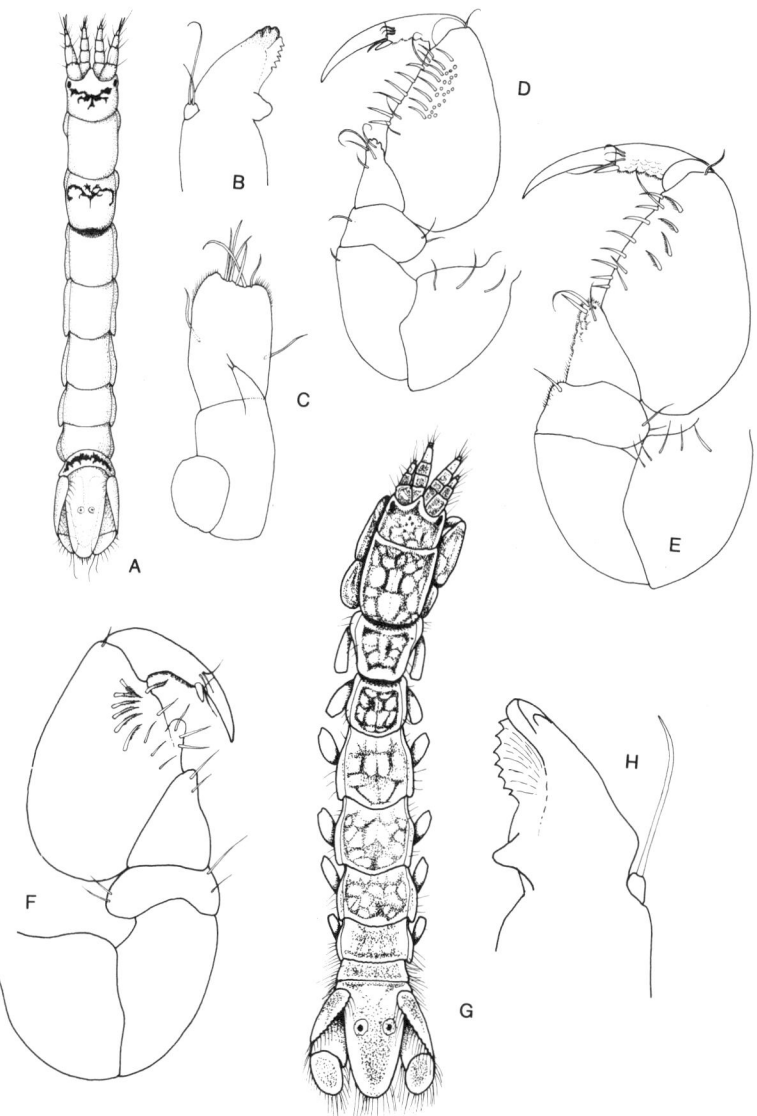

Figure 24. *Pendanthura hendleri:* A, ♀; B, mandible; C, maxilliped; D, pereopod 1, ♂; E, pereopod 1, ♀. *Pendanthura tanaiformis:* F, pereopod 1, ♀; G, ♀; H, mandible.

reticulation over entire dorsum. Mandibular palp with one seta. Pereopod 1, propodus expanded, palm with rounded lobe, mesial surface with six spines near palmar margin. ♂: 2.8 mm. Pigmentation as in ♀. Propodal palm with rounded lobe.

RECORDS   Bermuda; Carrie Bow Cay, Belize, intertidal to 1 m; Puerto Rico, intertidal; Cozumel, Mexico.

REMARKS   Kensley (1984) characterized this common species of the intertidal of the reef crest as stress-tolerant, breeding throughout the year.

Skuphonura Barnard, 1925

DIAGNOSIS   Antennal flagellum of single article. Cephalon with midventral toothlike process at base of mouthparts. Mandibular palp of three articles. Maxillipedal palp of three articles; endite lacking. Pereopod 1, propodus expanded. Pereopods 4–7, carpi triangular. Pleopod 1, exopod operculiform. Pleonites 1–5 fused; pleonite 6 dorsally demarked. Pleotelson with two basal statocysts.

Skuphonura laticeps Barnard, 1925
   Figure 25

DIAGNOSIS   ♂ 6.0 mm. Cephalon wider anteriorly than posteriorly, with anterolateral lobes extending well beyond rostrum. Antennular flagellum of two articles. Pereonite 1 with strong midventral forwardly directed tooth. Pereopod 1, carpus with posterodistal angle produced into triangular lobe; propodus expanded, palmar margin proximally convex, numerous setae on mesial surface. Pleopod 2, copulatory stylet of endopod reaching beyond rami. Pleonite 6 dorsally demarked, posterior margin middorsally incised. Uropodal exopod ovate, with distal notch; endopod length slightly more than basal width. Pleotelson widest at midlength, posteriorly narrowly rounded, with broadly rounded longitudinal raised area.

RECORDS   St. Thomas, U.S. Virgin Islands, 8–40 m.

Family Hyssuridae Wägele, 1981

DIAGNOSIS   Maxillipedal palp usually of five articles; endite present. Pereopods 1–3 subsimilar, often all three pairs subchelate. Pleopod 1 similar to

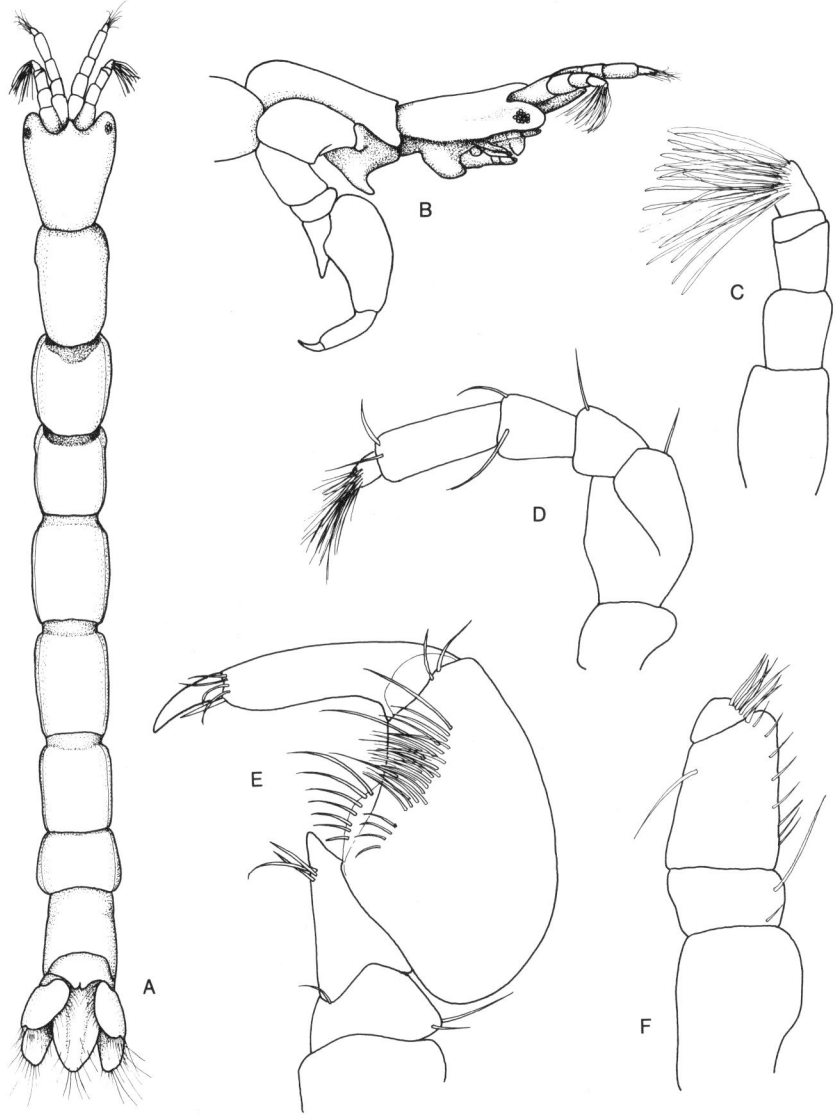

Figure 25. *Skuphonura laticeps:* A, ♂; B, cephalon and pereonite 1, lateral view; C, antennule, ♂; D, pereopod 1, ♂; E, antenna; F, maxilliped.

following pleopods, not operculiform. Pleonites 1–5 freely articulating, relatively elongate. Pleotelson lacking statocysts.

Key to genera of Hyssuridae

1. Pleotelson not covered by uropodal exopods; maxillipedal rami basally free .... *Kupellonura*
   Pleotelson covered completely by uropodal exopods; maxillipedal rami basally fused .... *Xenanthura*

Kupellonura Barnard, 1925

DIAGNOSIS   Mandibular palp of three articles. Maxillipedal palp of five articles; large endite present. Pereopods 1–3 similar, propodi somewhat expanded. Pereopods 4–7, carpi triangular. Pleonites 1–5 elongate, free, subequal. Pleopods 1–5 similar.

Kupellonura imswe (Kensley, 1982)
   Figures 26, 27

DIAGNOSIS   ♀: 3.4 mm. Eyes present. Antennular flagellum of four articles. Antennal flagellum of seven articles. Maxillipedal endite reaching to base of article 3. Pereopods 1–3 similar; pereopod 1, carpus posterodistally acute; propodus expanded, palm straight, unarmed; carpus triangular, free anterior margin shorter than posterior. Pleonite 6 with middorsal incision in posterior margin. Uropodal exopod, outer margin serrate, distally narrowly rounded; endopod length twice greatest width. Telson widest at midlength, posterolateral margin serrate, apically broadly rounded. ♂: 2.7 mm. Eyes larger than in ♀. Mouthparts reduced. Pereopod 1, propodal palm armed with row of eight fringed spines.

RECORDS   Carrie Bow Cay, Twin Cays, Glover's Reef, Belize, 0–6.0 m; Montego Bay, Jamaica, 1 m.

Xenanthura Barnard, 1925

DIAGNOSIS   Mandibular palp of single article. Maxillipedal rami fused; articulations of palp articles obscure. Pereopods 1–3 similar. Pereopods 4–7, carpi triangular.

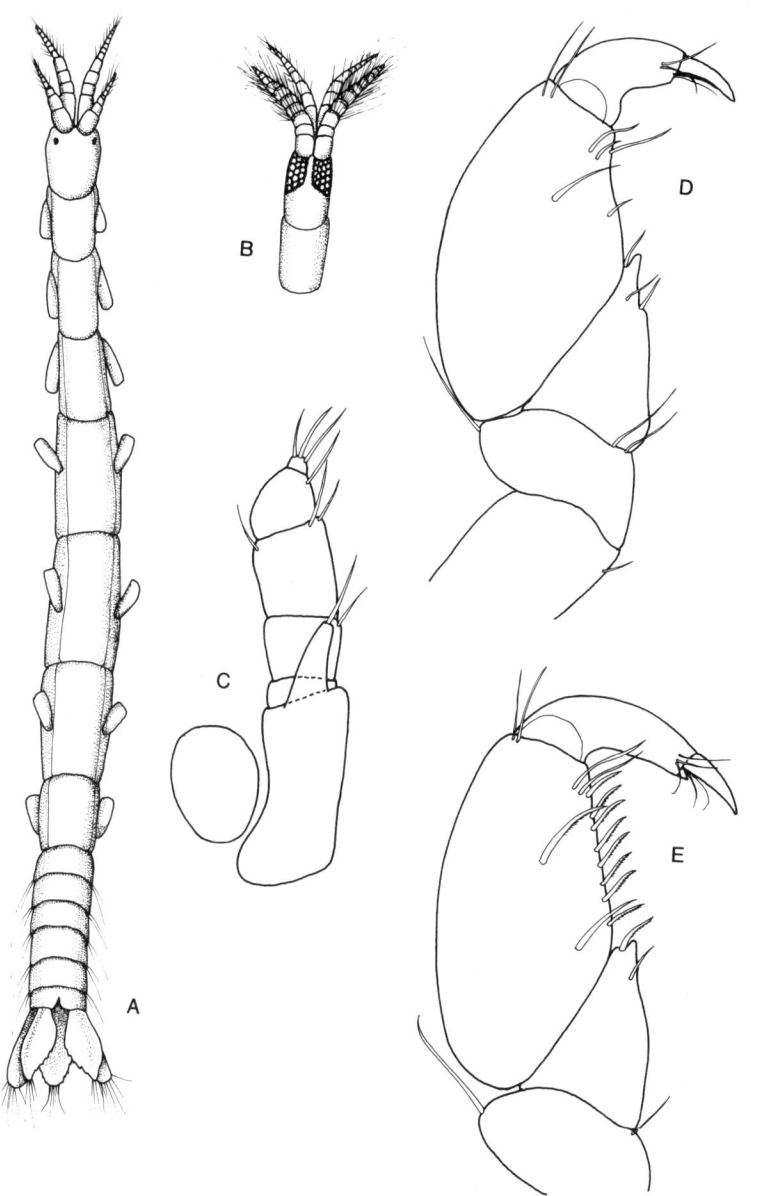

Figure 26. *Kupellonura imswe:* A, ♀; B, cephalon and pereonite 1, ♂; C, maxilliped; D, pereopod 1, ♀; E, pereopod 1, ♂.

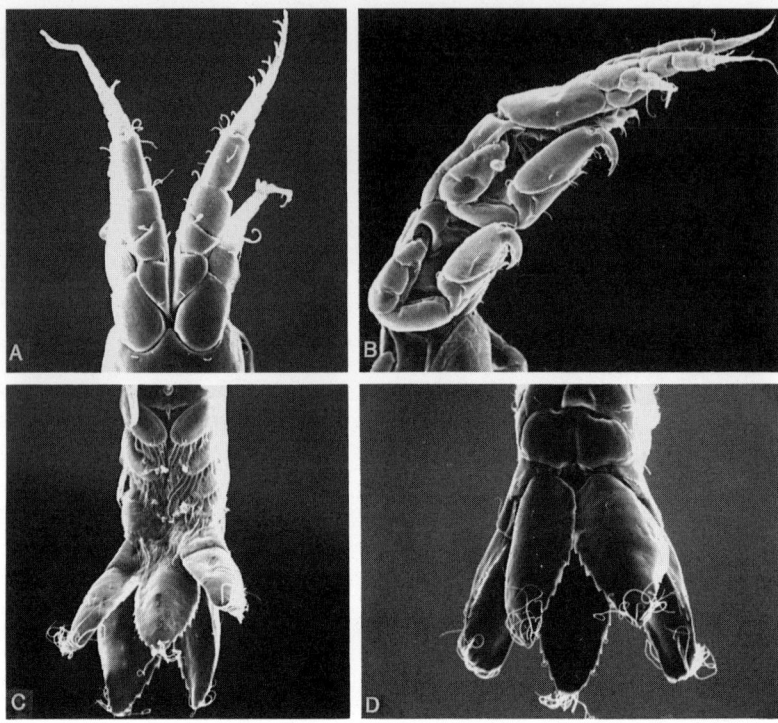

Figure 27. *Kupellonura imswe*: A, ♀ cephalon, dorsal view; B, posterior pleon, dorsal view; C, pleon, ventral view; D, cephalon and pereonites 1 and 2, lateral view.

Xenanthura brevitelson Barnard, 1925
Figure 28

DIAGNOSIS   ♀: 4.0 mm. Ommatidia of eyes in longitudinal row of three or four groups. Antennular flagellum of three articles. Antennal flagellum of three articles. Mandibular palp of single short article bearing single seta. Maxillipedal rami basally fused, palp of three or four obscurely separated articles. Pereopods 1–3 similar, carpus of pereopod 2 triangular, with strong triangular projecting lobe posterodistally; propodi expanded, palm incised into several rounded lobes. Uropodal exopods circular, overlapping dorsally and covering telson; endopod projecting beyond exopods ventrally, mesial margin with step, distally rounded. Telson shorter than uropodal exopod, posterior margin truncate to faintly concave. ♂: 3.5 mm. Antennular flagellum of seven articles.

RECORDS   Off Georgia, 20–145 m; off Florida, 8–10 m; Turks and Caicos Islands, 1 m; St.Thomas, U.S. Virgin Islands, 50–60 m; Gulf of Mexico.

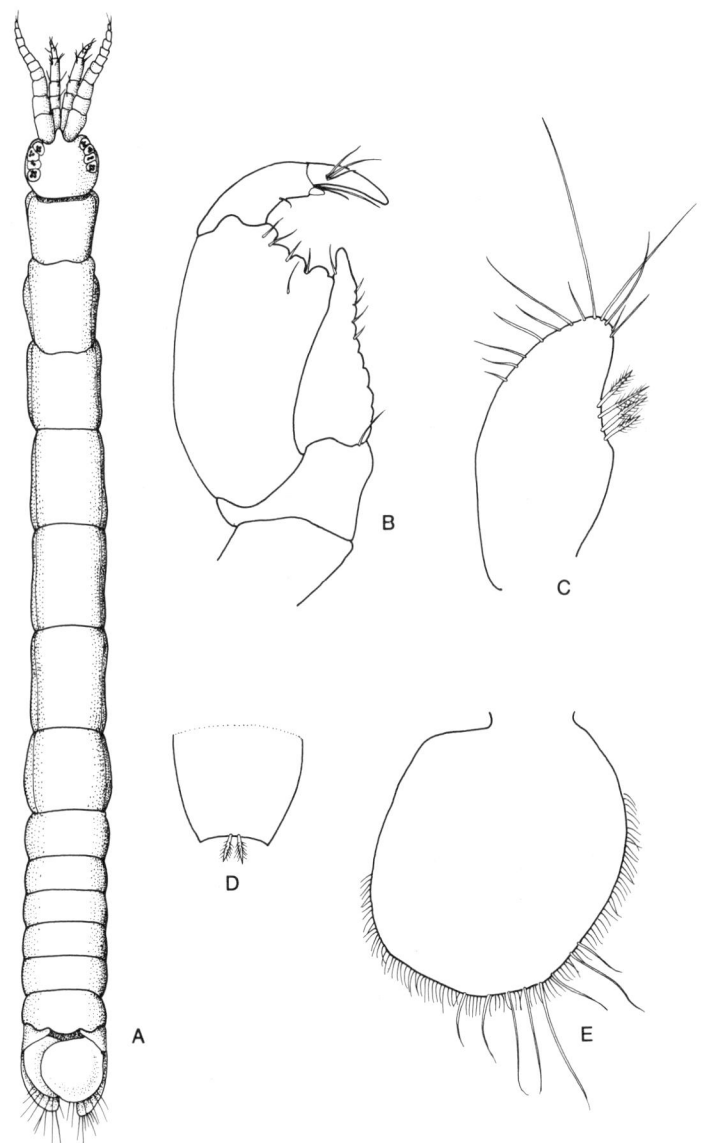

Figure 28. *Xenanthura brevitelson:* A, ♀; B, pereopod 1; C, uropodal endopod; D, telson; E, uropodal exopod.

Family Paranthuridae Menzies and Glynn, 1968

DIAGNOSIS   Mouthparts together forming somewhat elongate cone adapted for piercing and sucking. Mandible with styliform incisor, lacking lamina dentata and molar. Maxilla slender, styliform, distally serrate. Maxilliped elongate; number of palp articles usually reduced. Pereopod 1, or pereopods 1-3 subchelate. Pleopod 1 exopod operculiform. Pleonites short, free or fused. Pleotelson with single basal statocyst, or lacking statocyst.

Key to genera of Paranthuridae

1. Pereopod 7 lacking in adult 2
   Pereopod 7 present in adult 3
2. Eyes present .... *Colanthura*
   Eyes absent .... *Curassanthura*
3. Antennular and antennal flagellum of more than 10
      articles .... *Accalathura*
   Antennular and antennal flagella with fewer than 10 articles 4
4. Antennal flagellum a single (rarely two or 3) flattened article;
      maxillipedal palp of one or two articles .... *Paranthura*
   Antennal flagellum of seven articles; maxillipedal palp of three articles
      .... *Virganthura*

Accalathura Barnard, 1925

DIAGNOSIS   Eyes present. Antennular and antennal flagella multiarticulate, each of more than 10 articles. Mandibular palp of three articles. Maxillipedal

Key to species of *Accalathura*

1. Uropodal exopod elongate-narrow; endopod length twice basal width
      .... *crenulata*
   Uropodal exopod ovate, apically subacute; endopod length 1.5 times
      basal width .... *setosa*

palp of two articles, endite almost reaching end of palp. Pereopod 1 subchelate, propodus inflated, larger than pereopods 2 and 3. Pereopods 4–7, carpi with anterior and posterior margin subequal. Pleonites free, short. Pleopod 1, exopod operculiform. Telson with single statocyst.

Accalathura crenulata (Richardson, 1901)
Figure 29A–D

DIAGNOSIS ♀: 16.0 mm. Antennular flagellum of about 26 articles. Antennal flagellum of about 18 articles. Uropodal exopod narrow, parallel sided. Telson apically subacute. ♂: 15.0 mm. Pleopod 2, copulatory stylet of endopod apically acute, with subapical "heel."

RECORDS Off North Carolina, 30 m; off Georgia, 20 m; Cuba; Puerto Rico, intertidal; Cozumel, Mexico; Carrie Bow Cay, Twin Cays, Belize, intertidal to 6 m; west coast of Florida, Gulf of Mexico, 55 m.

Accalathura setosa Kensley, 1984
Figure 29E–H

DIAGNOSIS ♀: 8.5 mm. Antennular flagellum of 11 articles. Antennal flagellum of 13 articles. Uropodal rami, margins closely setose; exopod ovate, outer margin sinuate, apex subacute; endopod ovate, length 1.25 times greatest width. Telson apically rounded. ♂: 7.0 mm. Pleopod 2, copulatory stylet of endopod apically strongly bifid.

RECORDS Carrie Bow Cay, Belize, intertidal to 0.5 m.

Colanthura Richardson, 1902

DIAGNOSIS Integument with minute squamae. Mandible lacking palp. Maxillipedal palp articles fused except for minute terminal article. Pereopod 1 subchelate, propodus expanded. Pereopods 2 and 3 subchelate but smaller than pereopod 1. Pereopods 4–6, carpi rectangular. Pereonite 7 very short, pereopod 7 lacking. Pleotelson lacking statocyst.

Colanthura tenuis Richardson, 1902
Figure 30A–C

DIAGNOSIS ♀: 3.5 mm. Eyes present. Integument diffusely brown in color. Antennular flagellum of four articles. Antennal flagellum of single article.

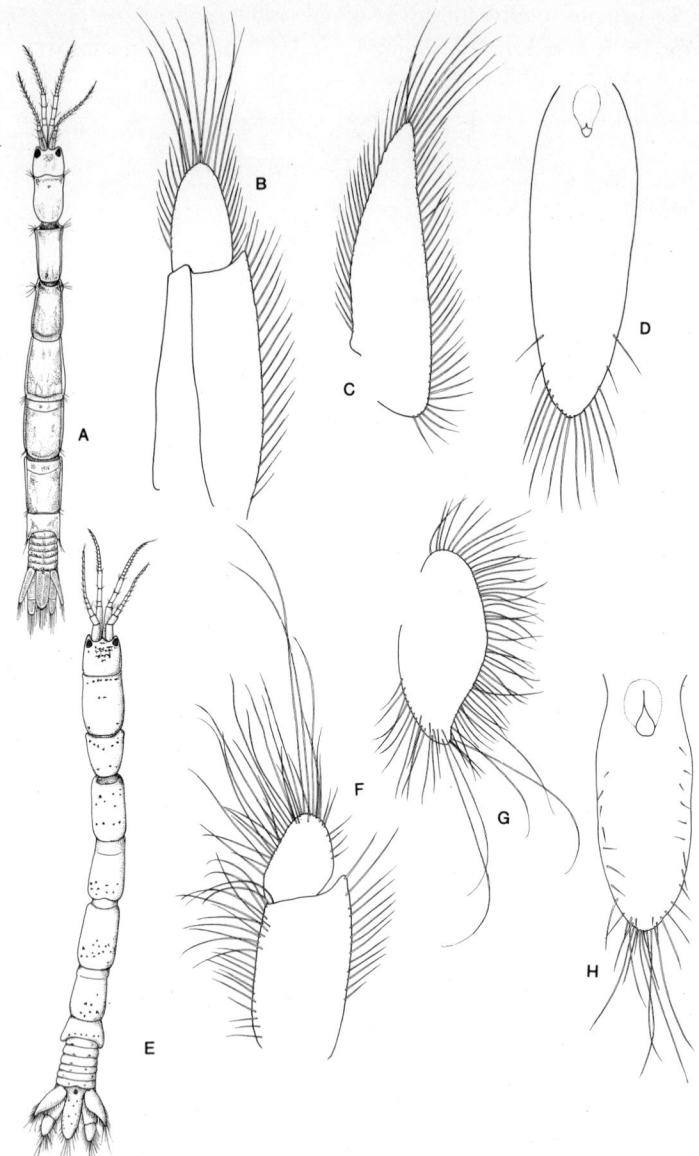

Figure 29. *Accalathura crenulata:* A, ♀; B, uropodal sympod and endopod; C, uropodal exopod; D, telson. *Accalathura setosa:* E, ♀; F, uropodal sympod and endopod; G, uropodal exopod; H, telson.

Pereopod 1, propodus with mesial surface bearing proximal row of six spines. Pleonites 1–5 short, fused, boundaries marked dorsally by folds. Telson posteriorly broadly rounded. ♂: 3.5 mm. Antennular flagellum of five articles.

RECORDS Bermuda, intertidal to 0.5 m.

Curassanthura Kensley, 1981

DIAGNOSIS Eyes lacking. Mandibular palp of three articles. Maxillipedal palp of five articles; short endite present. Pereopod 1 subchelate. Pereopods 2–6 similar, carpi rectangular. Pereopod 7 lacking. Pleonites 1–5 free; pleonite 6 dorsally demarked. Pleopod 1, exopod operculiform. Pleotelson with single statocyst. Interstitial littoral forms.

REMARKS Three species of this interstitial genus are known, two from the Caribbean, and one from the upper sublittoral gravels of a lava tunnel on Lanzarote, Canary Islands.

Key to species of *Curassanthura*

1. Telson with posterior fourth abruptly narrowed; uropodal exopod length 4.5 times greatest width .... *bermudensis*
   Telson tapering, but posterior fourth not abruptly narrowed; uropodal exopod length 2.5 times greatest width .... *halma*

Curassanthura bermudensis Wägele, 1985
   Figure 30G,H

DIAGNOSIS ♀ 3.0 mm. Pereopod 1 propodal length 2.5 times proximal width, palm with proximal strongly recurved hooklike tooth. Uropodal exopod slender, parallel sided, 4.5 times longer than greatest width. Telson constricted in posterior fourth.

RECORDS Church Cave, Bermuda, in shore sediments.

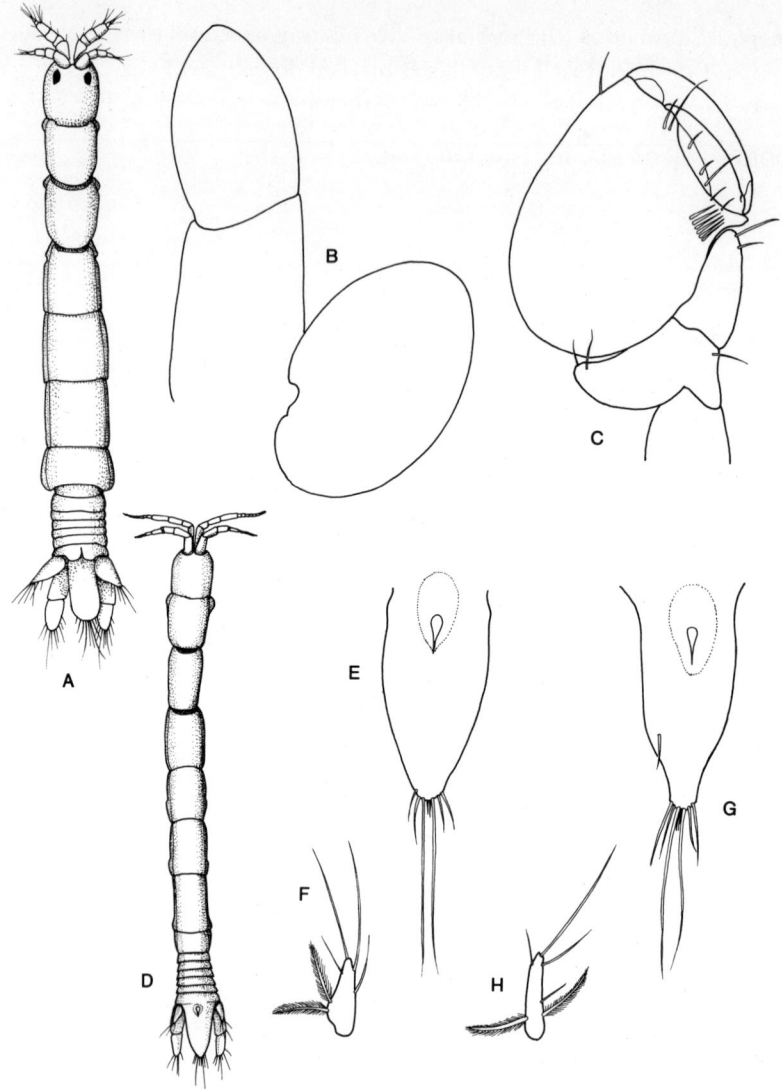

Figure 30. *Colanthura tenuis:* A, ♀; B, uropod; C, pereopod 1. *Curassanthura halma:* D, ♀; E, telson; F, uropodal exopod. *Curassanthura bermudensis:* G, telson; H, uropodal exopod (from Wägele, 1985).

Curassanthura halma Kensley, 1981
Figure 30D–F

DIAGNOSIS ♀ 2.3 mm. Pereopod 1, propodal length about 1.7 times proximal width, palm with eight fringed spines and basal tridentate lobe. Uropodal exopod triangular, length 2.5 times greatest width, shorter than sympod. Telson tapering but not abruptly narrowed in posterior fourth.

RECORDS   Curaçao, in shore sediments 1.5 m above tide line; Bonaire, in shore sediments above tide line.

Paranthura Bate and Westwood, 1868

DIAGNOSIS   Eyes present. Antennular flagellum shorter than peduncle. Antennal flagellum usually of single flattened setose article. Mandibular palp of three articles, article 3 with comb of spines. Maxillipedal palp of one or two articles; endite small to absent. Pereopod 1, propodus inflated, larger than that of pereopods 2 and 3. Pereopods 4–7, carpi rectangular. Pleonites short, more or less distinct. Pleopod 1, exopod operculiform. Telson lacking statocyst.

REMARKS   This is the largest of the paranthurid genera, with over 50 names in the literature. Many of these are poorly described. Species of *Paranthura* are common in the shallow waters of the temperate and tropical seas.

Key to species of *Paranthura*

1. Telson posteriorly truncate; uropodal exopod rectangular, margins serrate .... *infundibulata*
   Telson posteriorly rounded; uropodal exopox ovate, margins entire .. 2

2. Uropodal endopod longer than wide 3
   Uropodal endopod as long as wide .... *antillensis*

3. Uropodal exopod elongate-elliptical; telson parallel sided for half length
   .... *floridensis*
   Uropodal exopod ovate, outer margin sinuate; telson evenly elliptical
   .... *barnardi*

Paranthura antillensis Barnard, 1925
   Figure 31A–F

DIAGNOSIS   ♀ 5.1 mm. Antennular flagellum of four articles. Mandibular palp, article 3 with five spines. Maxillipedal palp of single article three times longer than basal width; short endite present. Pereopod 1, propodus ex-

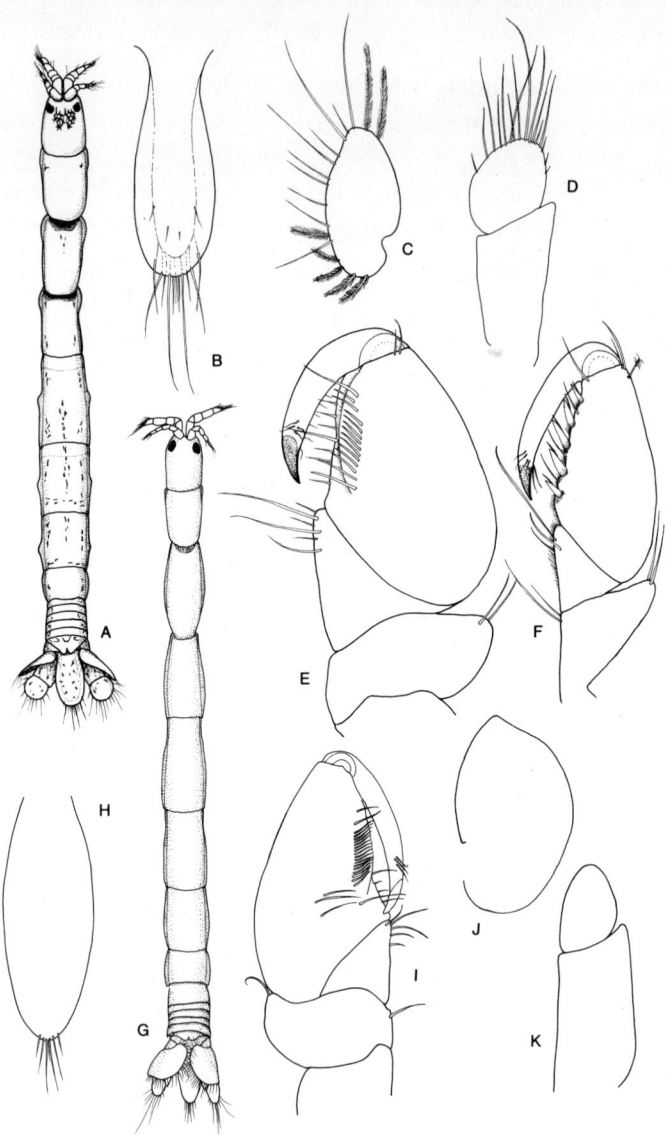

Figure 31. *Paranthura antillensis:* A, ♀; B, telson; C, uropodal exopod; D, uropodal sympod and endopod; E, pereopod 1, ♀; F, pereopod 2, ♀. *Paranthura barnardi:* G, ♀; H, telson; I, pereopod 1, ♀; J, uropodal exopod; K, uropodal sympod and endopod.

panded, palm bearing row of setae, mesial surface near palmar margin with convex flange and row of about 10 spines. Pereopod 2, propodal palm with five stout sensory setae. Pleonite 6 free, posterior margin bilobed. Uropodal exopod ovate, outer margin sinuous; endopod almost circular, as wide as long. Telson posteriorly rounded.

RECORDS St. Johns, St. James, U.S. Virgin Islands, 32 m; Carrie Bow Cay, Belize, intertidal to 1.5 m.

Paranthura barnardi Paul and Menzies, 1971
Figure 31G-K

DIAGNOSIS ♀ 6.0 mm. Mandibular palp article 3 with eight spines. Maxillipedal palp of single article, length four times basal width. Pereopod 1, propodal palm concave, with convex flange and row of about 17 spines on mesial surface. Uropodal exopod broadly ovate, apically subacute; endopod ovate, length about 1.5 times basal width. Telson evenly elliptical, apex evenly rounded.

RECORDS Off Venezuela, 95 m.

Paranthura floridensis Menzies and Kruczynski, 1983
Figure 32A-E

DIAGNOSIS ♀ 6.3 mm. Integument with sparse irregular pigmentation. Mandibular palp article 3 with eight spines. Maxillipedal palp of single article, length 3.5 times basal width. Pereopod 1, propodal palm with transparent flange and row of 10 setae on mesial surface. Uropodal exopod elongate-elliptical; endopod ovate, length 1.5 times greatest width. Telson posteriorly broadly rounded, parallel sided for about half its length.

RECORDS Off Sanibel Island, Florida, Gulf of Mexico, 73 m.

Paranthura infundibulata Richardson, 1902
Figure 32F-J

DIAGNOSIS ♀ 8.2 mm, ♂ 6.0 mm. Integument with red-brown pigmentation; broad irregular patch between eyes running onto bases of antennules; pereonites 1 and 2 with anterior patches, remainder of pereonites with posterior patches; strong patch on uropodal exopod, endopod, and telson. Mandibular palp article 3 with 11 or 12 spines. Maxillipedal palp of single article,

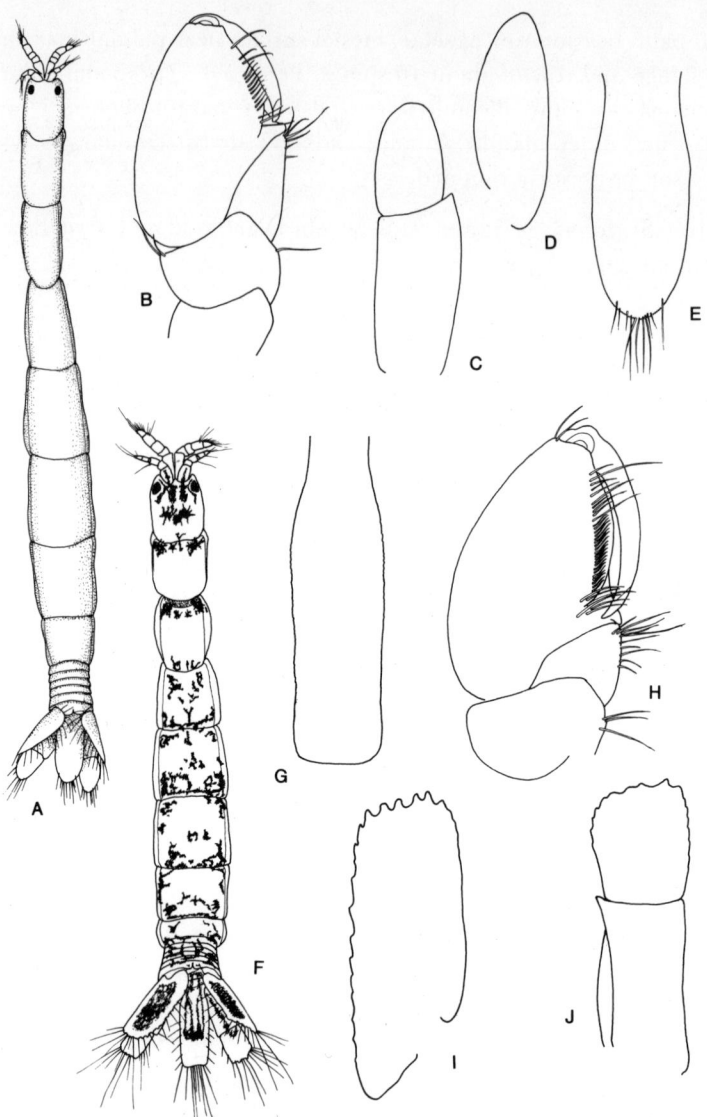

Figure 32. *Paranthura floridensis:* A, ♀; B, pereopod 1, ♀; C, uropodal sympod and endopod; D, uropodal exopod; E, telson. *Paranthura infundibulata:* F, ♀; G, telson; H, pereopod 1, ♀; I, uropodal exopod; J, uropodal sympod and endopod.

length 2.5 times basal width. Pereopod 1, propodus with convex flange and row of more than 20 setae. Uropodal exopod elongate-rectangular, mesial and distal margins serrate; endopod roughly square, margins serrate. Telson parallel sided, posterior margin truncate.

RECORDS   Bermuda, 11–12 m; Carrie Bow Cay, Belize, intertidal to 1 m; Cozumel, Mexico; Venezuela.

Virganthura Kensley, 1987b

DIAGNOSIS   Eyes present. Mandibular palp of three articles. Maxillipedal palp of three articles; endite present. Pereopods 1–3 subchelate, pereopod 1 larger than pereopods 2 and 3; pereopods 4–7, carpi rectangular. Pleopod 1, exopod operculiform. Pleonites 1–5 short, distinct; pleonite 6 dorsally demarked. Telson with single statocyst.

Virganthura crassa (Barnard, 1925)
Figure 33

DIAGNOSIS   ♀ 6.8 mm. Antennular flagellum of three articles. Antennal flagellum of seven articles. Maxillipedal endite reaching distal margin of first palp article. Pereopod 1, propodal palm slightly concave, bearing seven spines. Uropodal exopod ovate, outer margin sinuate; endopod distally rounded, articulating obliquely on sympod.

RECORDS   U.S. Virgin Islands, 30 m.

Suborder Asellota Latreille, 1803

DIAGNOSIS   Antennules uniramous. Antennae uniramous, with scale in some families. Mandible usually with palp, but palp lacking in some groups. Pereopod 1 usually subchelate, sexually dimorphic in some groups. Coxae small, sometimes not all visible in dorsal view. Pleon seldom of more than two free pleonites plus pleotelson. Pleopod 1 absent in ♀. One pair of pleopods in ♀, and one or two pairs of pleopods in ♂ forming operculum over remaining respiratory pleopods. Pleopod 2 in ♂ usually adapted for copulation. Uropods usually pedunculate, but peduncle may be reduced, biramous or uniramous, terminal or subterminal.

REMARKS   The suborder Asellota is usually divided into four superfamilies, based on pleopodal arrangement. The great majority of families, however,

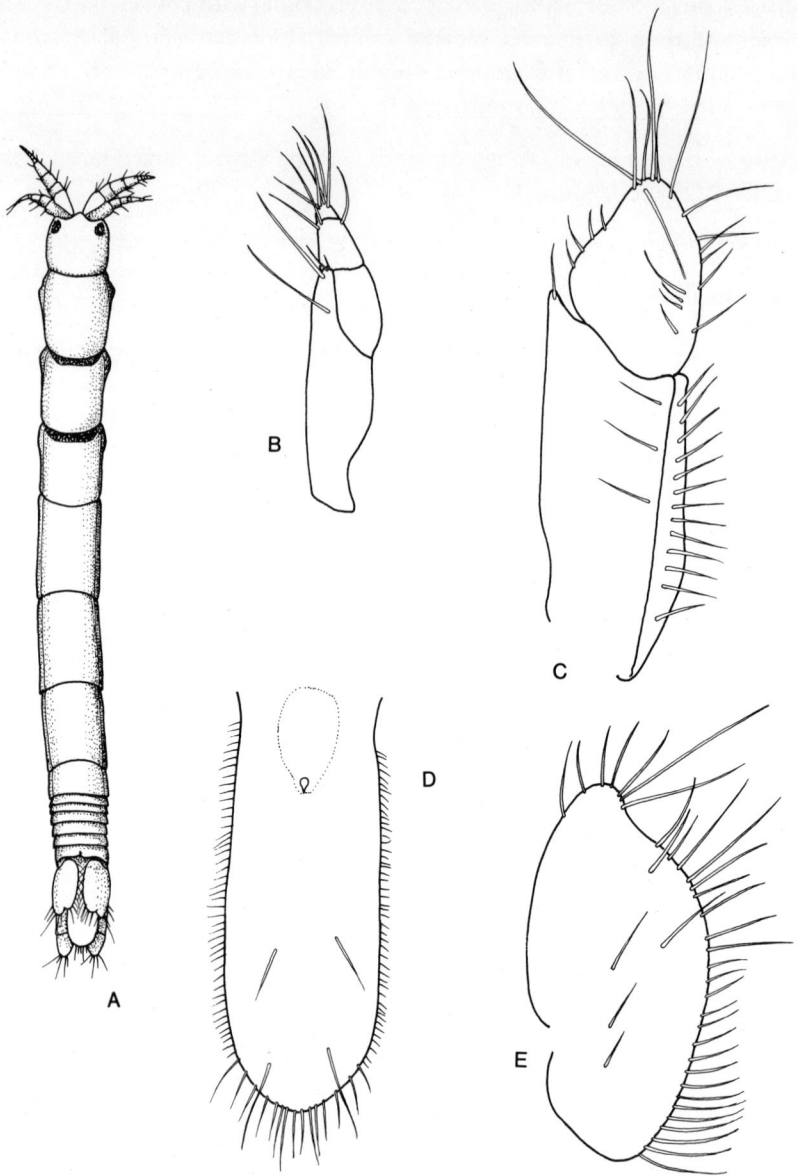

Figure 33. *Virganthura crassa*: A, ♀; B, maxilliped; C, uropodal endopod and sympod; D, telson; E, uropodal exopod.

belong to the superfamily Janiroidea, considered to be the most advanced. In place of a key to the four superfamilies, the chart shown in Figure 34 illustrates diagrammatically the arrangement of the pleopods in these groups.

Superfamily Aselloidea Rafinesque-Schmaltz, 1815

DIAGNOSIS ♂: Pleopod 1, peduncles separate, rami uniarticulate; pleopod 2 having biarticulate exopod and copulatory endopod; pleopod 3 biramous, opercular. ♀: Pleopod 1 absent; pleopod 2 absent, or of single article; pleopod 3 biramous, opercular.

Family Atlantasellidae Sket, 1979

DIAGNOSIS Body resembling that of a sphaeromatid isopod. Pleon consisting of two free pleonites plus pleotelson. ♂: Pleopod 1 of two articles; pleopod 2 with sympod, small exopod, and uniarticulate copulatory endopod. ♀: Pleopods 1 and 2 absent; pleopod 3 operculate but rami not fused.

Atlantasellus Sket, 1979

DIAGNOSIS Eyes lacking. Antennule flattened, short and broad. Antenna elongate, peduncle of four (?five) articles, flagellum of four articles. Mandibular palp of three articles; with incisor and lacinia, molar replaced by brushlike process. Maxillipedal palp of five articles of similar width and broad endite. Pereopod 1 subchelate; pereopods 2–7 ambulatory, slender, dactyli biunguiculate. Uropods uniarticulate, vestigial.

Atlantasellus cavernicolus Sket, 1979
  Figure 35

DIAGNOSIS ♂ 1.1 mm. Cephalon with strong triangular rostrum, equal to pereonites 1–3 combined in middorsal length. Pleotelson basally broad, tapering to trilobed posterior margin, uropods inserted in incisions between median and lateral lobes.

RECORDS Walsingham Cave, Bermuda.

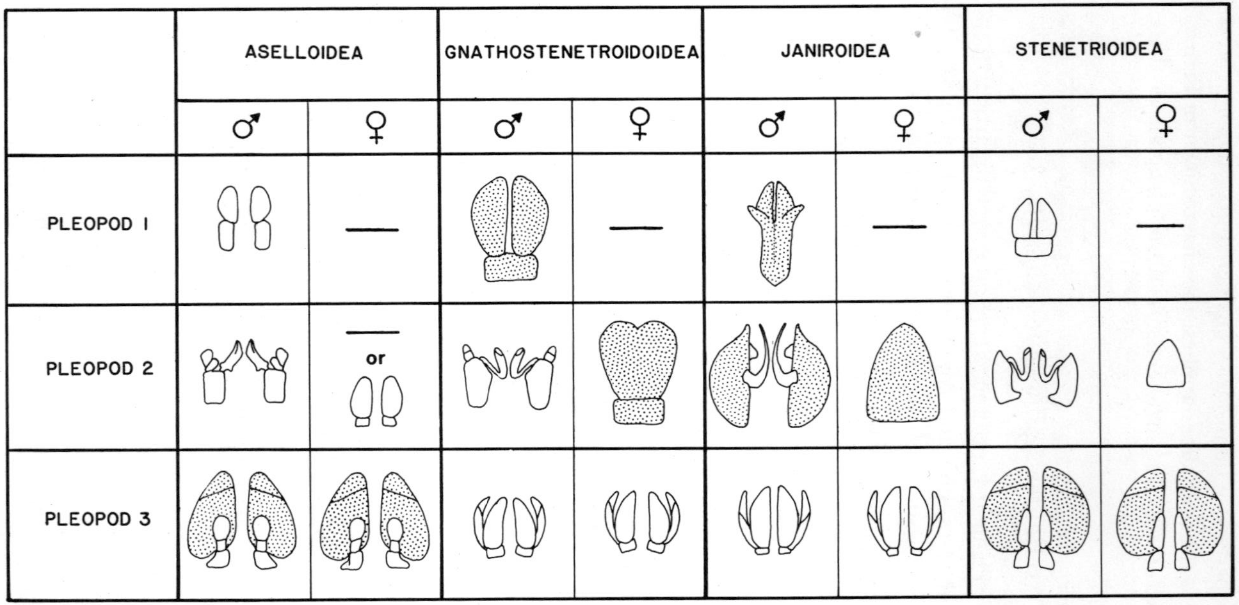

Figure 34. Comparison of pleopods 1–3 in the four superfamilies of the suborder Asellota.

Superfamily Gnathostenetroidoidea Kussakin, 1967

DIAGNOSIS ♂: Pleopod 1 peduncles fused, uniarticulate rami separate, opercular; pleopod 2 small, with biarticulate exopod and copulatory endopod; pleopod 3, broad endopod biarticulate, exopod slender, uniarticulate. ♀: Pleopod 1 absent; pleopod 2 rami fused to form operculum; pleopod 3 as in ♂.

Family Gnathostenetroididae Kussakin, 1967

DIAGNOSIS Uropods with short sympod, rami relatively well developed. ♂: Pleopod 1, protopodites short, fused, rami separate, together forming operculum covering remaining pleopods. Pleopod 2, separate, much smaller than pleopod 1. Pleopod 3, exopod elongate, slender; endopod uniarticulate, broad. ♀: Pleopod 2, rami fused to form operculum covering remaining pleopods.

Key to genera of Gnathostenetroididae

1. Eyes absent. Rostrum anteriorly rounded/truncate ...... *Neostenetroides*
   Eyes present. Rostrum anteriorly bilobed .... *Gnathostenetroides*

Gnathostenetroides Amar, 1957

DIAGNOSIS Eyes present. Rostrum well developed, anteriorly bilobed. Pereonite 1, especially in ♂, larger than following pereopods. Pleon having one very short free pleonite. Pleotelson with lateral margins subparallel, ending in strong tooth posteriorly; posterolateral margins sinuous, rounded. Mandible in ♂ with strong projection arising proximolateral to incisor, visible dorsally; in ♀, projection short, not visible dorsally.

Gnathostenetroides pugio Hooker, 1985
   Figure 36A–C

DIAGNOSIS ♂ 3.2 mm, ♀ 1.6 mm. Eyes of five ommatidia each. ♂: Pereopod 1, merus with anterior margin strongly produced into narrow triangular process; propodus broad, palm demarked by strong denticulate spine and

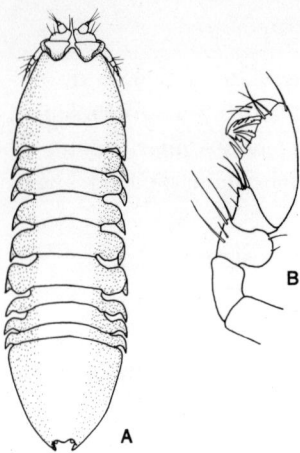

Figure 35. *Atlantasellus cavernicolus:* A, ♀; B, pereopod 1. (From Sket, 1979).

bearing row of nine more slender spines, posterior margin setose. ♀: Operculum with distal incision acute.

RECORDS   Florida Middlegrounds, Gulf of Mexico, 55 m.

REMARKS   The type of the genus, and only other species known, *G. laodicense* Amar, 1957, was recorded from the Mediterranean Sea. This species is included here, as several of the species first recorded from the Florida Middlegrounds (Hooker, 1985) have since been recorded from the Caribbean.

Neostenetroides Carpenter and Magniez, 1982

DIAGNOSIS   Pleonites 1 and 2 very short. Operculum of ♀ subcircular. Pleopod 2 in ♂ with elongate protopodite, copulatory organ prolonged by hyaline tonguelike structure. Pleopod 4, exopod ovate, wider than endopod.

Neostenetroides stocki Carpenter and Magniez, 1982
    Figure 36D,E

DIAGNOSIS   ♂ 1.46 mm, ♀ 1.87 mm. Eyes absent. Rostrum well developed, anteriorly rounded/truncate. Pleotelson wider than long, lateral margins entire, posterior margin regularly convex. Pereopod 1, propodus elongate, widening distally, palm poorly demarked, with few spines; dactylus stout, overlapping palm. Uropods unknown.

RECORDS   San Salvador, Bahamas, from Dixon Hill Lighthouse Cave.

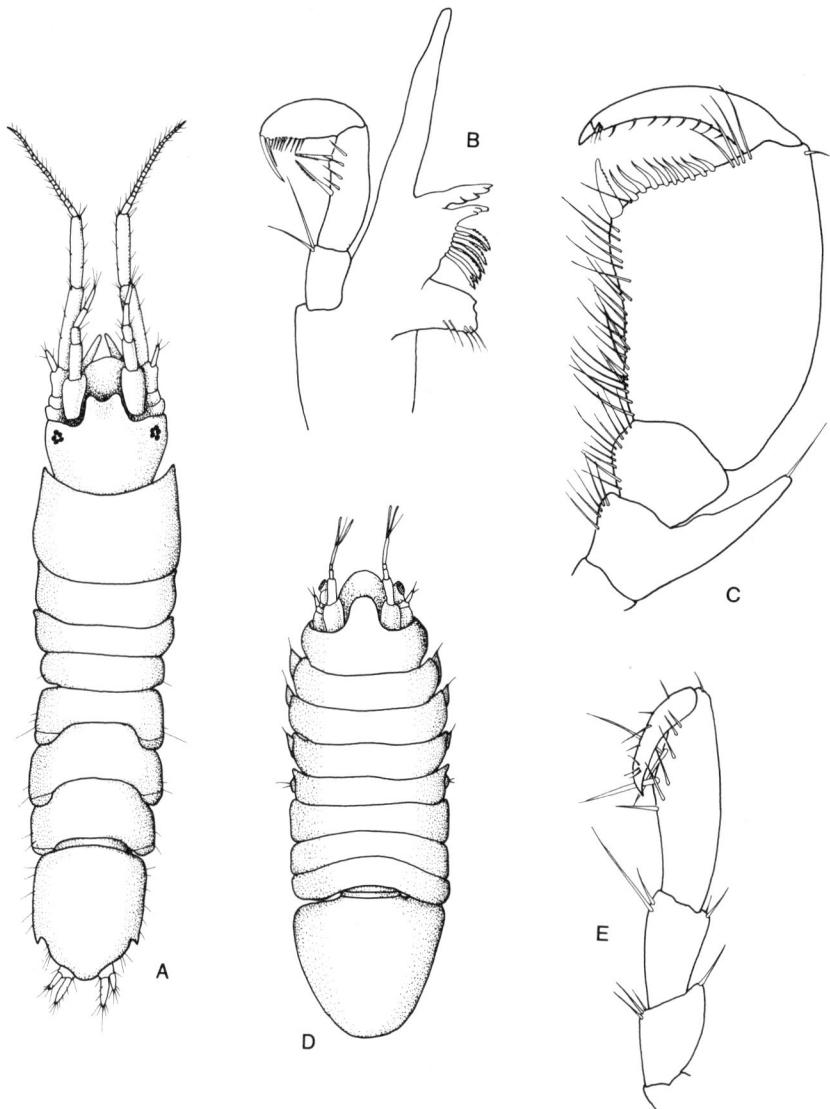

Figure 36. *Gnathostenetroides pugio:* A, ♂; B, mandible; C, pereopod 1. *Neostenetroides stocki* (from Carpenter and Magniez, 1982): D, ♀; E, pereopod 1.

Superfamily Janiroidea Sars, 1899

DIAGNOSIS ♂: Pleopod 1 with elongate peduncle, occasionally fused; pleopod 2 with short exopod and copulatory usually elongate endopod, pleopods 1 and 2 together forming operculum; pleopod 3 endopod uniarticulate, exopod biarticulate. ♀: Pleopod 1 absent; pleopod 2, rami fused to form operculum; pleopod 3 as in ♂.

Key to families of Janiroidea

1. Eyes (if present) on lateral processes of cephalon 2
   Eyes dorsolateral on cephalon, not on lateral processes 6
2. Uropods with large, easily visible sympod and rami 3
   Uropods with sympod minute or absent, rami short 5
3. Pleon posteriorly markedly produced; some pereonites produced laterally into fingerlike processes .... **Pleurocopidae**
   Pleon posteriorly rounded, barely produced; pereonites not laterally produced into prominent fingerlike processes 4
4. Pereopod 1 subchelate .... **Santiidae**
   Pereopod 1 ambulatory, biunguiculate .... *Mexicope*
5. Pleopod 1 ♂ distally sagittate; anus covered by pleopod 1 ♂, or pleopod 2 ♀ .... **Paramunnidae**
   Pleopod 1 ♂ distally truncate; anus exposed .... **Munnidae**
6. Uropodal rami minute, smaller than squat sympod ...... **Joeropsidae**
   Uropodal rami elongate, sympod variable, generally elongate 7
7. Eyes lacking; pereopods all similar, ambulatory .... **Microparasellidae**
   Eyes usually present; pereopod 1 prehensile, subchelate, pereopods 2–7 ambulatory .... **Janiridae**

Family Incertae Sedis

Mexicope Hooker, 1985

DIAGNOSIS Eyes on short lateral lobes of cephalon. Scale present on antennal peduncle. Coxae short, visible in dorsal view on pereonites 2–7. Pereonites laterally acute, not markedly produced. Pereopod 1 ambulatory, dactylus biunguiculate. Pleon consisting of one short pleonite plus broad pleotelson. Uropods large, with elongate peduncle, endopod, and exopod.

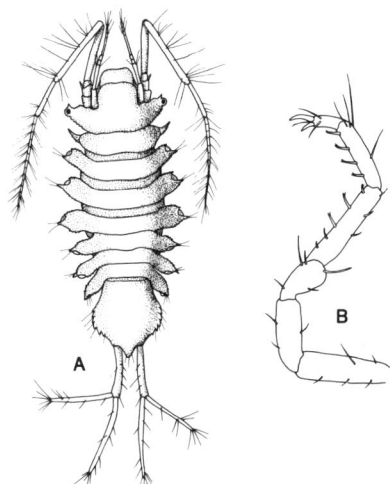

Figure 37. *Mexicope kensleyi*: A, ♀; B, pereopod 1.

REMARKS   The inability to place *Mexicope* in a family reflects the fact that the arrangement of the families and genera of the Janiroidea is still unsettled. While having affinities with the Pleurocopidae and the Janiridae, placement in either of these would require redefinition of both families, making the Janiridae even more of a hodgepodge of phylogenetically unrelated genera.

Mexicope kensleyi Hooker, 1985
Figure 37

DIAGNOSIS   ♂ 1.7 mm, ♀ 2.9 mm. Antennae slightly longer than body. Pereonites and pleon laterally finely serrate. Pleotelsonic lobe between uropodal bases narrowly rounded, barely produced.

RECORDS   Turks and Caicos Islands, 1 m; Florida Middlegrounds, Gulf of Mexico, 30 m.

Family Janiridae Sars, 1899

DIAGNOSIS   Antennae longer than antennules. Mandibles with palp and well-developed molar. Maxilliped with articles 1–3 at least as wide as endite, markedly broader than articles 4 and 5. Pereopod 1 prehensile, subchelate, sexually dimorphic, larger in ♂ than in ♀. Pereopods 2–7 ambulatory, dactyli biunguiculate. Coxae visible at least on three posterior pereonites. Pleon consisting of single free pleonite (often very narrow, short and inconspicuous) plus pleotelson. Uropods with well-developed sympod; usually biramous.

Carpias Richardson, 1902

DIAGNOSIS   Cephalon with dorsolateral eyes, lacking rostrum. Coxae visible dorsally on pereonites 1–7. Pleon with one short pleonite lacking free lateral margins, plus broad pleotelson. Antennules and antennae well developed, latter with scale. Articles 2 and 3 of maxillipedal palp expanded. Pereopod 1 sexually dimorphic, often relatively enormous and/or elongate in male, carpochelate, carpus expanded, propodus variously armed or expanded, dac-

Key to species of *Carpias* (based on pereopod 1 of mature ♂)

1. Propodus distally acute 2
   Propodus distally broad 4
2. Propodus apically acute, with proximal tooth .... *minutus*
   Propodus lacking teeth 3
3. Carpus with two strong distal teeth; propodus (when chela closed) reaching to merus .... *algicola*
   Carpus with two strong and one small teeth; propodus (when chela closed) reaching proximal half of carpus .... *serricaudus*
4. Propodus with distinct teeth 5
   Propodus lacking distinct teeth 6
5. Carpus distally with broadly rounded area; merus elongate-slender, length about four times greater than width .... *bermudensis*
   Carpus lacking broadly rounded area; merus short, broader than long .... *punctatus*
6. Carpus with two distal teeth; dactylus minute .... *triton*
   Carpus with three distal teeth; dactylus small, but not obsolete 7
7. Propodus distally truncate; carpus with middle tooth of palm distally faintly bilobed .... *brachydactylus*
   Propodus distally faintly bilobed; carpus with middle tooth of palm distally narrowly rounded .... *harrietae*

tylus reduced or rudimentary, bearing two claws (biunguiculate). Pereopods 2–7 similar, ambulatory, dactylus with three claws (triunguiculate). Uropods often longer than pleotelson, with relatively elongate sympod and rami.

REMARKS This genus has been, and continues to be, a source of taxonomic problems. Several authors (e.g., Pires, 1980) have separated the species into the genera *Carpias* and *Bagatus*; others (e.g., Bowman and Morris, 1979) have synonymized them. In part, this uncertainty reflects the general uncertainty of the state of taxonomy in the family Janiridae. In this work, the genus *Carpias* is used to contain all the species described under the names *Carpias* and *Bagatus*.

While these tiny asellotes are frequently extremely abundant in certain habitats (e.g., in reef-crest algal turfs; Kensley, 1983), difficulty is experienced in identifying specimens other than mature males. The first pereopod of the mature male is the feature best used for species separation, but variation with maturity and geographic locality have not been investigated. With more detailed work, some species will undoubtedly be synonymized.

Carpias algicola (Miller, 1941)
Figure 38A,B

DIAGNOSIS ♂ 2.9 mm, ovigerous ♀ 2.0 mm. Frontal margin straight. Pereopod 1 in ♂, carpus distally not much broader than proximally, with two teeth. Propodus reaching back to merus in mature male. Pleopod 1 in ♂, rami with outer lobe distally acute but not produced. Uropod longer than pleotelson. Pigment in scattered red-brown chromatophores.

RECORDS Looe Key, Florida, 1–1.5 m; Yucatan, Mexico; Carrie Bow Cay, Belize, 0–2 m; Puerto Rico; Jamaica; Venezuela.

Indo-west Pacific.

Carpias bermudensis Richardson, 1902
Figure 38C,D

DIAGNOSIS ♂ 2.7 mm. Frontal margin straight. Pereopod 1 in adult ♂ almost 1.6 times body length; carpus distally broadened, with rounded posterior area, palm with tooth at outer distal angle, larger tooth at midlength followed by deep notch; propodus with two teeth on flexor margin, widening to truncate distal margin. Pleopod 1 ♂, outer distal lobe narrowly acute, somewhat produced. Operculum of ♀ with distal margin emarginate.

RECORDS   Bermuda; eastern and southern coasts of Florida, 1.5–15 m.

Carpias brachydactylus Pires, 1982
    Figure 38E

DIAGNOSIS   ♂ 1.6 mm. Pereopod 1 ♂, carpus distally between two and three times wider than proximal width, with strong triangular outer tooth defining palm, middle tooth apically faintly bifid, inner tooth rounded; propodus widening to distal truncate margin, overreaching carpal palm by short distance, with very low tubercle at about midlength of flexor margin. Pleopod 1 ♂, outer distal lobe slightly produced, rounded. Operculum of ♀ wider than long, distal margin broadly emarginate.

RECORDS   Puerto Rico, 1.5 m.

Carpias harrietae Pires, 1981
    Figure 38F,G

DIAGNOSIS   ♂ 2.3 mm. Closely resembling *C. brachydactylus*. Pereopod 1 ♂, carpus with posterodistal area somewhat expanded, with strong tooth defining outer margin of palm, middle tooth rounded, inner tooth distally faintly bifid; propodus widening to shallowly bilobed distal margin. Pleopod 1 ♂, inner distal lobe rounded, outer lobe narrowly acute. Operculum of ♀ with mediodistal margin gently concave. Uropod about twice length of pleotelson.

RECORDS   Biscayne Bay, Florida, intertidal to 2 m.

Carpias minutus (Richardson, 1902)
    Figure 39A

DIAGNOSIS   ♂ 1.9 mm, ovigerous ♀ 1.8 mm. Pereopod 1 ♂, carpus distally widening, palm defined by strong triangular tooth, two inner teeth of palm separated by rounded notch; propodus with strong proximal tooth on flexor margin, distally produced into small triangular lobe below dactylus. Pleopod 1 ♂, outer distal lobe narrowly triangular and produced well beyond inner lobe. Uropod subequal to pleotelson in length.

RECORDS   Bermuda, on *Sargassum*.

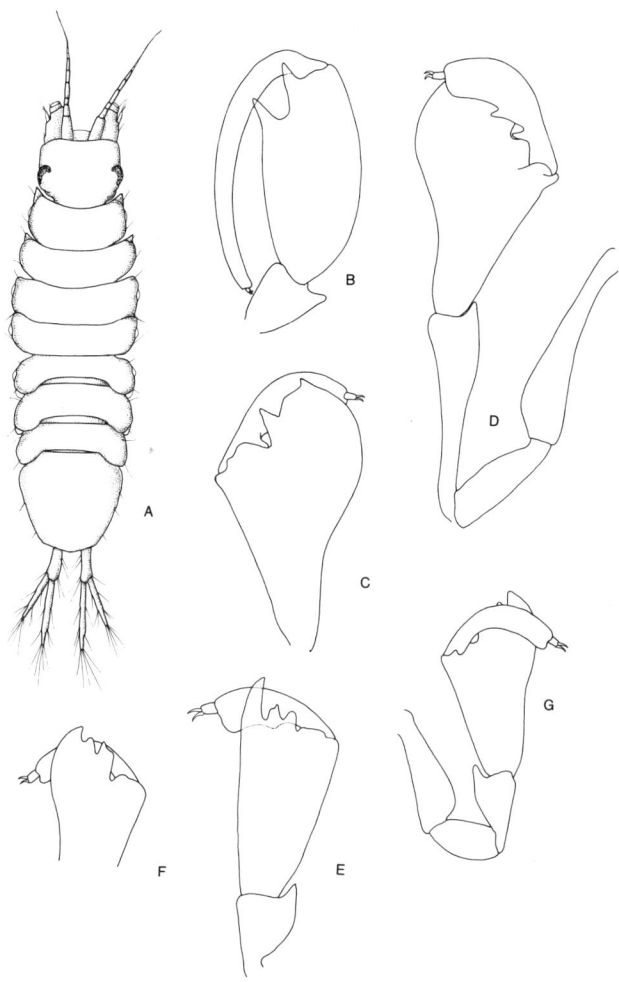

Figure 38. *Carpias algicola: A*, ♀; *B*, pereopod 1, ♂. *Carpias bermudensis: C, D*, pereopod 1, ♂. *Carpias brachydactylus: E*, pereopod 1, ♂. *Carpias harrietae: F, G*, pereopod 1, ♂.

Carpias punctatus (Kensley, 1984)
Figure 39 B,C

DIAGNOSIS ♂ 2.2 mm, ♀ 2.8 mm. Dorsal integument with patchy reticulate pattern of large dark brown chromatophores. Frontal margin faintly convex.

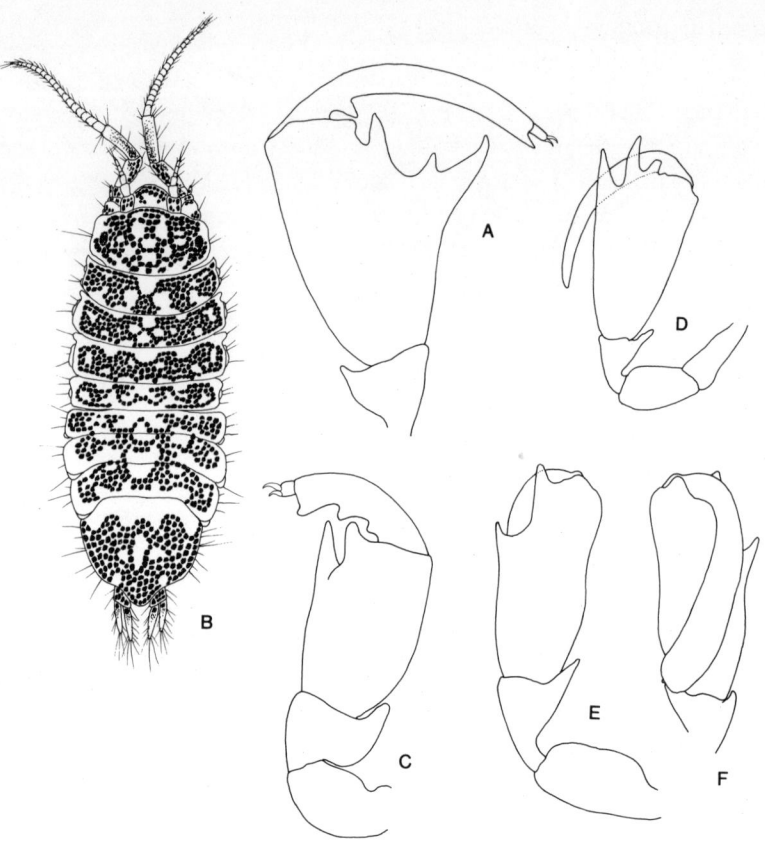

Figure 39. *Carpias minutus:* A, pereopod 1, ♂. *Carpias punctatus:* B, ♂; C, pereopod 1, ♂. *Carpias serricaudus:* D, pereopod 1, ♂. *Carpias triton:* E, F, pereopod 1, ♂.

Pereopod 1 ♂, distal two-thirds parallel sided, with strong acute tooth defining palm, and second rounded tooth; propodus with three lobe-teeth on flexor surface, overreaching carpus by a third of its length. Pleopod 1 ♂, outer distal lobe narrowly triangular, reaching well beyond inner lobe. Operculum of ♀ with distal margin shallowly concave. Uropod half length of pleotelson; latter with posterior margin a broadly rounded lobe between uropodal bases.

RECORDS   Carrie Bow Cay, Belize, intertidal to 15.2 m.

Carpias serricaudus Menzies and Glynn, 1968
Figure 39D

DIAGNOSIS ♂ 1.6 mm, ♀ 1.5 mm. Pereopod 1 ♂, palm of carpus with two strong outer teeth and one short inner tooth; propodus reaching back to proximal half of carpus, tapering distally; dactylus obsolete. Pleopod 1 ♂, outer distal lobe acute, reaching well beyond inner lobe. Pleotelsonic margins very faintly serrate. Uropod about 0.7 times length of pleotelson.

RECORDS   Turks and Caicos Islands, 1 m; Puerto Rico, intertidal to 1.5 m.

Carpias triton Pires, 1982
Figure 39E,F

DIAGNOSIS ♂ 2.3 mm. Very similar to *C. algicola*. Pereopod 1 ♂, carpus with two strong basally broad distal teeth; propodus extending back to merus in adult ♂, widening to broadly rounded distal margin; dactylus minute. Pleopod 1 ♂, outer distal lobe narrowly acute, reaching well beyond inner rounded lobe. Uropod about 1.5 times pleotelson length.

RECORDS   Carrie Bow Cay, Belize, intertidal reef crest.

Family Joeropsidae Nordenstam, 1933

DIAGNOSIS   Cephalon free, with distinct rostrum. Molar process of mandible reduced. Maxillipedal palp articles all of similar width. Antenna short, peduncle dilated, flagellum reduced. Pereonites similar, wider than long. Pereopods similar, biunguiculate. Uropods having short squat sympod and very reduced rami; inserted into submedian posterior notches of pleotelson.

Joeropsis Koehler, 1885

DIAGNOSIS   Dorsolateral eyes present. Antennule, basal article widest and longest, often with transparent distal dentition. Antenna, peduncular articles 3–5 somewhat dilated, article 2 often with fringe of transparent scales; flagellum of about six articles, together shorter than peduncle article 5. Pereonites similar, generally subequal in length and width. Pleotelson of single shield-shaped segment. Uropodal sympod usually with mesiodistal angle acute; rami reduced.

Key to species of *Joeropsis*

1. Lateral margins of cephalon serrate; rostrum triangular ..... *personatus*
   Lateral margins of cephalon entire; rostrum not triangular 2
2. Body glabrous; strong band of pigment on cephalon and pereonite 4
   .... *bifasciatus*
   Body setose; pigment in reticulation over entire body 3
3. Rostrum evenly convex; outer uropodal ramus longer than inner
   .... *rathbunae*
   Rostrum anteriorly shallowly notched; outer uropodal ramus shorter than inner .... *coralicola*

Joeropsis bifasciatus Kensley, 1984
   Figure 40A–F

DIAGNOSIS ♂ 2.5 mm, ♀ 2.4 mm. Body glabrous. Lateral margins of cephalon entire. Rostrum semicircular, with marginal flange of transparent teeth. Antennal flagellum of eight articles. Lateral margins of pleotelson serrate. Apex of ♀ operculum blunt. Broad band of pigment on cephalon between eyes and almost reaching posterior margin; broad band of pigment on pereonite 4.

RECORDS Carrie Bow Cay, Belize, 1–6 m, often on *Agaricia* sp. and *Porites* sp. corals, and *Halimeda* sp. alga; Anguilla.

Joeropsis coralicola Schultz and McCloskey, 1967
   Figure 40G

DIAGNOSIS ♂ 2.0 mm, ovigerous ♀ 1.9 mm. Body setose. Lateral margins of cephalon entire. Rostrum anteriorly notched. Antennal flagellum of five articles. Lateral margins of pleotelson serrate. Apex of ♀ operculum acute. Outer uropodal ramus shorter than inner. Pigment spread as reticulation over entire body.

RECORDS Off North Carolina, on coral *Oculina arbuscula;* Florida Middlegrounds, Gulf of Mexico, from sponge *Agelas* sp. and coral *Madracis* sp., 25–33 m.

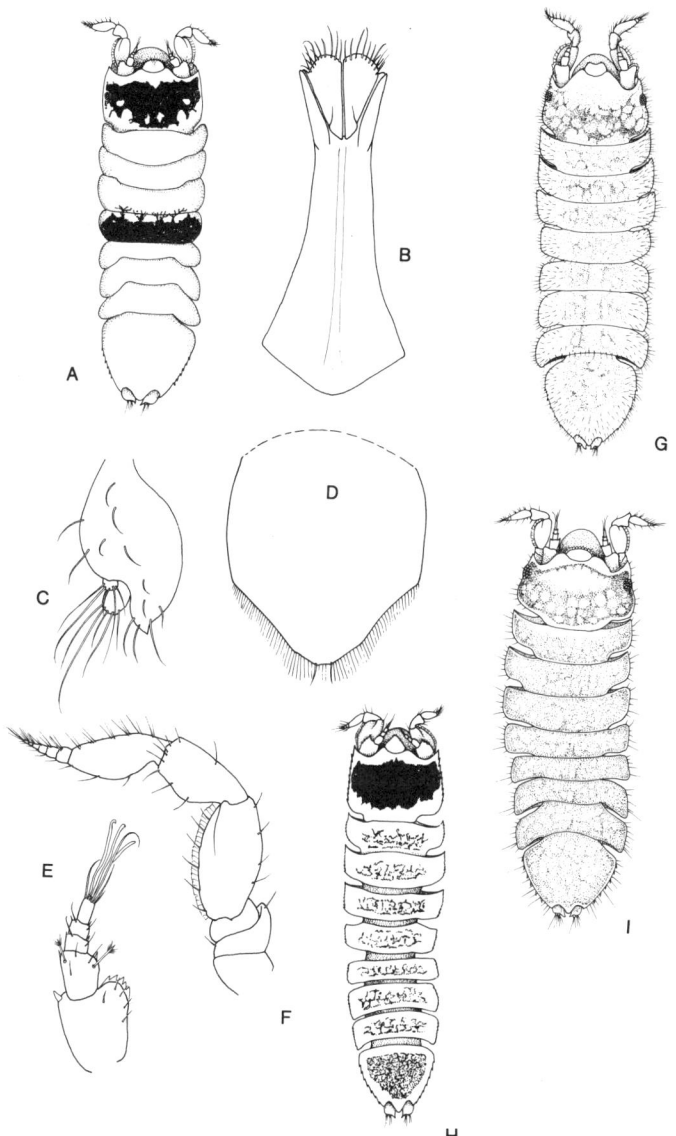

Figure 40. *Joeropsis bifasciatus: A*, ♂; *B*, pleopod 1, ♂; *C*, uropod; *D*, operculum, ♀; *E*, antennule; *F*, antenna. *Joeropsis coralicola: G*, ♂. *Joeropsis personatus: H*, ♂. *Joeropsis rathbunae: I*, ♂.

Joeropsis personatus Kensley, 1984
Figure 40H

DIAGNOSIS ♂ 2.2 mm, ♀ 2.0 mm. Body glabrous. Lateral margin of cephalon serrate. Rostrum triangular. Lateral margins of pleotelson serrate. Antennal flagellum of five articles. Apex of ♀ operculum acute. Outer uropodal ramus shorter than inner. Strong band of pigment on cephalon; rest of body with paler reticulation of pigment.

RECORDS Carrie Bow Cay, Belize, on *Porites* sp. and *Madracis* sp. corals, and on *Halimeda* sp. alga, 1–20 m.

Joeropsis rathbunae Richardson, 1902
Figure 40I

DIAGNOSIS ♂ 1.9 mm, ovigerous ♀ 1.6 mm. Body setose overall. Lateral margins of cephalon entire. Rostrum evenly convex with flange of transparent teeth. Antennal flagellum of three articles. Lateral margins of pleotelson serrate. Apex of ♀ operculum acute. Outer uropodal ramus longer than inner. Pigment in reticulation over entire body.

RECORDS Bermuda; Florida Keys; Turks and Caicos Islands; Puerto Rico; Gulf of Mexico; intertidal to 36 m.

Family Microparasellidae Karaman, 1933a

DIAGNOSIS Eyes lacking. Antennule much shorter than antenna. Antenna with scale. Pereopods all similar, with biunguiculate dactyli. Pleon of one free pleonite plus pleotelson. Uropods with well-developed sympod and rami.

Key to genera of Microparasellidae

1. ♂, pleopod 1 narrow, not overlapping external part of pleopod 2; maxillipedal palp of five articles .... *Microcharon*
   ♂, pleopod 1 broad, almost completely covering pleopod 2; maxillipedal palp of four articles, terminal article ending in pointed process .... *Angliera*

REMARKS All the members of this family are tiny (usually less than 2 mm), and most are interstitial in habit, being found in marine, brackish, and freshwater environments.

Angliera Chappuis and Delamare Deboutteville, 1955

DIAGNOSIS Mandibular palp with two proximal articles inflated, terminal article, slender hooklike, articles lacking setae and spines. Maxillipedal palp of four articles, articles 1 and 3 elongate, article 2 short, article 4 with terminal acute process. Pleopod 1 in ♂ forming broad lamella.

Angliera psamathus Kensley, 1984
Figure 41A–D

DIAGNOSIS ♂ 1.0 mm, ♀ 1.0 mm. Maxillipedal endite with seven setae on distal margin. Posterior four pairs of pereopods with claw on dactylus dorsal to unguis. Uropodal endopod subequal in length to sympod.

RECORDS Carrie Bow Cay, Belize, interstitial in intertidal sand bank.

REMARKS Two other species of *Angliera* have been recorded from the Caribbean area: *A. dubitans* Stock, 1977, from Bonaire, and *A. racovitzai* Coineau and Botosaneanu, 1973, from Cuba. The reader should refer to the original descriptions to distinguish the species, as differences are extremely subtle.

Microcharon Karaman, 1934

DIAGNOSIS Mandibular palp of three articles, two distal articles bearing spines and/or setae. Maxillipedal palp of five articles, articles 1, 2 and 3 expanded. ♂, pleopod 1 narrow, elongate, not obscuring pleopod 2.

REMARKS More than 20 species of *Microcharon* have been described worldwide. The genus is unusual in that the species have been found in true marine environments, in brackish habitats such as wells, and inland in freshwater.

Microcharon sabulum Kensley, 1984
Figure 41E–H

DIAGNOSIS ♂ 1.4 mm, ♀ 1.5 mm. Antennule of five articles. Inner ramus of maxilla 2 with pectinate spine. Pereopodal dactyli short, biunguiculate. En-

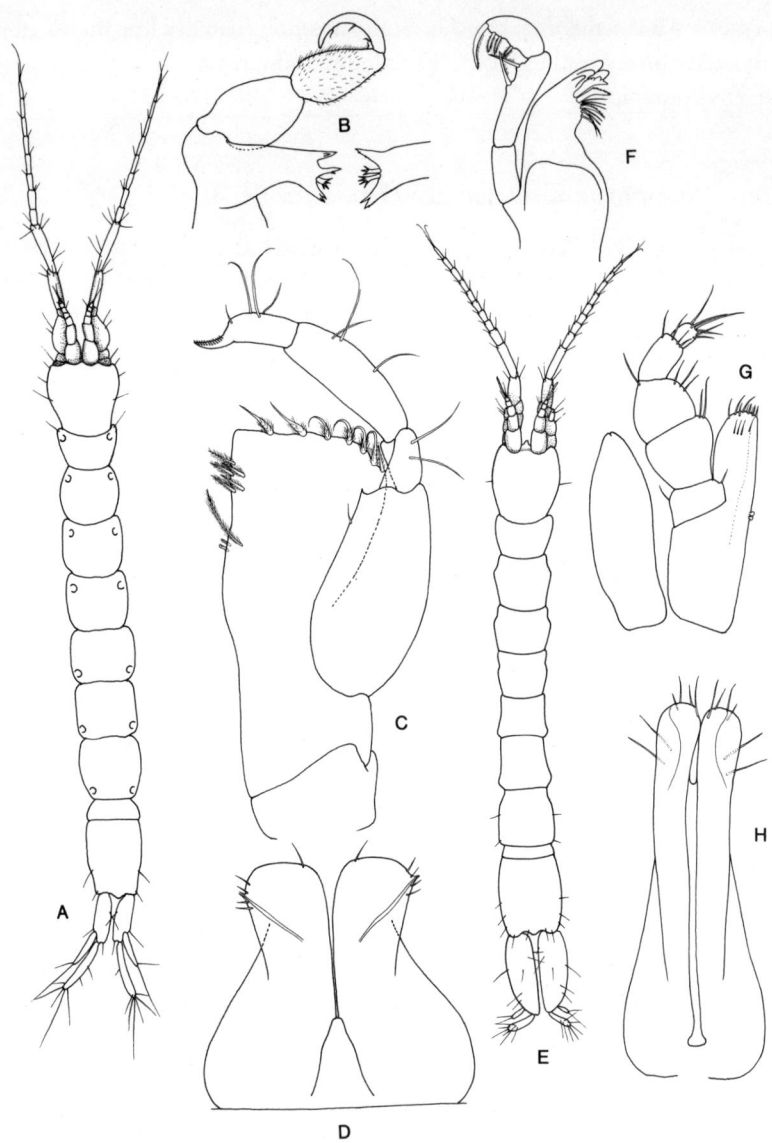

Figure 41. *Angliera psamathus:* A, ♂; B, right, and part of left, mandible; C, maxilliped; D, pleopod 1, ♂. *Microcharon sabulum:* E, ♂; F, mandible; G, maxilliped; H, pleopod 1, ♂.

dopod of pleopod 3 with three distal plumose setae. Uropodal sympod stout, longer than rami.

RECORDS   Carrie Bow Cay, Belize, interstitial in intertidal sand bank.

REMARKS   Two other species of *Microcharon* have been described from the Caribbean area: *M. phreaticus* Coineau and Botosaneanu, 1973, from interstitial freshwater on Cuba, and *M. herrerai* Stock, 1977, from freshwater wells on Bonaire. The reader should refer to the original descriptions to distinguish these species.

Family Munnidae Sars, 1899

DIAGNOSIS   Body ovate. Cephalon and all pereonites free; pleon narrower than rest of body, longer than broad. Eyes on lateral processes of cephalon. Mandible with molar and incisor present; palp present or absent. Maxillipedal palp articles 2 and 3 broader than remaining articles. Pereopod 1 prehensile; pereopods 2–7 ambulatory. Uropods tiny, without sympod. Anus exposed, not covered by pleopods.

REMARKS   Poore (1984) has provided the most useful and recent survey of the genera, and especially of *Munna*.

Key to genera of Munnidae

1. Mandibular palp present; pereopod 1 enormous in ♂, as compared with ♀ .... *Munna*
   Mandibular palp absent; pereopod 1 not sexually dimorphic
   .... *Uromunna*

Munna Krøyer, 1839

DIAGNOSIS   Body dorsally with numerous setae and/or articulating spines. Antennular flagellum with two distal articles each with single aesthetasc, terminal article minute. Mandibular molar strong, subcylindrical, distally truncate, with accessory setae; palp reaching beyond incisor, article 2 with few serrate spines. Pereopod 1 sexually dimorphic; pereopods 2–7 not (or barely) sexually dimorphic, dactyli with accessory claw. Pleopod 3, article 2 of exopod reaching well beyond endopod.

Munna petronastes Kensley, 1984
Figure 42A–D

DIAGNOSIS ♂ 1.1 mm, ♀ 1.0 mm. Pereopod 1 in ♂ enormously enlarged, carpochelate. Pleopod 1 in ♂ with distolateral angles projecting, acute. Body with anterodorsal U-shaped pigment band, and two converging bands on posterior pereon.

RECORDS Carrie Bow Cay, Belize, intertidal to 2 m, usually on corals.

Uromunna Menzies, 1962c

DIAGNOSIS Body with few if any dorsal setae, without articulating spines. Terminal antennular flagellar article not minute, bearing single aesthetasc. Mandibular molar strong, cylindrical, distally truncate, lacking accessory setae; palp present or absent. Pereopod 1 not sexually dimorphic, small. Pereopod 2 (rarely 2–7) sexually dimorphic, carpi and propodi broader in ♂ than in ♀. Pleopod 3, article 2 of exopod not reaching beyond endopod.

Key to species of *Uromunna*

1. Larger uropodal ramus parallel sided, about 3.5 times longer than basal width; inner uropodal ramus tiny, obscured by pleotelsonic margin .... *reynoldsi*
   Larger uropodal ramus tapering, about 1.5–2.0 times longer than basal width; inner uropodal ramus smaller than outer, but visible beyond pleotelsonic margin .... *caribea*

Uromunna caribea (Carvacho, 1977)
Figure 42E,F

DIAGNOSIS ♂ 1.5 mm, ♀ 1.5 mm. Propodus of pereopod 1 1.5–2 times longer than wide. Operculum of ♀ distally truncate. Shorter uropodal ramus visible beyond pleotelsonic margin, with single seta. Pigmentation in reticulation on cephalon and pereon; with six marginal patches on pleon.

RECORDS Turks and Caicos Islands, 1 m; Canal de la Belle Plaine, Guadeloupe, in water of 25‰.

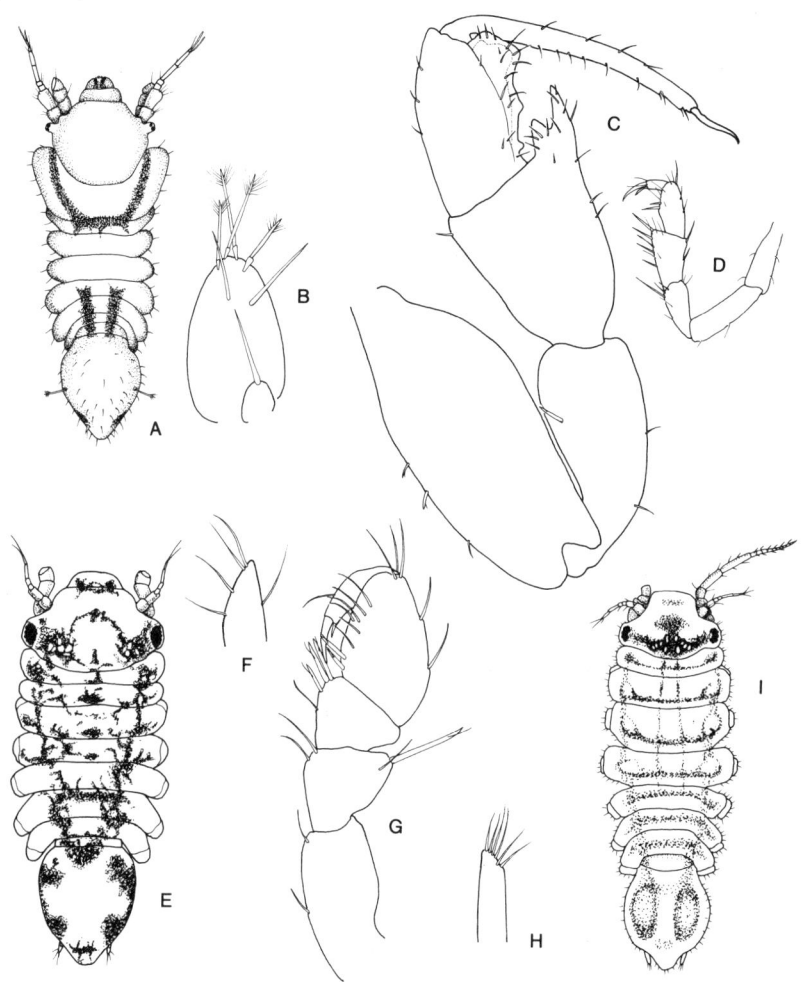

Figure 42. *Munna petronastes: A*, ♂; *B*, uropod; *C*, pereopod 1, ♂; *D*, pereopod 1 ♀ (*C* and *D* same scale). *Uromunna caribea: E*, ♂; *F*, larger uropodal ramus. *Uromunna reynoldsi: G*, pereopod 1, ♂; *H*, larger uropodal ramus; *I*, ♂.

Uromunna reynoldsi Frankenberg and Menzies, 1966
  Figure 42G–I

DIAGNOSIS ♂ 1.5 mm, ♀ 1.6 mm. Propodus of pereopod 1 two or three times longer than wide. Operculum of ♀ distally rounded. Shorter uropodal ramus obscured by pleotelsonic margin, with single seta. Pigmentation a broad patch between eyes on cephalon, lateral bands on pereon, and anterior and lateral patches on pleon.

RECORDS   Sapelo Island, Georgia, in tidal saltmarsh creek; Lake Ponchartrain, Louisiana; Atlantic and Pacific locks of Panama Canal.

Family Paramunnidae Vanhöffen, 1914

DIAGNOSIS   Body broad, ovate, often with laterally produced tergal or epimeral plates. Cephalon recessed into pereonite 1. Eyes, if present, on lateral projections of cephalon. Antennule short, usually of six articles, with single terminal aesthetasc. Antenna never longer than body. Mandibular palp present or absent. Pereopod 1 prehensile; pereopods 2–7 ambulatory. Pleopod 1 ♂ distally sagittate. Uropods with sympod minute or absent; rami tiny. Anus covered by pleopods.

Munnogonium George and Strömberg, 1968

DIAGNOSIS   Eyes present on short lateral processes of cephalon. Antennal peduncular scale present. Coxal plates visible on pereonites 2–7.

Munnogonium wilsoni Hooker, 1985
Figure 43A,B

DIAGNOSIS   ♂ 0.86 mm, ♀ 0.98 mm. Frontal margin of cephalon broadly rounded. Mandibular palp absent. Uropodal endopod twice length of exopod. Lateral margins of pleotelson to uropodal insertion serrate, posterior margin between uropodal insertions tapering to rounded apex.

RECORDS   Florida Middlegrounds, Gulf of Mexico, 55 m.

Family Pleurocopidae Fresi and Schiecke, 1972

DIAGNOSIS   Cephalon broader than long. Eyes (or at least ocular peduncles) present. Mandible with or without palp; molar truncate. Maxillipedal palp articles narrow, less than half width of endite. At least coxae of pereonites 5–7 dorsally visible. Pereopods 2–7 uni- or biunguiculate. Pleopod 1 in ♂ not sagittate. Uropod pedunculate, inserted laterally or slightly dorso- or ventrolaterally on pleotelson; biramous, or with one ramus fused with sympod.

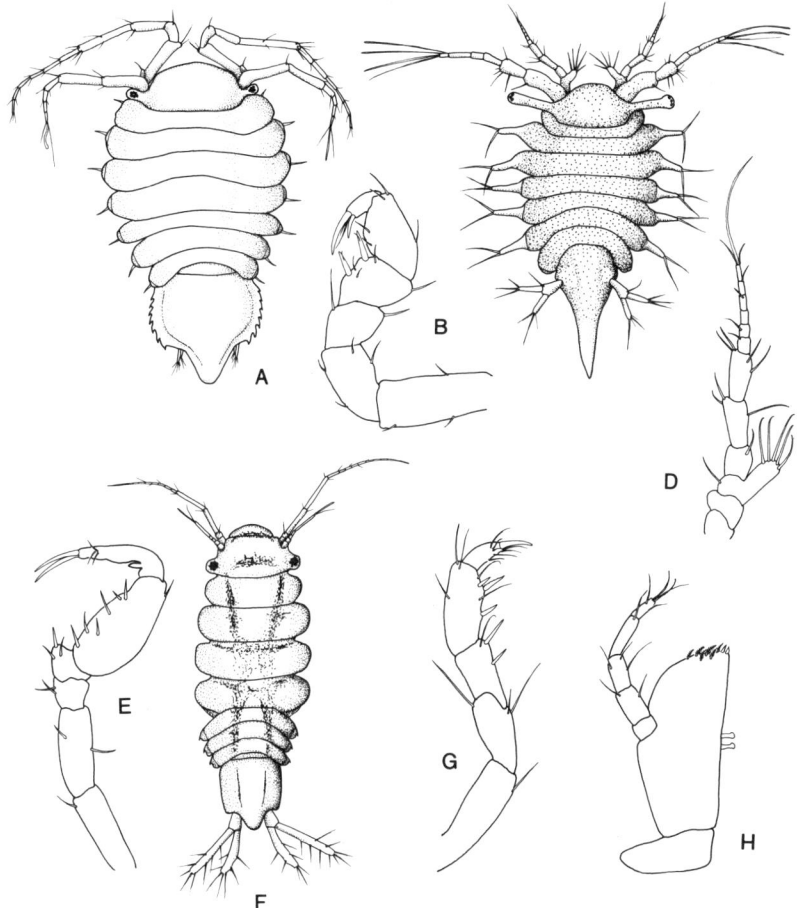

Figure 43. *Munnogonium wilsoni:* A, ♀; B, pereopod 1, ♂. *Pleurocope floridensis:* C, ♂; D, antennule ♂; E, pereopod 1 ♂. *Santia milleri:* F, ♂; G, pereopod 1, ♂; H, maxilliped.

Pleurocope Walker, 1901

DIAGNOSIS   Eyes present on lateral peduncle. Antennular peduncle of two articles; flagellum of four articles. Antennal peduncle of five (six) articles, scale lacking; flagellum of six or seven articles. Mandibular palp lacking. Pereopod 1 subchelate. Pereopods 2–7 uniunguiculate.

Figure 44. *Pleurocope floridensis:* adult in dorsal view.

Pleurocope floridensis Hooker, 1985
 Figure 43C–E, 44

DIAGNOSIS ♂ 1.15 mm, ♀ 0.96 mm. Body ovate, tapering posteriorly. Integument very finely tuberculate. Mesiodistal lobe on antennal peduncle article 3 bearing five distal setae. Pereon lacking long dorsal setae. Pereopod 1 subchelate, but almost carpochelate. Pleon consisting of single segment, posteriorly narrowly tapered and produced. Uropodal rami as long as sympod.

RECORDS Turks and Caicos Islands, 1 m; Carrie Bow Cay, Belize, 3–10 m; Florida Middlegrounds, Gulf of Mexico, 55 m.

Family Santiidae Wilson, 1987

DIAGNOSIS Antennular flagellum with at most, three articles, antennular scale sometimes present. Pereopod 1 prehensile. Pereopods 2–7, dactyli biunguiculate. Coxae visible at least on pereonites 5–7. Pleon consisting of single short pleonite plus pleotelson. Uropods pedunculate, biramous, inserted dorsally or laterally. (One species of *Santia* possesses a uniramous uropod.)

Santia Sivertsen and Holthuis, 1980

DIAGNOSIS Cephalon about twice wider than long. Eyes present. Antennular peduncle of three articles. Pereonites laterally rounded, sometimes bearing short lateral spines.

Santia milleri (Menzies and Glynn, 1968)
Figure 43F–H

DIAGNOSIS ♂ and ♀ 1.0 mm. Eye on short lateral process of cephalon. Mandibular palp of three articles. Maxillipedal palp articles all of similar width. Pereopod 1 barely subchelate. Uropod with sympod well developed, rami prominent, well developed.

RECORDS Carrie Bow Cay, Belize, intertidal to 30 m; Puerto Rico, 1.5 m; San Salvador, Bahamas, 6 m; Turks and Caicos Islands, 1 m; Anguilla; Jamaica; Cozumel, Mexico; Gulf of Mexico. Brazil, 1–6 m.

Superfamily Stenetrioidea Hansen, 1905a

DIAGNOSIS ♂: Pleopod 1 small, peduncles fused, rami separate, uniarticulate; pleopod 2 small, copulatory; pleopod 3 biramous, opercular. ♀: Pleopod 1 absent; pleopod 2 rami and peduncle fused to form operculum; pleopod 3 as in ♂.

Family Stenetriidae Hansen, 1905a

DIAGNOSIS ♂: Pereopod 1 frequently much bigger than in ♀, with distinctive lobes and teeth. Pleopods 1 and 2 reduced; pleopod 1 protopodite short, fused, rami separate. Pleopod 2, endopod elongate, flexed, exopod short. Pleopod 3 exopod basally broad, distally narrowed; endopod broad, biarticulate. ♀: Pleopod 2, rami fused, short, covering base of pleopod 3. Uropod with short sympod, rami relatively well developed, styliform, of single article.

Key to genera of Stenetriidae

1. Rostrum narrowly triangular, spikelike; two free, very short pleonites anterior to pleotelson .... *Stenobermuda*
   Rostrum short, basally broad, anteriorly truncate or broadly rounded; two or three very short free pleonites anterior to pleotelson
   .... *Stenetrium*

Stenetrium Haswell, 1881

DIAGNOSIS Eyes present. Cephalon broader than long. Rostrum short, basally broad, anteriorly truncate or rounded. Pereonites 1–4 with anterolateral projections; pereonites 5–7 projecting posteriorly. Pleotelson with sharp tooth anterior to small lateral notch.

REMARKS If fresh material is not available, and color pattern is lost in preservation, mature male material is needed as identification is based on the structure of male pereopod 1.

Key to species of *Stenetrium*

1. Eyes of few (not more than 10) ommatidia, not reniform 2
   Eyes of many ommatidia, reniform 3

2. Eyes of four ommatidia; ♂ pereopod 1 small, propodus not unusually broad .... *minocule*
   Eyes of more than four ommatidia; ♂ pereopod 1, propodus broad, with wide palm .... *patulipalma*

3. Pleotelsonic margins serrate 4
   Pleotelsonic margins entire 5

4. Rostrum convex, with fine marginal teeth; pereopod 1 ♂, propodal palm with three straight teeth .... *bowmani*
   Rostrum truncate; pereopod 1 ♂, propodal palm with two teeth, outer tooth elongate, curved .... *serratum*

5. Pereopod 1 ♂, carpus produced, apically acute .... *stebbingi*
   Pereopod 1 ♂, carpus produced, apically rounded .... *spathulicarpus*

Stenetrium bowmani Kensley, 1984
Figure 45

DIAGNOSIS ♂ 5.0 mm, ♀ 5.2 mm. Rostrum convex, with tiny marginal teeth. Lateral lobes of cephalon acute, margins serrate. Color pattern in small scattered red-brown chromatophores; irregular unpigmented patches on cephalon. pereonite 4, and pleon; chalky-white bands on antennae and uropods. ♂: Pereopod 1 propodus broad, palm with three teeth, outermost

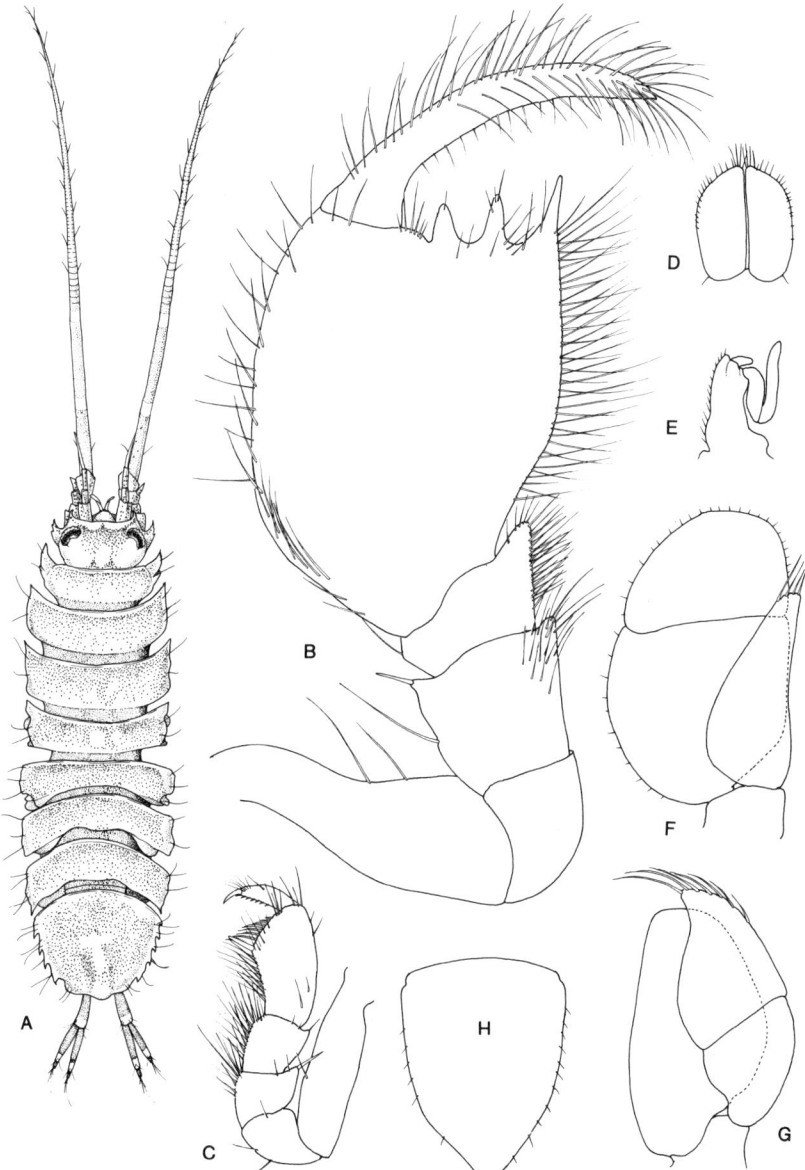

Figure 45. *Stenetrium bowmani*: *A*, ♂; *B*, pereopod 1, ♂; *C*, pereopod 1, ♀; *D*, pleopod 1, ♂; *E*, pleopod 2, ♂; *F*, pleopod 3; *G*, pleopod 4; *H*, operculum, ♀.

longest, slender. ♀: Pereopod 1 propodus with strong denticulate spine demarking palm, latter straight, with row of about seven slender spines.

RECORDS   Cozumel, Mexico; Carrie Bow Cay, Belize, 0.5–15.2 m, on algae and corals in reefcrest, and spur and groove zone of reef.

Stenetrium minocule Menzies and Glynn, 1968
Figure 46A–C

DIAGNOSIS   ♂ 2.8 mm, ♀ 3.7 mm. Eye of four ommatidia. Anterolateral lobes of cephalon blunt, barely produced. Rostrum poorly defined, truncate. ♂: Pereopod 1, carpus produced posterodistally into broadly rounded lobe; propodus broad, palm demarked by strong spine, with six low rounded teeth. ♀: Pereopod 1 propodus little broadened, palm demarked by strong denticulate spine, bearing several more spines.

RECORDS   Puerto Rico, intertidal to 3 m; Carrie Bow Cay, Belize, intertidal to 36 m, from rubble, algal turfs, and seagrass.

Stenetrium patulipalma Kensley, 1984
Figure 46D,E

DIAGNOSIS   ♂ 2.0 mm, ♀ 2.7 mm. Eyes of about 10 ommatidia in cluster. Rostrum poorly defined, truncate. Two basal articles of maxillipedal palp not enlarged. ♂: Pereopod 1 unknown. ♀: Pereopod 1 broadening to palm, and bearing row of about 12 small fringed spines. Color pattern: entire body with red-brown reticulation, dark transverse bars anteriorly on cephalon, pereonites 2 and 3, posteriorly on pereonites 4–7.

RECORDS   Carrie Bow Cay, Belize, 9.1–27.4 m; Barbados; Jamaica.

Stenetrium serratum Hansen, 1904
Figure 46F–H

DIAGNOSIS   ♂ 4.0 mm, ovigerous ♀ 4.9 mm. Rostrum truncate. Pereonites 1–5 with acute anterolateral angles. Pleotelsonic lateral margins with five teeth. ♂: Pereopod 1, propodus broad, palm with three teeth, outermost elongate, curved; dactylus reaching well beyond outermost palmar tooth. ♀: Pereopod 1, propodus much smaller than in ♂, palm bearing series of about nine fringed spines. Color: tiny red-brown chromatophores arranged in re-

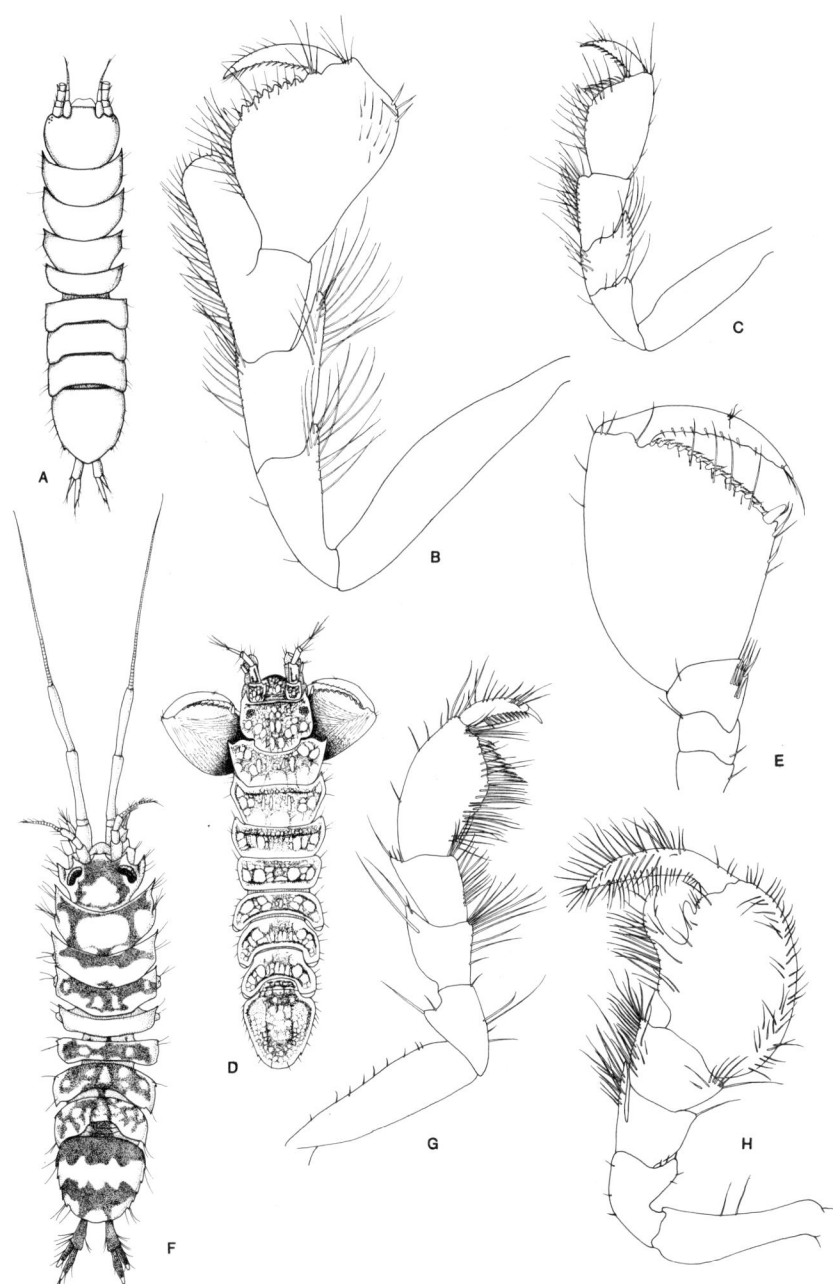

Figure 46. *Stenetrium minocule:* A, ♂; B, pereopod 1 ♂; C, pereopod 1 ♀. *Stenetrium patulipalma:* D, ♀; E, pereopod 1 ♀. *Stenetrium serratum:* F, ♂; G, pereopod 1 ♀; H, pereopod 1 ♂.

ticulate bands; distinctive open patches on cephalon and pereonite 1; pleon with two broad transverse bands.

RECORDS   Looe Key, Florida, 0.5–6 m; Turks and Caicos Islands, 1 m; Jamaica; Puerto Rico, intertidal to 3 m; St. Thomas, U.S. Virgin Islands; Carrie Bow Cay, Belize, intertidal to 15 m.

Stenetrium spathulicarpus Kensley, 1984
   Figure 47A–C

DIAGNOSIS   ♂ 4.1 mm, ♀ 4.1 mm. Rostrum truncate. ♂: Pereopod 1, merus and ischium each with setose fingerlike anterodistal projection; carpus with large spatulate and setose lobe almost reaching level of palm; propodal palm with large outer tooth and four or five low rounded teeth, broad band of setae near anterior margin; dactylus reaching slightly beyond palm, with band of setae along anterior margin. ♀: Propodal palm straight, with row of slender spines; band of setae in similar position as in ♂. Color: pigment in ill-defined and scattered reticulation; strong band on cephalon between eyes.

RECORDS   Carrie Bow Cay, Belize, intertidal to 36 m; Puerto Rico, intertidal.

Stenetrium stebbingi Richardson, 1902
   Figure 47 D–H

DIAGNOSIS   ♂ 4.8 mm, ovigerous ♀ 4.1 mm. Rostrum truncate. Ommatidia of eye more bunched than in *S. spathulicarpus*. ♂: Pereopod 1 variable according to maturity; carpus produced posterodistally into narrowly triangular, apically acute lobe; propodal palm poorly defined, with group of two to four teeth near dactylar articulation. ♀: Pereopod 1, palm straight, defined by strong outer tooth and bearing row of six or seven low rounded tubercles. Color: irregular reticulation of red-brown pigment, no strong band between eyes.

RECORDS   Bermuda, 0.5–4 m; Florida Keys, 18.3 m; Bahamas, 5 m; Turks and Caicos Islands, 1 m; Cuba; Jamaica; U.S. Virgin Islands, 50 m; Carrie Bow Cay, Belize, 0.5–36 m; Gulf of Mexico.

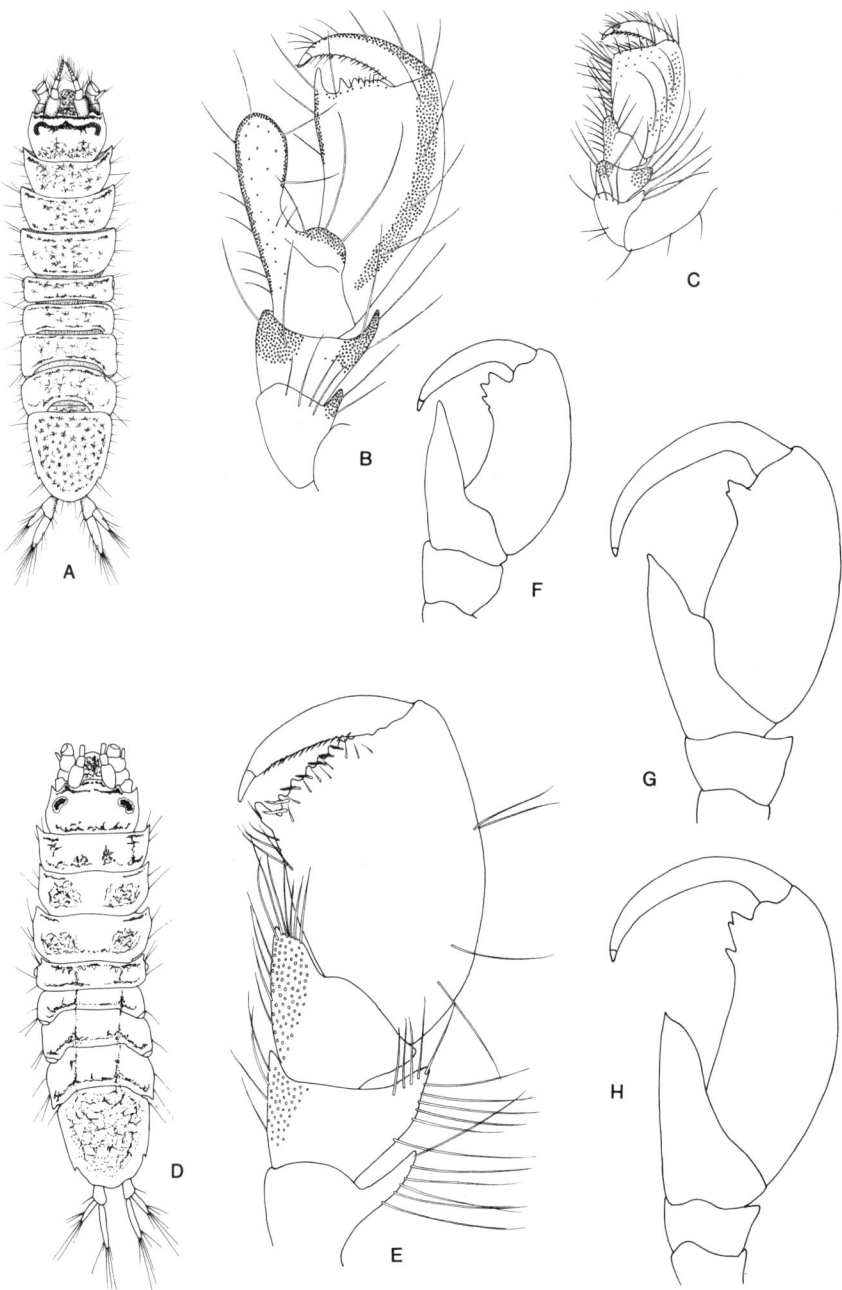

Figure 47. *Stenetrium spathulicarpus:* A, ♂; B, pereopod 1, ♂, many setae removed; C, pereopod 1, ♀. *Stenetrium stebbingi:* D, ♂; E, pereopod 1, ♀; F, G, H, variation in pereopod 1, ♂.

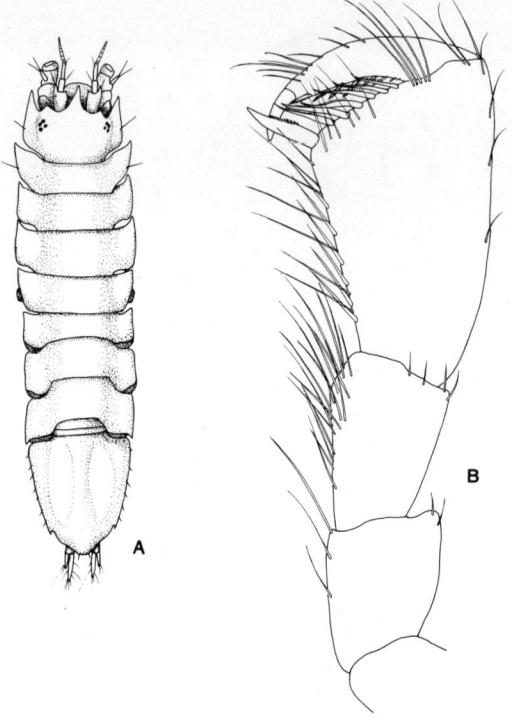

Figure 48. *Stenobermuda acutirostrata:* A, ♀; B, pereopod 1, ♀.

Stenobermuda Schultz, 1979

DIAGNOSIS Eyes of few ommatidia; rostrum narrow-based, elongate and spikelike. Pleon consisting of two free pleonites plus pleotelson, with posterolateral notch marked by tooth.

Stenobermuda acutirostrata Schultz, 1979
   Figure 48

DIAGNOSIS ♂ 4.8 mm. Eyes having five ommatidia. Spikelike rostrum reaching well beyond anterolateral angles of cephalon. Pereopod 1, propodus longer than wide, palm straight, bearing eight fringed spines, posterior margin with six spines and several setae.

RECORDS Off Bermuda, 90 m; Turks and Caicos, 1 m.

Suborder Epicaridea Latreille, 1831

DIAGNOSIS   Predominantly ectoparasites of marine crustaceans, feeding on blood. Eyes sessile, usually present in ♂, often reduced or lost in ♀. Antennae and antennules reduced; mouthparts reduced, forming a suctorial cone containing pair of piercing stylets formed from modified mandibles. Maxillae 1 and 2 reduced or lost. All mouthparts may be lost, and replaced by proboscis. ♂ small and isopodlike. ♀ undergo considerable distortion or reduction, often to unsegmented sacs of eggs in some forms. Ostegites usually retained. Two larval mancalike stages, epicaridium and cryptoniscium (Figure 49), characteristic of entire suborder.

REMARKS   The epicarideans are ectoparasites of other crustaceans, with the juveniles often using copepods as intermediate hosts. Sexual dimorphism is marked, the males being symmetrical with unambiguous segmentation, and considerably smaller than the often highly distorted females. In these, body segmentation is often obscured, with body segments often expanded on one side and reduced and compressed on the other. The marsupium of the ovigerous female, except in the Crytoniscidae and Entoniscidae, is made up of broadly lamellar oostegites, is relatively enormous and often obscures the rest of the body structure.

Crustacean hosts of the epicarideans are found in four classes: Ostracoda, Copepoda, Cirripedia, and Malacostraca, and in nine orders of the Malacostraca: Leptostraca, Stomatopoda, Mysidacea, Cumacea, Tanaidacea, Isopoda, Amphipoda, Euphausiacea, and Decapoda. The Epicaridea have been divided into two superfamilies, the Bopyroidea, containing families Bopyridae, Dajidae, and Entoniscidae, and the Cryptoniscoidea, containing the Crytoniscidae.

In the Bopyridae (Figure 50), the often asymmetrical adult female shows some segmentation. Seven pereopods may be present only on one side, their number being variable on the other. This is largest of the epicaridean families, containing over 400 species (Markham, 1974). Ten subfamilies have been recognized: the monotypic Entophilinae parasitizes galatheid crabs; six subfamilies, the Argeiinae, Bopyrinae, Bopyrophryxinae, Ioninae, Pseudioninae, and Orbioninae are all branchial parasites of decapod crustaceans; two subfamilies are abdominal parasites, the Phyllodurinae on callianassid mud-shrimps, and the Athelginae on hermit crabs; the Hemiarthrinae are known from the dorsal and ventral abdominal surfaces, and from the branchial chamber of caridean shrimps.

The Dajidae are ectoparasites pelagic mysidaceans, euphausiaceans, and decapod caridean shrimps. Adult females are often found unattached in plankton and pelagic samples. When attached, dajids are found on the

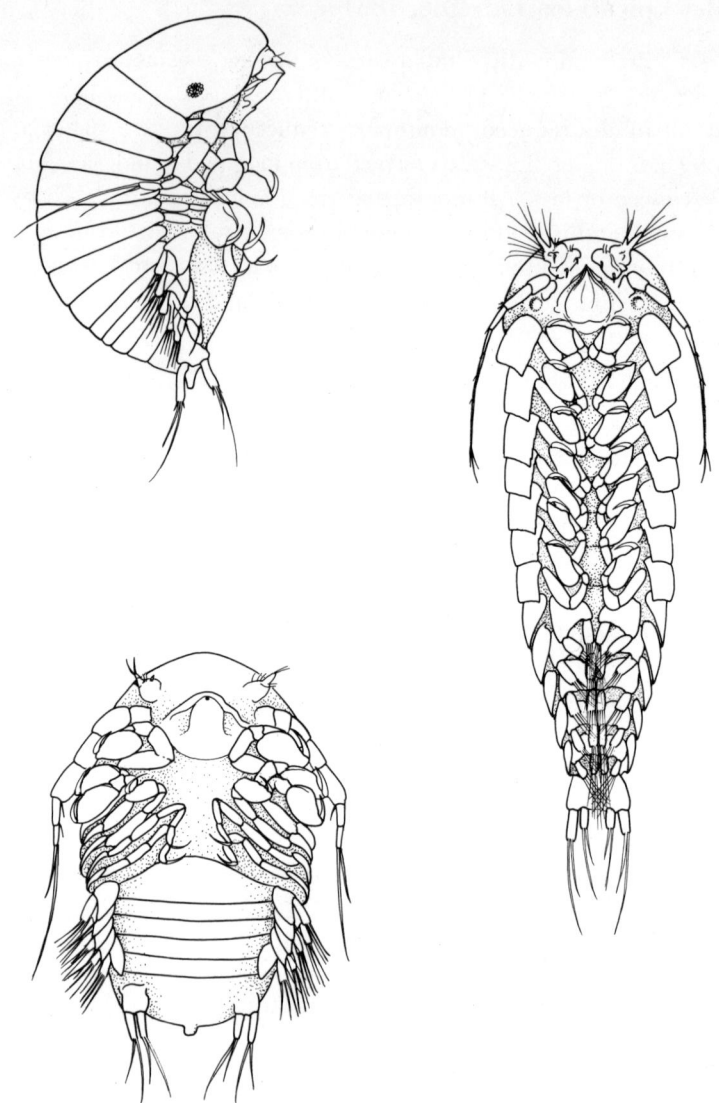

Figure 49. *A*, epicaridium larva, lateral view; *B*, epicaridium larva, ventral view; *C*, cryptoniscium larva, ventral view (from Bonnier, 1900).

EPICARIDEA 109

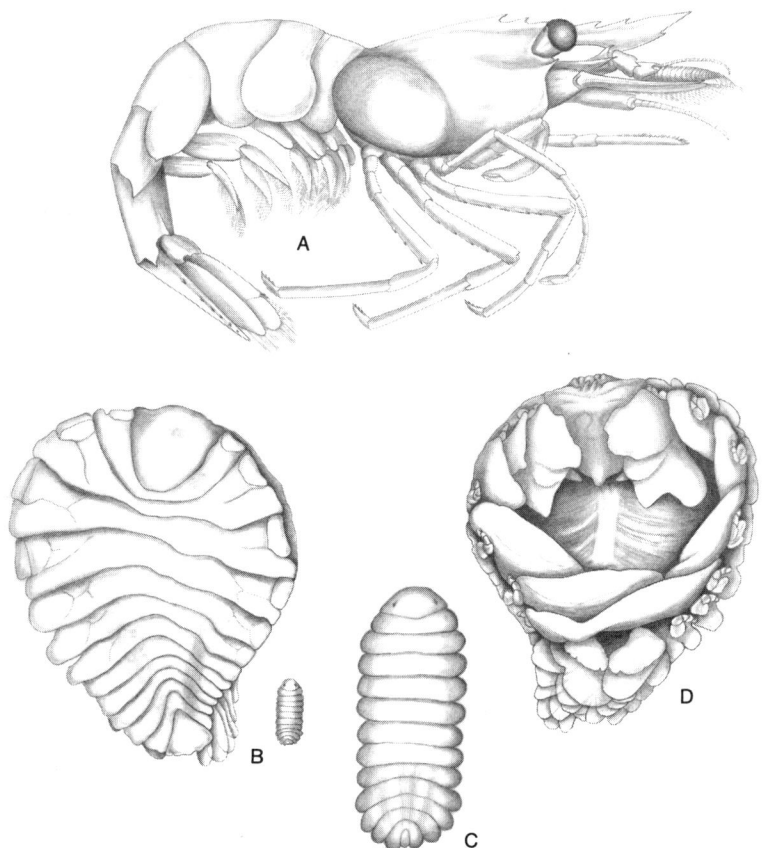

Figure 50. *A*, caridean shrimp with bopyrid parasite in branchial chamber. *Probopyrus pandalicola*: *B*, ♀ and ♂ in dorsal view, same scale; *C*, ♂ enlarged; *D*, ♀, ventral view, eggs removed from marsupium.

cephalothorax of the host, attached dorsally to the carapace, ventrally and laterally in the gill chambers and on the pereopods, or in the brood chambers.

The Entoniscidae are internal parasites of decapod crustaceans, being found in the visceral cavity, with the parasite's head in the position of the host's gonads or hepatopancreas. Veillet (1945) demonstrated that a pore to the host's branchial chamber connecting the parasite to the exterior is present only in hosts with mature parasites, to facilitate the release of epicaridium larvae.

The Cryptoniscidae are protandrous hermaphrodites. The female is

TABLE 2. CARIBBEAN EPICARIDEAN ISOPODS, THEIR HOSTS AND LOCALITIES

Achelion occidentalis Hartnoll, 1966
   Microphrys bicornutus (Latreille)
   Stenorhynchus seticornis (Herbst)
   Jamaica
Aporobopyrina anomala Markham, 1973
   Munida valida Smith
   Florida Keys; off Colombia; Gulf of Mexico
Aporobopyrus curtatus (Richardson, 1904)
   Petrochirus diogenes (Linnaeus)
   Petrolisthes armatus (Gibbes)
   Petrolisthes galathinus (Bosc)
   Petrolisthes marginatus Stimpson
   Porcellana sayana (Leach)
   Florida Keys; U.S. Virgin Islands; North Carolina
Argeia atlantica Markham, 1977
   Sclerocrangon jacqueti (A. Milne Edwards)
   Bahamas; Newfoundland
Astalione cruciaria Markham, 1975b
   Clastotoechus vanderhorsti (Schmitt)
   U.S. Virgin Islands
Asymmetrione clibanarii Markham, 1975d
   Clibanarius tricolor (Gibbes)
   Florida; Bahamas; Ascension Island
Asymmetrione desultor Markham, 1975d
   Pagurus bonairensis Schmitt
   Pagurus longicarpus Say
   Pagurus provenzanoi Forest and de Saint Laurent
   Pylopagurus sp.
   North Carolina; Florida Keys; Curaçao; Bonaire
Azygopleon schmitti (Pearse, 1932)
   Synalpheus brooksi Coutière
   Synalpheus hemphilli Coutière
   Synalpheus longicarpus Coutière
   Synalpheus mcclendoni Coutière
   Synalpheus pectiniger Coutière
   North Carolina to Florida; Bahamas; Hispaniola; Jamaica; Bonaire; Curaçao; Belize; Gulf of Mexico
Balanopleon tortuganus Markham, 1973
   Munida simplex Benedict
   Tortuga Island
Bopyrella harmopleon Bowman, 1956
   Synalpheus brevicarpus (Herrick)
   Synalpheus fritzmuelleri Coutière
   Synalpheus hemphilli Coutière
   Synalpheus minus Say
   Venezuela; Brazil
Bopyrina abbreviata Richardson, 1904
   Hippolyte curacaoensis Schmitt
   Hippolyte pleuracanthus (Stimpson)
   Hippolyte zostericola (Smith)
   North Carolina to Florida; Belize; West Indies; Gulf of Mexico
Bopyrinella thorii (Richardson, 1904)
   Thor floridanus Kingsley
   Florida; Curaçao; Yucatan Peninsula, Mexico
Bopyrione synalphei Bourdon and Markham, 1980
   Synalpheus goodei Coutière
   Synalpheus bousfieldi Chace
   Synalpheus brevicarpus (Herrick)
   Synalpheus pectiniger Coutière
   Florida; Haiti; Curaçao; Gulf of Mexico
Bopyrissa wolffi Markham, 1978
   Clibanarius tricolor (Gibbes)
   Clibanarius vittatus (Bosc)
   Bermuda; North Carolina to Florida; Bahamas; Puerto Rico; Gulf of Mexico
Cabirops sp.
   Synsynella deformans Hay
   Bermuda
Cancricepon choprae (Nierstrasz and Brender à Brandis, 1925)

Domecia acanthophora (Desbonne and Schramm)
Domecia hispida Eydoux and Souleyet
Eriphia gonagra (Fabricius)
Hexapanopeus angustifrons (Benedict and Rathbun)
Micropanope barbadensis Rathbun
Neopanope packardii (Kingsley)
Neopanope texana sayi (Smith)
Panopeus herbstii H. Milne Edwards
Panoplax depressa Stimpson
Paraliomera dispar (Stimpson)
Rithropanopeus harrisii (Gould)
Carolinas to Florida; Bermuda; Curaçao; Gulf of Mexico
Cancrion carolinus Pearse and Walker, 1939
Panopeus herbstii H. Milne Edwards
North Carolina; Bahamas
Dactylokepon caribaeus Markham, 1975c
Iliacantha liodactyla Rathbun
Iliacantha subglobosa Stimpson
Dominican Republic; Costa Rica–Panama
Dicropleon periclimenis Markham, 1972a
Periclimenes americanus Kingsley
St. Lucia Island
Diplophryxus sp. (see Markham, 1985)
Alpheus formosus Gibbes
Georgia; Florida; Yucatan, Mexico
Eophrixus subcaudalis (Hay, 1917)
Synalpheus brooksi Coutière
Synalpheus goodei Coutière
Synalpheus hemphilli Coutière
Synalpheus longicarpus (Herrick)
Synalpheus mcclendoni Coutière
Synalpheus pandionis Coutière
Synalpheus pectiniger Coutière
North Carolina to Florida; Yucatan Pensinsula, Mexico; Belize; Hispaniola; Curaçao

Gigantione mortenseni Adkison, 1984b
Dromidia antillensis Stimpson
Hypoconcha sabulosa (Herbst)
Hypoconcha spinosissima Rathbun
Florida; Haiti; Yucatan, Mexico; U.S. Virgin Islands; Gulf of Mexico
Hemiarthrus synalphei (Pearse, 1950)
Synalpheus fritzmuelleri Coutière
Synalpheus hemphilli Coutière
Synalpheus longicarpus (Herrick)
North Carolina to Florida; Haiti; Gulf of Mexico
Leidya bimini Pearse, 1951
Cyclograpsus interger (H. Milne Edwards)
Pachygrapsus transversus (Gibbes)
Sesarma ricordi H. Milne Edwards
Bermuda; Florida Keys; Bahamas; U.S. Virgin Islands; Jamaica; Panama
Leidya distorta (Leidy, 1855)
Uca pugilator (Bosc)
Uca spp.
New Jersey to Florida; Guadeloupe; Trinidad
Loki circumsaltanus Markham, 1972a
Thor floridanus Kingsley
Thor manningi Chace
Southern Florida; U.S. Virgin Islands; Belize
Metaphrixus carolii Nierstrasz and Brender à Brandis, 1931
Hippolyte pleuracanthus Stimpson
Southern Florida; U.S. Virgin Islands
Munidion cubense Bourdon, 1972
Munida flinti Benedict
Munida stimpsoni A. Milne Edwards
Cuba; Venezuela
Munidion irritans Boone, 1927
Munida irrasa A. Milne Edwards
Florida Keys; Belize

(*continued*)

TABLE 2. (*Continued*)

Munidion longipedis Markham, 1975a	*Parathelges piriformis* Markham, 1972b
Munida longipes A. Milne Edwards	*Paguristes oxyophthalmus* Holthuis
Munida schroederi Chace	*Pagurus brevidactylus* (Stimpson)
East coast of Florida; Florida Keys; Cuba; Gulf of Mexico	*Pagurus provenzanoi* Forest and de Saint Laurent
Parabopyrella lata (Nierstrasz and Brender à Brandis, 1929)	Bermuda; Bahamas; Colombia
Alpheus normanni Kingsley;	*Parathelges tumidipes* Markham, 1972b
Upogebia affinis (Say)	*Allodardanus bredini* Haig and Provenzano
Florida; U.S. Virgin Islands; Brazil	*Dardanus fucosus* Biffar and Provenzano
Parabopyrella mortenseni (Nierstrasz and Brender à Brandis, 1929)	Bermuda; Jamaica
Lysmata rathbunae Chace	*Pleurocrypta floridana* Markham, 1974
Lysmata wurdemanni (Gibbes)	*Galathea rostrata* A. Milne Edwards
Florida; U.S. Virgin Islands; Venezuela	Alligator Reef, Florida
Parabopyrella richardsonae (Nierstrasz and Brender à Brandis, 1929)	*Pleurocryptella fimbriata* Markham, 1973
Alpheus formosus Gibbes	*Munida constricta* A. Milne Edwards
Alpheus heterochaelis (Say)	*Munida miles* A. Milne Edwards
U.S. Virgin Islands; Gulf of Mexico	Western Caribbean; Cuba
Parabopyrella thomasi (Nierstrasz and Brender à Brandis, 1929)	*Probopyria alphei* (Richardson, 1900a)
Tozeuma carolinense Kingsley	*Alpheus armillatus* H. Milne Edwards
U.S. Virgin Islands	*Alpheus heterochaelis* Say
Parapagurion imbricata Markham, 1978	*Alpheus normanni* Kingsley
Paguristes tortugae Schmitt	*Alpheus viridari* (Armstrong)
Parapagurus sp.	North Carolina to Florida; Antilles; Brazil; Gulf of Mexico
Cuba; Colombia	*Probopyrinella latreuticola* (Gissler, 1882)
Parathelges foliatus Markham, 1972b	*Latreutes fucorum* (Fabricius)
Clibanarius vittatus (Bosc)	Bermuda; Sargasso Sea to Azores; North Carolina to Florida; Bahamas; Antilles; Gulf of Mexico
Pagurus brevidactylus (Stimpson)	
Barbados; Curaçao; Trinidad	*Probopyrus pandalicola* (Packard, 1879)
Parathelges occidentalis Markham, 1972b	*Macrobrachium acanthurus* (Wiegmann)
Clibanarius tricolor (Gibbes)	*Macrobrachium amazonicum* (Heller)
Iridopagurus sp.	*Macrobrachium bonelli* (Nobili)
Pylopagurus corallinus (Benedict)	*Macrobrachium carcinus* (Linnaeus)
North Carolina; Florida Keys; Bahamas; Venezuela	*Macrobrachium faustinum* (de Saussure)
	Macrobrachium ohione (Smith)
	Macrobrachium olfersii (Wiegmann)
	Macrobrachium surinamicum Holthuis

Palaemon northropi (Rankin)
Palaemon pandaliformis (Stimpson)
Palaemonetes exilipes Stimpson
Palaemonetes intermedius Holthuis
Palaemonetes kadiakensis Rathbun
Palaemonetes paludosus (Gibbes)
Palaemonetes pugio Holthuis
Palaemonetes vulgaris (Say)
Periclimenes americanus (Kingsley)
New Hampshire to Florida; Caribbean to Brazil; Pacific Panama
Pseudasymmetrione sp. (see Adkison and Heard, 1978)
  Iridopagurus iris (A. Milne Edwards)
  Venezuela
Pseudione affinis (Sars, 1882)
  Pandalus annulicornis Leach
  Pandalus bonnieri Caullery
  Pandalus leptorhynchus Kinahan
  Pandalus montagui Leach
  Plesionika antiguai Zariquiey
  Plesionika edwardsi (Brandt)
  Plesionika ensis (A. Milne Edwards)
  Plesionika heterocarpus (Costa)
  Plesionika martia (A. Milne Edwards)
  Bermuda; Northeastern Atlantic; South Africa; Java
Schizobopyrina urocaridis (Richardson, 1904)
  Periclimenes longicaudatus (Stimpson)
  Pontonia margarita Smith
  North Carolina to Florida; Belize; Gulf of Mexico
Stegias clibanarii Richardson, 1904
  Clibanarius tricolor (Gibbes)
  Bermuda

Stegophryxus hyptius Thompson, 1902
  Iridopagurus sp.
  Pagurus annulipes (Stimpson)
  Pagurus bonairensis Schmitt
  Pagurus brevidactylus (Stimpson)
  Pagurus longicarpus Say
  Pagurus provenzanoi Forest and de Saint Laurent
  Massachusetts to Florida; Curaçao
Synalpheion giardi Coutière, 1908
  Synalpheus longicarpus Herrick
  Yucatan, Mexico
Synsynella choprae (Pearse, 1932)
  Synalpheus brooksi Coutière
  Synalpheus longicarpus (Herrick)
  Synalpheus minus (Say)
  Synalpheus pandionis Coutière
  Bermuda; North Carolina to Florida; Bahamas; Haiti; U.S. Virgin Islands; Gulf of Mexico
Synsynella deformans Hay, 1917
  Synalpheus brooksi Coutière
  Synalpheus longicarpus (Herrick)
  Synalpheus pectiniger Coutière
  Bermuda; Carolinas to West Indies; Gulf of Mexico
Urobopyrus processae Richardson, 1904
  Ambidexter symmetricus Manning and Chace
  Processa acutirostris Nouvel and Holthuis
  Processa canaliculata Leach
  Processa edulis (Risso)
  Processa fimbriata Manning and Chace
  Processa tenuipes Manning and Chace
  Caribbean; Gulf of Mexico; Brazil; Mediterranean; Eastern Atlantic

reduced to a simple or lobed sac, generally without appendages. The broodpouch is formed by invagination of the ventral body wall. The eggs are released by the bursting of the sac. Cryptoniscids have been recorded as parasites of ostracods, cirripedes, mysidaceans, amphipods, isopods, and cumaceans. The majority feed on blood, but the females of some forms have been reported to be egg predators.

Given the highly variable morphology of the epicarideans, and the necessity of examining large series of specimens, keys are not provided and species are not treated individually here. As there is a degree of genus- and species-specificity for the hosts, Table 2 is provided to give a clue to the possible identity of a specimen. The student is then advised to consult one of the detailed works on the group. The most useful single work on the speciose Bopyridae for the area covered here is Markham (1985).

Suborder Flabellifera Sars, 1882

DIAGNOSIS Eyes usually well developed, reduced or absent in cave forms. Antennules and antennae uniramous; antennal peduncle of five or six articles. Mandible usually strong, adapted for cutting and grinding, occasionally for piercing; lacinia mobilis, spine-row, and molar usually present, although latter sometimes reduced; usually with triarticulate palp. Maxilla 1 biramous, sometimes adapted for piercing; maxilla 2 biramous, outer ramus

Key to families of Flabellifera

1. Pleon consisting of four or five free pleonites plus pleotelson 3
   Pleon consisting of not more than three free pleonites plus pleotelson 2

2. Pleon consisting of one or two free pleonites plus pleotelson; body usually dorsally strongly convex; pleopods subequal
   .... Sphaeromatidae
   Pleon consisting of three free pleonites plus pleotelson; body strongly depressed; pleopods 1–3 small, natatory, pleopods 4 and 5 large and broadly ovate .... Serolidae

3. Uropodal rami flattened, generally not reduced 4
   Uropodal rami reduced, exopod often hooklike .... **Limnoriidae**

4. Pereopods 4–7 prehensile, with dactyli longer than propodi; antennae reduced, with no clear distinction between peduncle and flagellum .... Cymothoidae
   Pereopods 4–7 ambulatory, with dactyli shorter than propodi; antennae normal, peduncle and flagellum clearly distinguished 5

5. Maxilliped bearing distal recurved hooks; pereopods 1–3 strongly prehensile .... Aegidae
   Maxilliped lacking distal recurved hooks; pereopods 1–3 ambulatory or at most weakly prehensile 6

6. Maxilliped lacking, or with very reduced endite; maxilla 1 a strongly falcate hook .... Corallanidae
   Maxilliped with strong endite; maxilla 1 not strongly falcate 7

7. Mandibular incisor distally narrowed, lacinia lacking; maxilla 1 slender and elongate, with 3–5 distal hooked spines .... Tridentellidae
   Mandibular incisor distally broad, cusped; maxilla 1 relatively broad, with several distal spines and setae .... Cirolanidae

usually consisting of two lobes. Pereopods generally ambulatory, sometimes prehensile; pereopods 1 and 2 subchelate only in Serolidae, ancinine Sphaeromatidae, and some Cirolanidae; posterior pereopods sometimes secondarily natatory in some cirolanids. Pleon consisting of as many as five free pleonites plus pleotelson, but pleonites variously fused in several families. Five pairs of pleopods usually present. Uropods lateral, usually forming tailfan with pleotelson.

REMARKS   This suborder contains a large group of diverse families, largely held together by primitive features such as the tailfan structure. Future work will undoubtedly show the Flabellifera to be an artificial polyphyletic group.

Family Aegidae Leach, 1815

DIAGNOSIS   Dorsal integument usually unornamented. Coxae distinct on pereonites 2–7. Eyes usually present, large, often almost, or complete contiguous. Mandible lacking lacinia mobilis, spine-row, and molar. Maxilla 1 slender, with apical spines. Maxilla 2 with two terminal unequal lobes bearing apical spines. Maxillipedal palp of two, three, or five articles. Pereopods 1–3 prehensile, with dactyli strongly curved; pereopods 4–7 ambulatory.

Pleopods biramous, bearing plumose marginal setae. Uropods forming tailfan with pleotelson. Pleon of four or five free pleonites plus pleotelson.

REMARKS   Although these large isopods (up to 60 mm) are often referred to as fish parasites, Brusca (1983) prefers the term "carnivorous scavengers and micropredators," as they attach to fish hosts infrequently and only long enough to feed. When feeding, they engorge themselves on the host's blood. Aegids show almost no host- (or rather prey-) specificity, being opportunistic feeders, and are most frequently captured by bottom trawls on the ocean bed. In ovigerous females, the maxillipedal articles become expanded and, along with the anterior oostegites, cover the buccal field, thereby making feeding impossible.

Key to genera and subgenera of Aegidae

1. Maxillipedal palp of two or three articles; frontal lamina small, narrow
 .... *Rocinela*
 Maxillipedal palp of five articles; frontal lamina large, broad 2
2. Antennular peduncle articles 1 and 2 expanded; cephalon lacking true rostrum, not completely separating antennular bases .... *Aega (Aega)*
 Antennular peduncle articles 1 and 2 not expanded; cephalon with true rostrum completely separating antennular bases ... *Aega (Rhamphion)*

Aega Leach, 1815

DIAGNOSIS   Eyes large, contiguous or separate. Cephalon with or without true rostrum. Frontal lamina broad, separating bases of antennae. Mandibular palp article 2 elongate. Maxilla 1 bearing strong apical and subapical spines. Maxilla 2 of two usually unequal lobes bearing stout spines. Maxillipedal palp of four or five articles, terminal article often small, with setae or recurved spines; article 4 with stout recurved spines; endite small, seldom reaching beyond palp article 2. Pleon not much narrower than pereon.

REMARKS   Brusca (1983) published a useful account of the genus *Aega* in the Eastern Pacific.

Key to species of *Aega (Aega)*

1. Eyes contiguous .... *deshaysiana*
   Eyes separate .... *ecarinata*

Key to species of *Aega (Rhamphion)*

1. Posterior margin of pleotelson distinctly dentate .... *dentata*
   Posterior margin of pleotelson at most faintly crenulate .... *tenuipes*

Aega (Aega) deshaysiana (H. Milne Edwards, 1840)
   Figure 51A

DIAGNOSIS ♀ 18.0 mm. Eyes large, contiguous. Cephalon with frontal margin acute to subacute. Frontal lamina large, shield shaped. Antennular peduncle articles 1 and 2 not expanded; flagellum of more than 15 articles. Uropodal endopod with deep notch in lateral (outer) margin. Pleotelson with basal width subequal to middorsal length, triangular, lateral margins faintly to markedly convex, tapering to narrowly rounded to subacute apex.

RECORDS Cuba; Yucatan Peninsula, Mexico; Gulf of Mexico.
Azores; Cape Verde Islands; Tristan da Cunha; Mediterranean; St. Paul and Amsterdam Islands; Seychelles; east coast of South Africa; Philippines; Japan; Hawaii; northeast Australia; Tasmania; Cocos Islands; Costa Rica.

REMARKS This species is more familiarly known in the Caribbean region as *Aega antillensis* Schioedte and Meinert, 1879.

Aega (Aega) ecarinata Richardson, 1898
   Figure 51B

DIAGNOSIS ♀ 21.0 mm. Eyes well separated. Articles 1 and 2 of antennular peduncle expanded. Propodus of pereopod 3 with posterodistal lobe. Uropo-

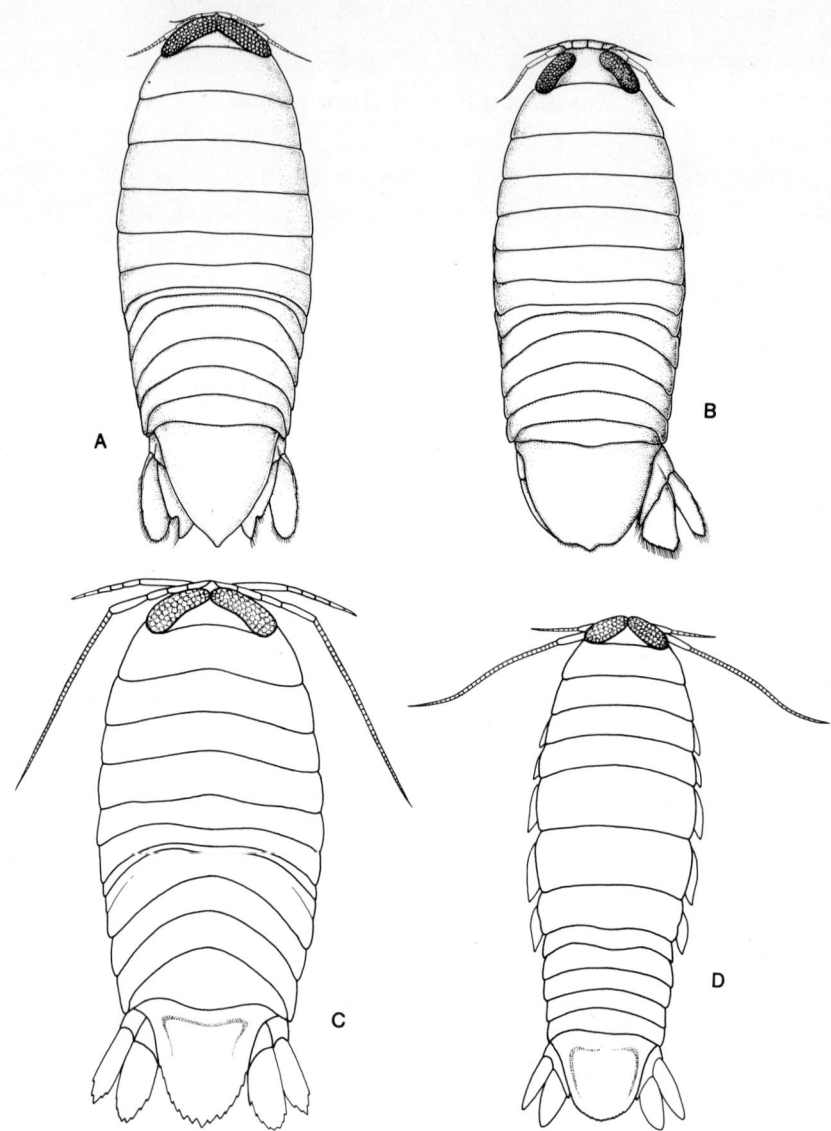

Figure 51. A, *Aega (Aega) deshayesiana;* B, *Aega (Aega) ecarinata;* C, *Aega (Rhamphion) dentata;* D, *Aega (Rhamphion) tenuipes.*

dal exopod narrower than endopod; latter distally truncate, lacking marginal notch. Pleotelson dorsally smooth, with posterior margin broadly trilobed.

RECORDS   Bahamas, 776 m; Puerto Rico; Gulf of Mexico, 176 m.

Aega (Rhamphion) dentata Schioedte and Meinert, 1879
Figure 51C

DIAGNOSIS ♀ 7.5 mm. Eyes large, just contiguous in midline. Frontal lamina distally acute. Uropodal exopod shorter than and half width of endopod; latter with lateral margin entire. Pleotelson with two obscure dorsal depressions anteriorly; posterior margin crenulate with seven teeth.

RECORDS Cuba.

Aega (Rhamphion) tenuipes Schioedte and Meinert, 1879
Figure 51D

DIAGNOSIS ♀ 11.5 mm. Eyes large, contiguous. Frontal lamina distally broadly rounded. Uropodal exopod shorter and narrower than endopod; latter with entire lateral margin. Pleotelson dorsally smooth; posterior margin evenly and broadly convex, obscurely crenulate.

RECORDS Cuba.

Rocinela Leach, 1818

DIAGNOSIS Cephalon with short rostrum sometimes covering antennular bases. Eyes well developed. Frontal lamina small, often indistinct. Mandibular palp of three articles, article 1 elongate. Maxillipedal palp of two or three articles. Pereopods 1–3 usually with spine-bearing expanded lobe on posterior margin of propodi.

Key to species of *Rocinela*

1. Eyes contiguous .... *oculata*
   Eyes not contiguous 2
2. Cephalon produced anteriorly into broadly rounded rostrum .. *cubensis*
   Cephalon lacking obvious broadly rounded rostrum 3
3. Eyes well separated, cephalon anteriorly broadly triangular .... *signata*
   Eyes barely separate, cephalon anteriorly narrowly triangular *insularis*

Rocinela cubensis Richardson, 1898
   Figure 52A

DIAGNOSIS ♂ 16 mm. Cephalon with two small tubercles between well-separated eyes; rostrum broadly rounded, extending anteriorly, very obvious. Flagellum of antenna with about 15 articles. Propodi of pereopods 1–3 with two spines. Pleotelson basally wider than middorsal length, lateral margins convex, apex rounded.

RECORDS Off Cuba, 290 m.

Rocinela insularis Schioedte and Meinert, 1879
   Figure 52B

DIAGNOSIS Ovigerous ♀ 24.5 mm. Eyes medially barely separated but not contiguous. Flagellum of antenna of more than 12 articles. Propodus of pereopods 1–3 with two to four spines on posterior lobe. Uropodal endopod barely reaching beyond pleotelsonic apex. Pleotelson basally slightly wider than middorsal length, lateral margins convex, apex rounded.

RECORDS Florida Keys; West Indies; between Mississippi delta and west coast of Florida, Gulf of Mexico; 550 m.

Rocinela oculata Harger, 1883
   Figure 52C

DIAGNOSIS ♀ 21.0 mm. Eyes contiguous. Cephalon with rostrum truncate in dorsal view. Antennal flagellum with 12 articles. Propodi of pereopods 1–3 with six to eight spines on lobed posterior margin. Pleotelson with basal width subequal to middorsal length; posterior margin broadly rounded.

RECORDS Off Georgia; Gulf Stream off Florida, 360–400 m; Puerto Rico, 84 m; Gulf of Mexico, 380–750 m.
   New South Wales, Queensland, Australia, 450–630 m.

Rocinela signata Schioedte and Meinert, 1879
   Figure 52D

DIAGNOSIS Ovigerous ♀ 13.0–15.0 mm. Cephalon anteriorly broadly triangular, produced over bases of antennules. Eyes widely separate. Flagellum of antenna with 10 or 11 articles. Pereopods 1–3, propodi unarmed or with

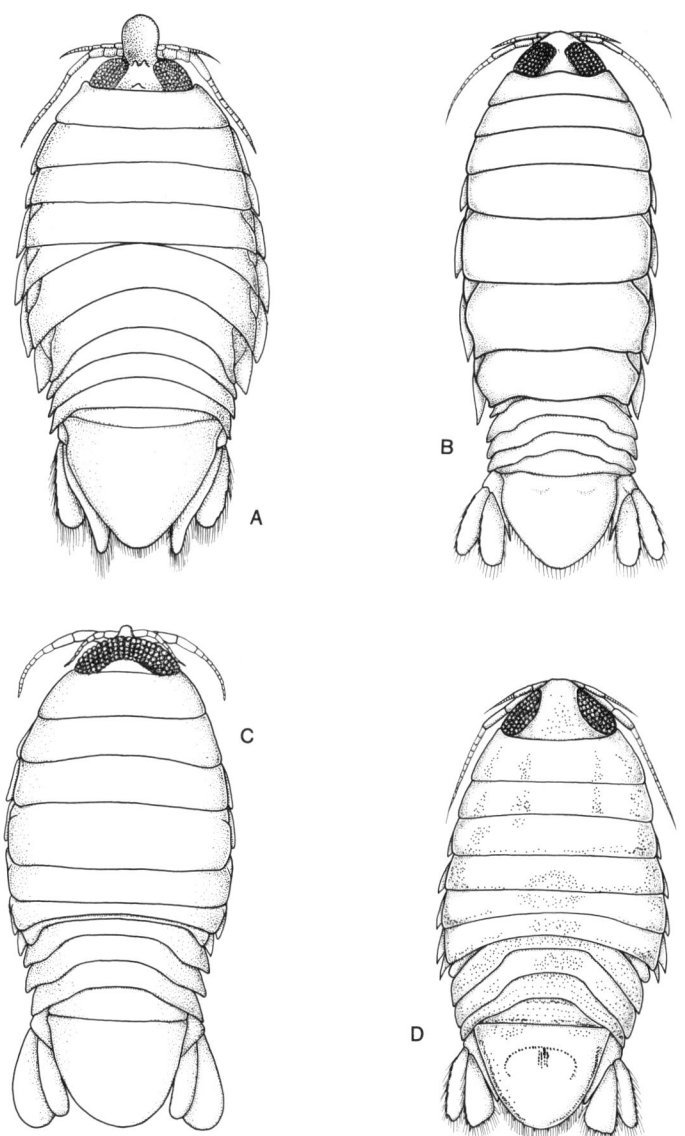

Figure 52. *A, Rocinela cubensis; B, Rocinela insularis; C, Rocinela oculata; D, Rocinela signata.*

single spine on posterior margin. Pleotelson with posterior margin evenly and broadly rounded; usually with inverted W-shaped band of pigment.

RECORDS   Florida Keys, shallow infratidal–4 m; Tortugas, from gills of jewfish *Epinephelus itajara,* mutton snapper *Lutjanus analis;* U.S. Virgin Islands, on mutton snapper *Lutjanus analis,* on yellowfin grouper *Mycteroperca venenosa;* Bahamas, on sheepshead *Archosargus probatocephalus,* on mutton snapper *Lutjanus analis,* on blackfin snapper *Lutjanus buccanella,* on queen triggerfish *Balistes vetula,* on saucereye porgy *Calamus calamus;* Jamaica, on French grunt *Haemulon flavolineatum,* hogfish *Lachnolaimus maximus,* on parrotfish *Sparisoma viride;* Haiti; Yucatan Peninsula, Mexico, 60–93 m, on gills of tiger shark *Galeocerdo cuvieri;* Puerto Rico, in gill slits of southern stingray *Dasyatis americana,* in gill slits of nurse shark *Ginglymostoma cirratum;* Jamaica; Carrie Bow Cay and Blue Ground Range, Belize, 0.5–2 m, on jolthead porgy *Calamus bajonado* and sheepshead porgy *Calamus penna,* on peacock flounder *Bothus lunatus,* on queen triggerfish *Balistes vetula,* on *Caranx* sp., on barracuda *Sphyraena barracuda,* on hogfish *Lachnolaimus maximus,* on mutton snapper *Lutjanus analis;* Venezuela, on *Orthopristis ruber,* on *Haemulon steindachneri;* Surinam, on gills of sheepshead porgy *Calamus penna;* Gulf of Mexico off Florida, shallow infratidal–55 m, on red grouper *Epinephelus morio,* on *Lutjanus blackfordi,* on black grouper *Mycteroperca bonaci,* on clearnose skate *Raja eglanteria.*

Pacific records: Southern California and Gulf of California; Socorro Island; Panama; Costa Rica.

REMARKS   While often taken from fish hosts (sometimes in the gill chamber), this species is equally frequently found freeliving in shallow water over sand and coral rubble. The species will also attach itself to humans, inflicting a sharp bite as it tries to feed.

Family Cirolanidae Dana, 1852

DIAGNOSIS   Eyes when present, relatively small, lateral. Frontal lamina present. Mandible with tridentate incisor, lacinia mobilis, blade- or sawlike molar, and palp. Maxillipedal palp of five articles, endite present. Coxal plates present on pereonites 2–7, distinctly separated by suture from tergite. Pereopods generally ambulatory, although anterior three pairs prehensile in some genera, and posterior four pairs natatory in some genera. Pleon of five free pleonites plus pleotelson in most genera; pleonite 5 with free lateral margins or overlapped by pleonite 4. Pleopods membranous, lacking ridges or folds. Uropods situated at anterolateral angles of pleotelson, freely articulating, both rami well developed, mobile.

REMARKS Of the many recent publications on the cirolanids, the most comprehensive is that of Bruce (1986) on the cirolanids of Australia. Botosaneanu, Bruce, and Notenboom (1986) tabulate all the known troglobitic cirolanids of the world.

Key to subfamilies of Cirolanidae

1. Clypeus projecting; pleonite 5 with free lateral margins (except in
   Xylolana) .... **Eurydicinae**
   Clypeus flattened, not projecting; pleonite 5 lacking free lateral margin,
   overlapped by pleonite 4 2

2. Pereopods 1–3 with ischium and merus not anterodistally produced;
   antennal peduncular articles 4 and 5 subequal; secondary unguis
   present on pereopodal dactyli .... **Cirolaninae**
   Pereopods 1–3 with ischium and merus anterodistally produced;
   antennal peducular articles 3 and 4 subequal; no secondary unguis
   on pereopodal dactyli .... **Conilerinae**

Subfamily Cirolaninae Dana, 1852

DIAGNOSIS Frontal lamina short, flat. Clypeus flattened, not projecting. Antennal peduncular articles 4 and 5 subequal, longer than articles 1–3. Pereopods with secondary unguis on dactyli. Penes reduced or absent. Pleonite 5 always overlapped by pleonite 4. Pleopod 2 in ♂ with copulatory stylet articulating basally.

Key to genera of Cirolaninae

1. Pleopods having accessory branchial filaments .... *Bathynomus*
   Pleopods lacking accessory branchial filaments 2

2. Pleopod 1 operculiform 7
   Pleopod 1 not operculiform 3

3. Pleopods 3–5, endopods lacking, or with very few, marginal setae ... 4
   Only endopod of pleopod 5 lacking marginal setae .... *Cirolana*

(continued)

Key to genera of Cirolaninae (*Continued*)

4. Merus of pereopod 1 posterodistally produced; merus of pereopods 2 and 3 anterodistally produced .... *Bahalana*
   Meri of pereopods 1–3 not markedly produced 5

5. Animal able to conglobate .... *Creaseriella*
   Animal unable to conglobate 6

6. Mandibular palp directed anteriorly .... *Anopsilana*
   Mandibular palp directed posteriorly .... *Haptolana*

7. Pleopod 1, exopod longer and broader than endopod ..... *Oncilorpheus*
   Pleopod 1, endopod longer and broader than exopod ...... *Calyptolana*

Anopsilana Paulian and Delamare Deboutteville, 1956

DIAGNOSIS Body unable to conglobate. Eyes present or absent. Frontal lamina well developed, as long as broad, or longer than broad, anteriorly

Key to species of *Anopsilana*

1. Estuarine-brackish water species; integument pigmented when alive 2
   Cave species; lacking integumental pigment 3

2. Frontal lamina distally rounded, projecting .... *browni*
   Frontal lamina distally acute, not projecting .... *jonesi*

3. Posterior margin of pleotelson with 10 or more spines 4
   Posterior margin of pleotelson with less than 10 spines 5

4. Posterior margin of pleotelson with 10 spines; found in cave on Cuba
   .... *cubensis*
   Posterior margin of pleotelson with 10–12 spines; found in cave on Haiti
   .... *acanthura*

5. Posterior margin of pleotelson with eight spines; found in cave on Grand Cayman .... *crenata*
   Posterior margin of pleotelson with four spines; found in cave on Haiti
   .... *radicicola*

somewhat expanded. Antennular peduncle of two articles. Maxillipedal endite with two coupling hooks. Pereopod 1 prehensile, pereopods 2–3 weakly prehensile, pereopods 4–7 ambulatory. Pleopod 2 ♂, copulatory stylet articulating at base of endopod. Pleopods 3–5, exopods biarticulate, endopods lacking marginal setae. Uropodal sympod produced along mesial margin of endopod.

Anopsilana acanthura (Notenboom, 1981)
Figure 53A,B

DIAGNOSIS ♂ 7.0 mm. Lacking eyes and integumental pigment. Frontal lamina anteriorly rounded. Posterior margin of pleotelson with 10–12 spines.

RECORDS Well at Marigot, Haiti.

Anopsilana browni (Van Name, 1936)
Figure 53C,D

DIAGNOSIS ♂ 11.1 mm, ♀ 10.0 mm. Eyes well developed, pigmented. Dorsal integument strongly pigmented with red-brown chromatophores. Frontal lamina as wide as long, anteriorly rounded. Cephalon with two fused middorsal tubercles near posterior margin. Pereonites and pleonites each with row of tubercles near posterior margin. Pleotelson triangular, with scattered dorsal tubercles, apex rounded, with eight spines.

RECORDS River in Santa Clara Province, Cuba (freshwater); Sittee River and Salt Creek, Stann Creek District, Belize (brackish water).
Golfo de Nicoya, Pacific Costa Rica, in red mangroves.

Anopsilana crenata Bowman and Franz, 1982
Figure 53E,F

DIAGNOSIS ♂ 6.2 mm. Lacking eyes and integumental pigmentation. Frontal lamina longer than wide, anteriorly rounded. Posterior margin of pleotelson with eight spines.

RECORDS West Bay Cave, Grand Cayman Island.

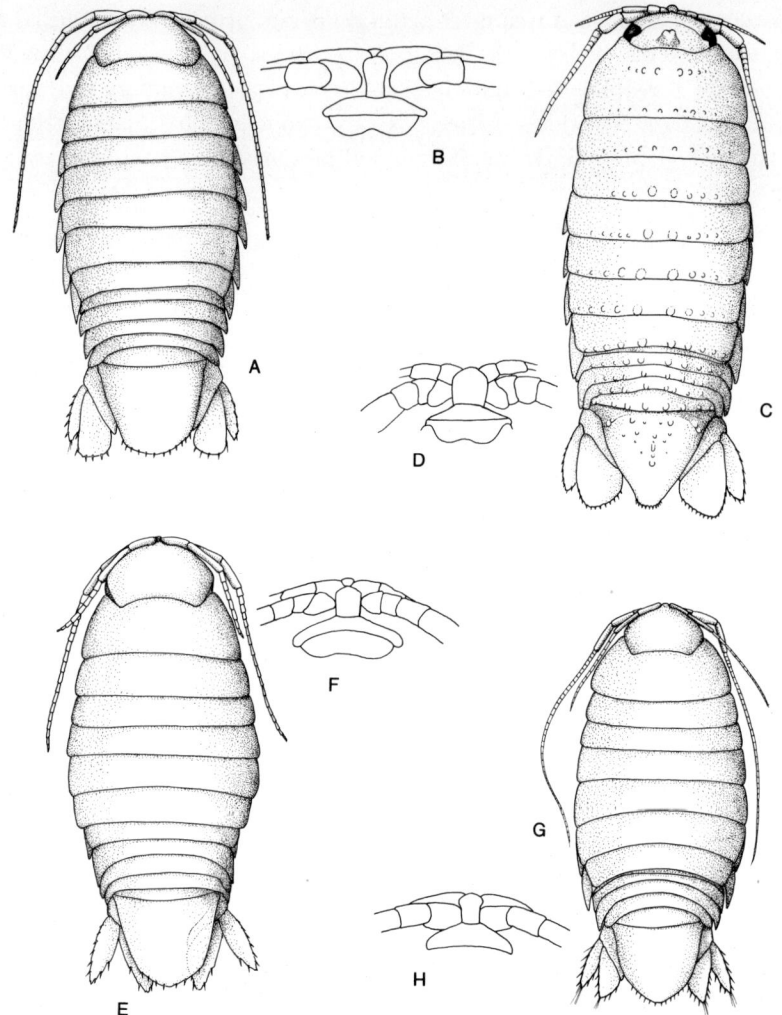

Figure 53. *Anopsilana acanthura:* A, ♀; B, anterior cephalon. *Anopsilana browni:* C, ♀; D, anterior cephalon. *Anopsilana crenata:* E, ♀; F, anterior cephalon. *Anopsilana cubensis:* G, ♀; H, anterior cephalon.

Anopsilana cubensis (Hay, 1903)
Figure 53G,H

DIAGNOSIS ♀ 7.0 mm. Lacking eyes and integumental pigmentation. Frontal lamina longer than wide, anteriorly expanded, rounded. Posterior margin of pleotelson with 10 spines.

RECORDS   Caves in provinces of Pinar del Río, La Habana, Matanzas, and on Isla de Pinos, Cuba.

Anopsilana jonesi Kensley, 1987
    Figure 54A–D

DIAGNOSIS   ♂ 7.2 mm, ovigerous ♀ 5.9 mm. Eyes well developed and pigmented. Dorsal integument strongly pigmented with almost solid central area on pereonites 1–7. ♂ cephalon with three low tubercles near posterior margin; pereonite 1 with four to six low tubercles. ♀ lacking tubercles on cephalon or pereonite 1. Pereonites 2–7 with low submedian longitudinal ridges near posterior margin. Frontal lamina narrow, pentagonal, anteriorly acute, not projecting. Posterior margin of pleotelson with 9 or 10 spines.

RECORDS   Salt Creek, and Sittee River, Stann Creek District, Belize, in estuarine mangroves.

Anopsilana radicicola (Notenboom, 1981)
    Figure 54E,F

DIAGNOSIS   ♂ 6.3 mm, ♀ 6.5 mm. Lacking eyes and integumental pigmentation. Frontal lamina longer than wide, anteriorly expanded and rounded. Posterior margin of pleotelson with four spines.

RECORDS   Source Débarasse, a spring near Jérémie, Haiti.

Bahalana Carpenter, 1981

DIAGNOSIS   Eyes lacking. Frontal lamina basally triangular, anteriorly narrowed into poorly developed carina. Pereopods 1–3 prehensile, dactyli and propodi relatively elongate; pereopod 1 carpus small, almost concealed; merus with strong posterodistal extension almost reaching dactylar base and armed with spines. Pereopods 2 and 3, meri with elongate anterodistal extension, meri and carpi with shorter posterodistal extensions. Pereopods 4–7 slender, ambulatory. Pleopod 2 in ♂ with copulatory stylet articulating basally on endopod. Pleopods 3–5, exopods biarticulate, endopods with few distal marginal setae, or lacking setae. Pleonite 5 with free lateral margin, hardly overlapped by pleonite 4.

Key to species of *Bahalana*

1. Pleopods 3–4, endopods lacking setae; maxillipedal endite with one coupling hook .... *cardiopus*
   Pleopods 3–4, endopods with few setae; maxillipedal endite with one or two coupling hooks 2

2. Antennular peduncle, article 3 longest; maxillipedal endite with two coupling hooks .... *geracei*
   Antennular peduncle, article 2 longest; maxillipedal endite with one coupling hook .... *mayana*

Bahalana cardiopus Notenboom, 1981
   Figure 55A,B

DIAGNOSIS   ♀ 10.0 mm. Maxillipedal endite with single coupling hook. Pereopod 1, meral projection bearing five distal spines. Pleopods 3–5, endopods lacking marginal setae. Uropodal exopod half width of endopod, four spines on outer margin.

RECORDS   Mount Misery Cave, Mayaguana Island, Bahamas.

Bahalana geracei Carpenter, 1981
   Figure 55C–G

DIAGNOSIS   ♀ 15.0 mm, ♂ 8 mm. Maxillipedal endite with two coupling hooks. Pereopod 1, meral projection bearing seven distal spines. Pleopods 3–5, endopods with 9–13 distal marginal setae; pleopod 5 endopod with four distal setae. Uropodal endopod bearing few spines at outer distal margin, margin not serrate, lacking distinct apex; exopod half width of endopod, with four spines on outer margin.

RECORDS   Lighthouse Cave, San Salvador Island, Bahamas.

Bahalana mayana Bowman, 1987
   Figure 55H

DIAGNOSIS   ♀ 8.4 mm, ♂ 10.0 mm. Clypeus acutely pointed. Antennular peduncle, article 2 longest. Maxillipedal endite with one coupling hook. Per-

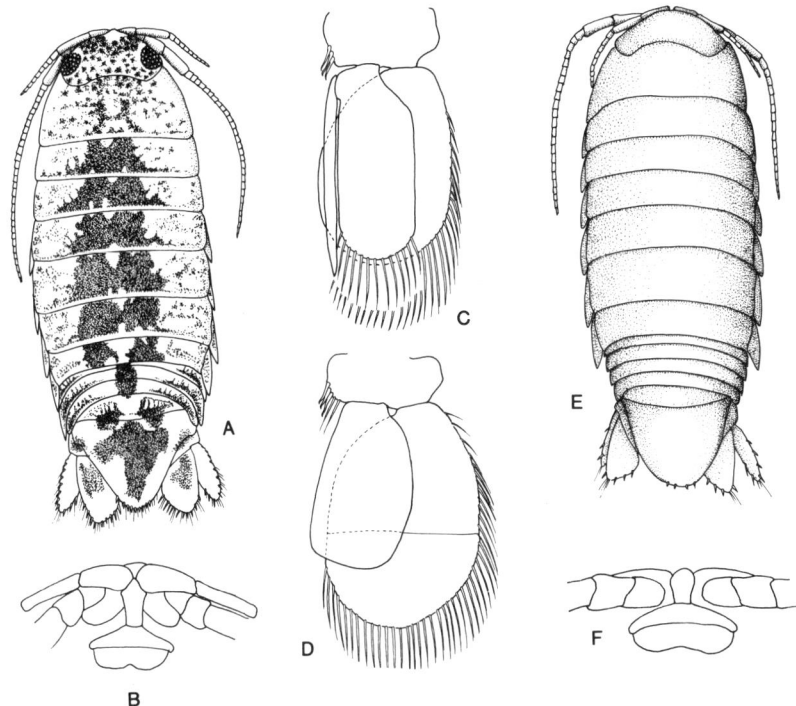

Figure 54. *Anopsilana jonesi*: A, ♂; B, anterior cephalon; C, pleopod 2, ♂; D, pleopod 3. *Anopsilana radicicola*: E, ♂; F, anterior cephalon.

eopod 1, meral projection rudimentary. Pleopods 3–4, endopods with few marginal setae, pleopod 5 lacking marginal setae. Uropodal exopod narrow, ⅓ width of endopod; endopod with distal margin slightly concave.

RECORDS Anchialine caves on Cozumel Island and at Tulum, Yucatan Peninsula, Mexico.

Bathynomus A. Milne Edwards, 1879

DIAGNOSIS Frontal lamina triangular; clypeus projecting anteriorly. Antennal peduncle with articles 3 and 4 subequal, article 5 longest. Maxillipedal endite with four to seven coupling hooks. Pereopods 1–3 with anterodistal margins of ischia and meri produced. Pleopods with all rami bearing marginal setae; endopods bearing accessory gills at bases. Posterior margin of pleotelson dentate.

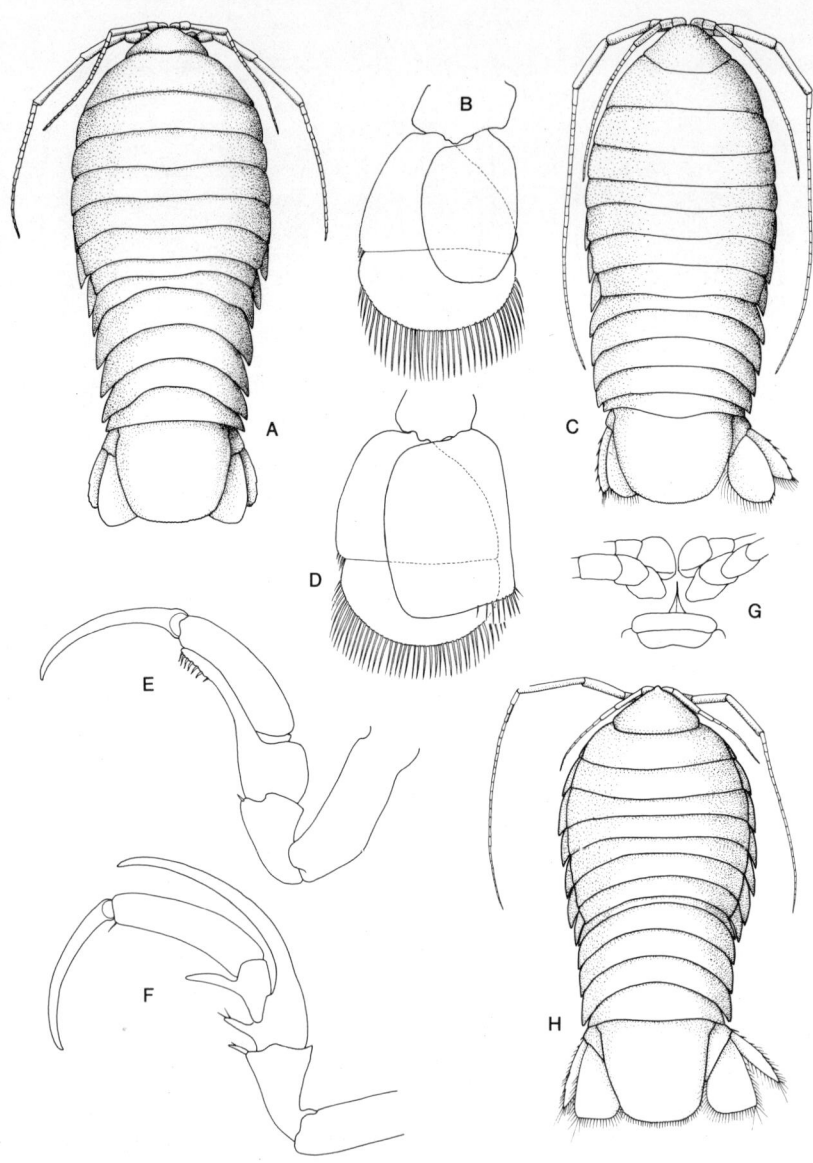

Figure 55. *Bahalana cardiopus:* A, ♀; B, pleopod 3. *Bahalana geracei:* C, ♀; D, pleopod 3; E, pereopod 1; F, pereopod 2; G, anterior cephalon. *Bahalana mayana:* H, ♀.

 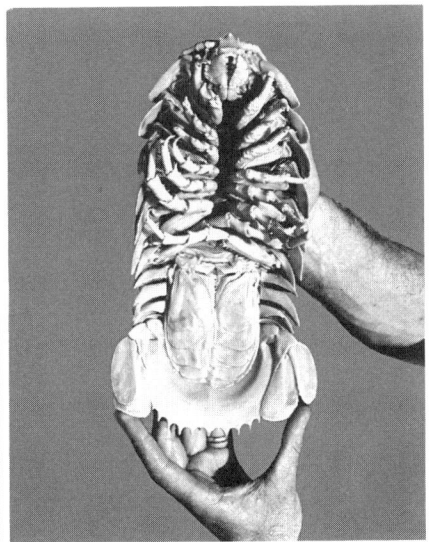

Figure 56. *Bathynomus giganteus*.

Bathynomus giganteus A. Milne Edwards, 1879
Figure 56

DIAGNOSIS   Up to 280 mm. Large pigmented eyes present, not visible dorsally. Antennule with small exopod distally on peduncular article 3. Pleopods with marginal setae on all rami; pleonites 3 and 4 with epimera produced posteriorly. Posterior margin of pleotelson with median tooth and five or six teeth on each side.

RECORDS   Gulf of Mexico; Caribbean; Bahamas; Florida Keys; 360–2300 m.

REMARKS   Gut-content analysis of these deep-water giants has shown them to be scavengers, commonly feeding on dead fish, cephalopods, crabs, and polychaete worms.

Calyptolana Bruce, 1985

DIAGNOSIS   All pereopods ambulatory; each dactylus with secondary unguis. Pleopod 1 operculate, longer than following pleopods. All pleopodal rami except endopod of pleopod 5 with marginal setae.

Calyptolana hancocki Bruce, 1985
Figure 57

DIAGNOSIS ♀ 3.0 mm. Body dorsally strongly convex. Cephalon with small rostrum curving ventrally to meet frontal lamina; latter pentagonal. Eyes small, well pigmented. Coxae of pereonites barely produced. Pleonite 5 lacking free lateral margins. Pleotelson with broadly rounded posterior margin. Uropodal sympod produced along mesial margin of endopod; exopod slightly less than half length of endopod; latter distally rounded.

RECORDS Dominican Republic; Aruba Island, Netherlands Antilles, 43.2 m.

Cirolana Leach, 1818

DIAGNOSIS Frontal lamina usually twice as long as wide, not projecting; clypeus flat. Mandible with strong incisor, dentate molar, palp of three articles. Pereopods all ambulatory. Pleon consisting of five free pleonites plus pleotelson, pleonite 5 overlapped laterally by pleonite 4. All pleopodal rami

Key to species of *Cirolana*

1. Pleotelson posteriorly very broad to subtruncate 2
   Pleotelson posteriorly narrowed 4

2. Uropodal endopod broad, distally rounded, lacking marginal spines
   .... *obtruncata*
   Uropodal endopod broad, distal margin having apical tooth or
   angle 3

3. Posterior margin of pleotelson bearing spines; uropodal endopod not
   dentate .... *minuta*
   Posterior margin of pleotelson faintly crenulate, lacking spines;
   uropodal endopod strongly dentate .... *crenulitelson*

4. Uropodal endopod evenly tapering to acute apex; uropodal exopod
   length about four times greatest width .... *albidoida*
   Uropodal endopod, outer margin convex; uropodal exopod length less
   than four times greatest width .... *parva*

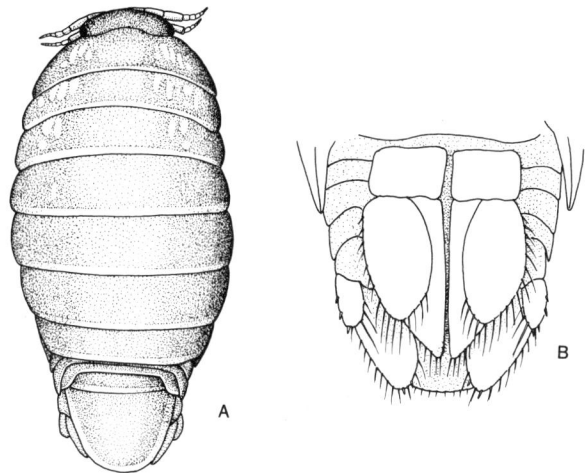

Figure 57. *Calyptolana hancocki*: A, ♀; B, ventral pleon.

bearing marginal setae except endopod of pleopod 5. Copulatory stylet on endopod of ♂ pleopod 2 inserted proximally.

Cirolana albidoida Kensley and Schotte, 1987
Figure 58A–C

DIAGNOSIS ♂ 7.8 mm. Integument sparsely pitted. Antenna reaching posteriorly to pereonite 3. Uropodal endopod triangular, evenly tapering, margins serrate; exopod length about four times greatest width, apically acute. Pleopod 1 ♂, endopod markedly narrowed in distal half. Pleopod 2 ♂, copulatory stylet reaching beyond rami by about half its length. Pleotelson with sides gently convex, tapering to rounded posterior margin bearing eight spines and three or four small serrations anterior to first spine.

RECORDS Off Lucaya, Grand Bahama, 180–220 m.

Cirolana crenulitelson Kensley and Schotte, 1987
Figure 58D–F

DIAGNOSIS ♂ 7.0 mm, ♀ 7.0 mm. Antenna reaching posteriorly to anterior of pereonite 3. Pleopod 2 ♂, copulatory stylet reaching by ⅕ its length be-

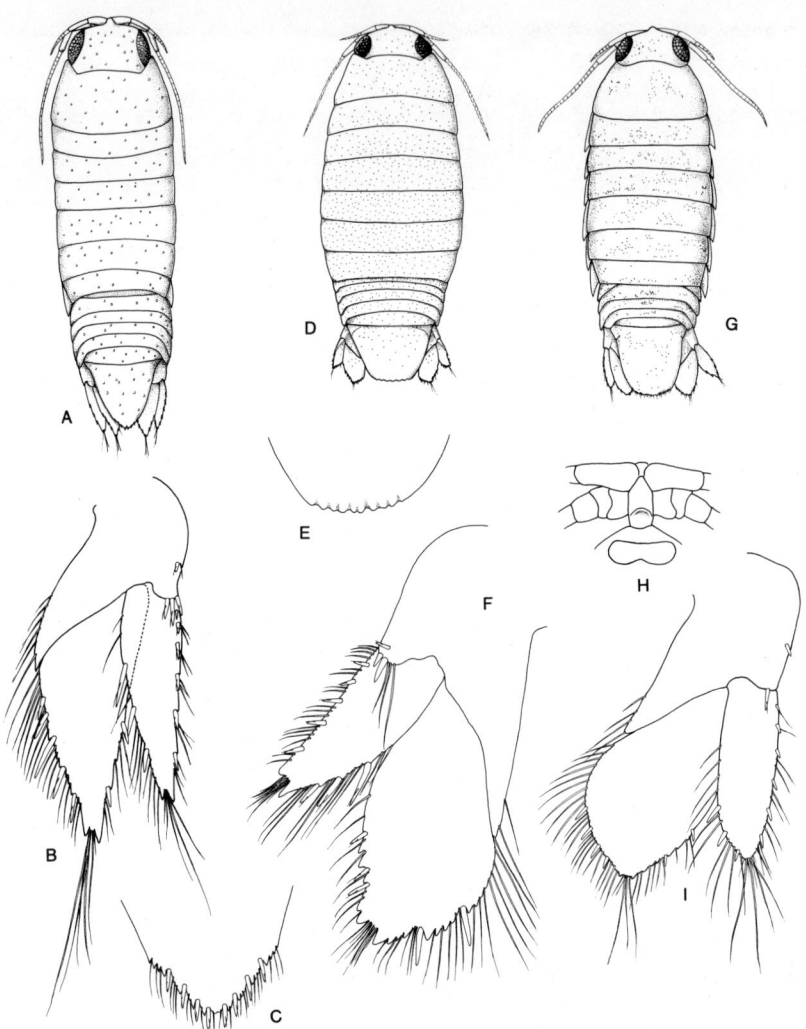

Figure 58. *Cirolana albidoida*: A, ♂; B, uropod; C, pleotelsonic apex. *Cirolana crenulitelson*: D, ♀; E, pleotelsonic apex; F, uropod. *Cirolana minuta*: G, ♂; H, anterior cephalon; I, uropod.

yond rami. Uropodal endopod with mesial margin broadly convex, serrate, apically acute; exopod about 2.5 times longer than wide, mesial margin weakly convex, apically acute. Posterior margin of pleotelson truncate, faintly crenulate, lacking spines.

RECORDS   Carrie Bow Cay, Belize, 36 m.

Cirolana minuta Hansen, 1890
Figure 58G–I

DIAGNOSIS ♂ 8.9 mm, ♀ 5.0 mm. Antenna reaching posteriorly to pereonite 3. Pleopod 2 ♂, copulatory stylet reaching by about ⅙ of its length beyond rami. Uropodal endopod, mesial margin broadly convex, with apical angle of about 90°; exopod about three times longer than wide, parallel sided for proximal two-thirds, with marginal spines, apically narrowly rounded. Pleotelson with posterior margin subtruncate to broadly rounded, with about eight marginal spines.

RECORDS St. Thomas, U.S. Virgin Islands; Tobago; 180–220 m.

Cirolana obtruncata Richardson, 1901
Figure 59A,B

DIAGNOSIS ♂ 11.3 mm, ♀ 11.0 mm. Antenna reaching posteriorly to middle of pereonite 3. Pleopod 2 ♂, copulatory stylet just reaching to distal margin of rami. Uropodal endopod distally broadly rounded, margin with rounded teeth; exopod 2.5 times longer than wide, margin with rounded teeth, apically obscurely subacute. Posterior margin of pleotelson subtruncate, with about eight spines.

RECORDS Jamaica; Puerto Rico; Cozumel, Mexico; Gulf of Mexico.

Cirolana parva Hansen, 1890
Figures 59C–E, 60

DIAGNOSIS ♂ 6.9 mm, ♀ 7.9 mm. Cephalon with furrow between eyes having middorsal posterior deflection. Antenna reaching posteriorly to pereonite 4. Pleopod 1 ♂, endopod narrowed in distal third. Pleopod 2 ♂, copulatory stylet reaching by ¼ its length beyond rami. Uropodal endopod with mesial margin convex, apically acute; exopod 2.5 times longer than wide, apically acute. Pleotelson evenly tapering to angled posterior margin, with seven or eight spines.

RECORDS North and South Carolina; Turks and Caicos Islands; St. Thomas and St. Croix, U.S. Virgin Islands; Andros Island, Bahamas; Puerto Rico; Jamaica; Florida Keys; Dry Tortugas; Barbados; Carrie Bow Cay, Belize; Cozumel, Mexico; Panama; Gulf of Mexico; intertidal to 55 m.

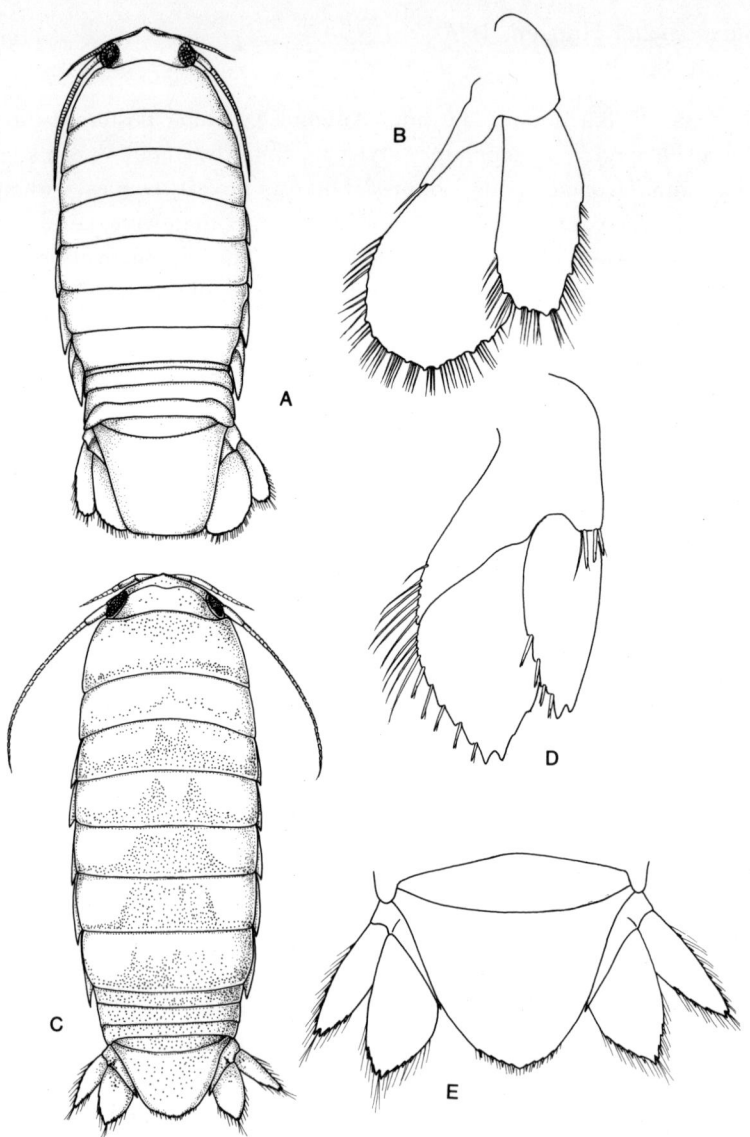

Figure 59. *Cirolana obtruncata:* A, ♀; B, uropod. *Cirolana parva:* C, ♀; D, uropod; E, pleotelson and uropods.

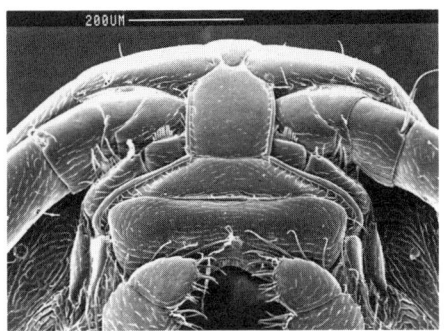

Figure 60. *Cirolana parva*, anteroventral cephalon.

Creaseriella Rioja, 1953

DIAGNOSIS Antennular peduncle of two articles (articles 1 and 2 fused). Antennal peduncle of five articles. Pereopods all ambulatory. Penial rami fused to form short stout process. Pleopod 2 in ♂ with copulatory stylet inserted at base of endopod. Pleopods 3–5, endopods lacking marginal setae. Pleonite 5 lacking free lateral margins.

Creaseriella anops (Creaser, 1936)
Figure 61A,B

DIAGNOSIS ♂ 15.5 mm. Animal able to conglobate. Eyes lacking. Frontal lamina pentagonal, longer than wide, with transverse ridge at widest point. Maxillipedal endite with four or five coupling hooks. Pleotelson wider than long, posterior margin very broadly rounded and finely crenulate. Both uropodal rami distally rounded, margins bearing spines and setae, sympod produced along mesial margin of endopod.

RECORDS Several caves and cenotes on the Yucatan Peninsula, Mexico.

Haptolana Bowman, 1966

DIAGNOSIS Mandibular palp directed posteriorly. Antennular peduncle of two articles, basal article expanded. Pereopods all prehensile, with dactyli closing in propodal groove. Pleopods 3–5, exopods with partial suture and marginal setae; endopods undivided, lacking marginal setae. Pleonite 4 overlapping pleonite 5 laterally.

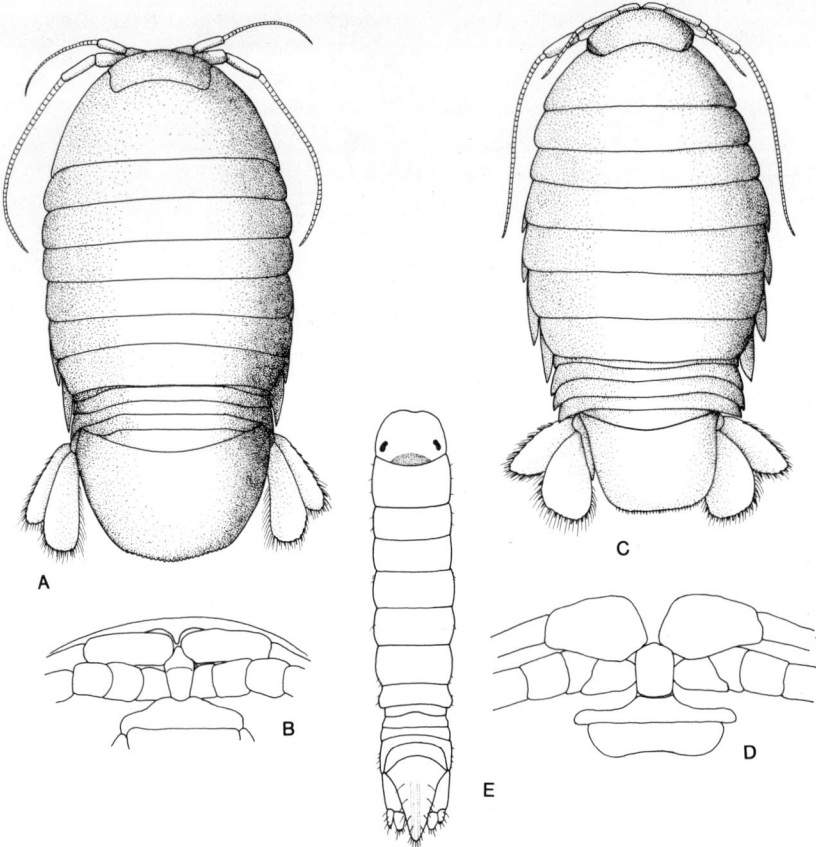

Figure 61. *Creaseriella anops:* A, adult; B, frontal lamina. *Haptolana trichostoma:* C, adult; D, frontal lamina. *Oncilorpheus stebbingi:* E, juvenile (from Paul and Menzies, 1971).

Haptolana trichostoma Bowman, 1966
    Figure 61C,D

DIAGNOSIS ♂ 13.8 mm. Eyes lacking. Frontal lamina broad, pentagonal, with anterior angle very broad. Pleotelson wider than long, roughly rectangular, posterior margin faintly crenulate, with spine between and setae on crenulations. Uropodal endopod distally broad, exopod distally narrowly rounded, ²/₃ width of endopod, both rami with marginal setae and spines.

RECORDS Cave in Camaguey Province, Cuba.

REMARKS A second species of *Haptolana*, *H. somala* Messana and Chelazzi, has been described from northern Somalia in Africa.

Oncilorpheus Paul and Menzies, 1971

DIAGNOSIS Frontal lamina projecting ventrally. Pleopod 1, exopod indurate, opercular; endopod membranous, less than half width of exopod. Uropodal sympod longer than rami, slightly produced along mesial margin of endopod; rami inserted subapically.

Oncilorpheus stebbingi Paul and Menzies, 1971
Figure 61E

DIAGNOSIS ♀ 11.0 mm. Body narrow, about five times longer than wide. Pleotelson triangular, bearing faint middorsal longitudinal ridge, apex narrowly rounded.

RECORDS Off Venezuela, 73 m.

Subfamily Conilerinae, new name

DIAGNOSIS Clypeus flattened. Frontal lamina flat, narrow. Antennal peduncular articles 3 and 4 subequal. Pereopods 1–3, ischium and merus produced anterodistally. Pereopods lacking secondary unguis on dactyli. Natatory setae present on pereopods 4–7.

REMARKS In a discussion of the Cirolanidae, Botosaneanu, Bruce, and Notenboom (1986:412) refer to the subfamilies Eurydicinae and Cirolaninae but place the Conilera group under the heading "Unnamed Subfamily." For consistency, the Conilera group is here recognized as a subfamily.

Key to genera of Conilerinae

1. Uropodal endopod with distal notch; pereopods 4–7 natatory, ischium, merus, and carpus flattened .... *Politolana*
   Uropodal endopod lacking distal notch; pereopods 4–7 with basis flattened and expanded, bearing natatory setae .... *Natatolana*

Natatolana Bruce, 1981

DIAGNOSIS Frontal lamina narrow; clypeus flat. Pereopods 1–3 bearing long setae. Pleopod 2 ♂, copulatory stylet articulating basally on endopod.

Natatolana gracilis (Hansen, 1890)
Figure 62

DIAGNOSIS ♂ 8.0 mm. Antennule short, not reaching distal end of antennal peduncle. Antenna reaching posteriorly to pereonite 4. Pleopod 2, copulatory stylet of endopod cylindrical, bowed, distally rounded. Pleotelson with obtuse apex, with slight transverse indentation near anterior margin.

RECORDS Probably St. Thomas, U.S. Virgin Islands; off Sombrero Light, Florida, 100–120 m.
Northern Brazil, 7–85 m.

REMARKS Hansen (1890) indicated some uncertainty about the exact type locality, but thought it likely to have been St. Thomas. Koening (1972) recorded this species from several localities off northern Brazil, but did not illustrate her material.

Politolana Bruce, 1981

DIAGNOSIS Frontal lamina slender, flattened; clypeus flattened. Antennal peduncular articles 3–5 subequal. Pereopods 1–3 with ischium and merus anterodistally produced. Pereopods 4–7 with ischium, merus, and carpus flattened. ♂ pleopod 2 with copulatory stylet inserted basally on endopod. Endopod of pleopod 5 lacking marginal setae. Pleonite 5 overlapped laterally by pleonite 4. Uropodal endopod with distal emargination; exopod slender, elongate; sympod produced along mesial margin of endopod.

Key to species of *Politolana*

1. Uropodal endopod broad distal to emargination, margin obliquely truncate; coxae of pereonites 2–6 with impressed line ...... *impressa*
   Uropodal endopod distal to emargination somewhat narrowed, margin evenly convex; coxae of pereonites lacking impressed line ..... *polita*

Politolana impressa (Harger, 1883)
Figure 63A,B

DIAGNOSIS ♂ and ♀ up to 27 mm. Frontal lamina slightly expanded anteriorly. Coxae of pereonites 2–6 with impressed oblique line. Uropodal endo-

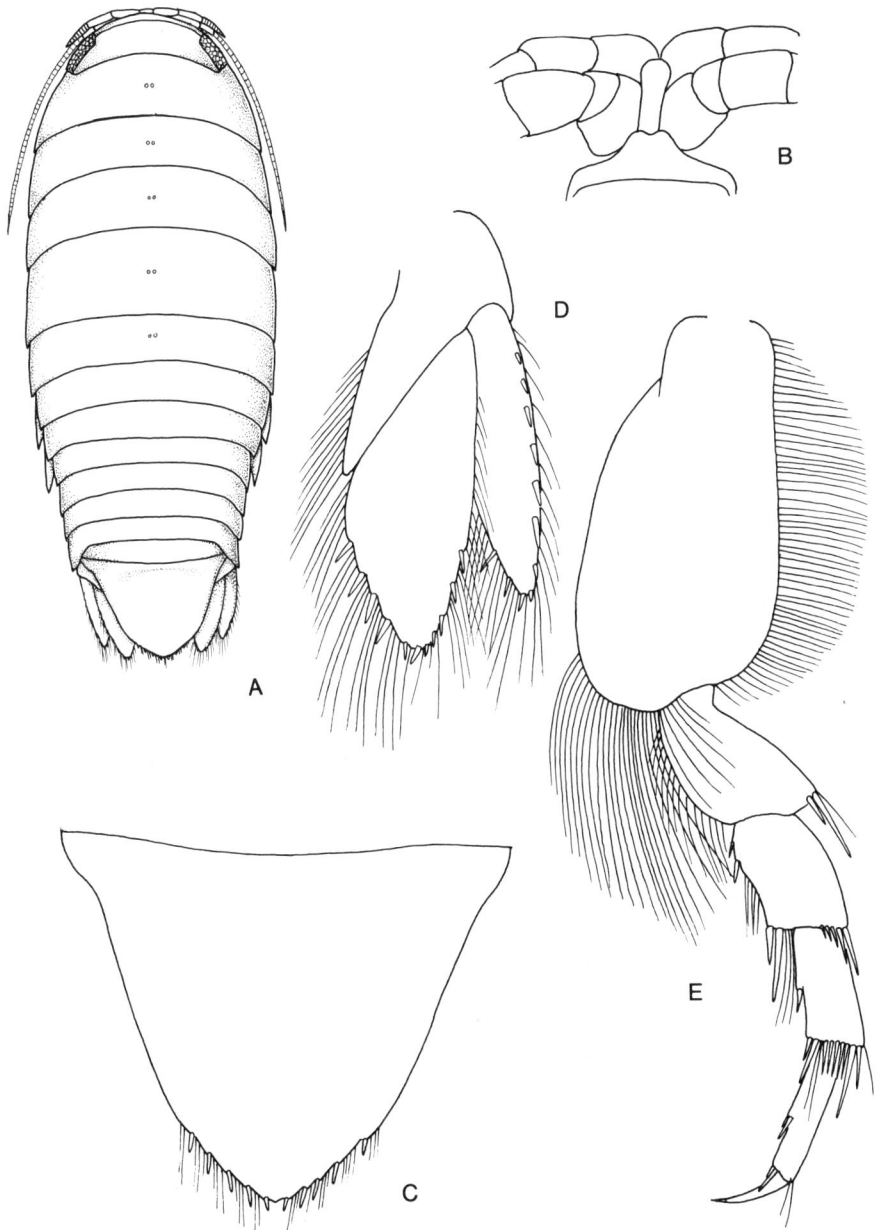

Figure 62. *Natatolana gracilis*: A, ♀; B, frontal lamina; C, pleotelson; D, uropod; E, pereopod 7.

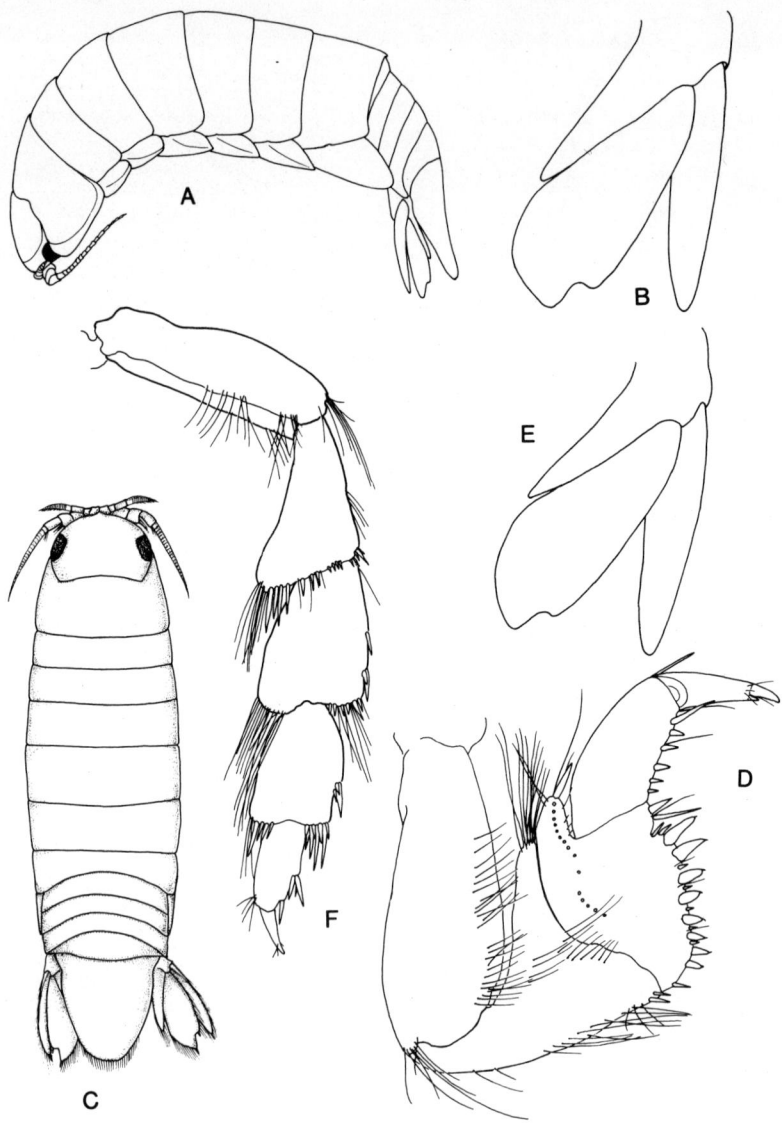

Figure 63. *Politolana impressa:* A, ♂, lateral view; B, uropod, (setae and spines omitted). *Politolana polita:* C, ♂; D, pereopod 1; E, uropod, setae and spines omitted; F, pereopod 1.

pod broad distal to emargination, margin subtruncate. Pleotelson posteriorly broadly rounded.

RECORDS   Massachusetts to Palm Beach, Florida, 32–650 m.

Figure 64. *Politolana polita*, anteroventral cephalon.

Politolana polita (Stimpson, 1853)
Figures 63C–F, 64

DIAGNOSIS ♂ 27.0 mm, ♀ 29.0 mm. Antennule barely reaching distal end of antennal peduncle. Frontal lamina basally slender, anteriorly expanded, just visible in dorsal view. Coxae lacking impressed oblique line. Uropodal endopod distal to emargination narrowed, margin convex. Pleotelson posteriorly broadly rounded.

RECORDS Bay of Fundy, Canada, to Florida Keys, 2–600 m; Gulf of Mexico.

Subfamily Eurydicinae Stebbing, 1905

DIAGNOSIS Clypeus projecting. Pleonite 5 with free lateral margins (except in *Xylolana*). Penes prominent. Pleopod 2 of ♂ with copulatory stylet articulating subbasally, medially, or subapically.

Key to genera of Eurydicinae

1. Uropodal sympod not produced along mesial margin of endopod
   Eurydice
   Uropodal sympod produced along mesial margin of endopod 2
2. Rostrum prominent, fused with frontal lamina, separating antennal
   bases 5
   Rostrum not prominent 3

(continued)

Key to genera of Eurydicinae (*Continued*)

3. Pleon of five free pleonites plus pleotelson 4
   Pleon of three free pleonites plus pleotelson .... *Colopisthus*

4. Endopods of pleopods 3–5 lacking marginal setae, or with no more than three marginal setae; copulatory stylet of endopod of pleopod 2 in ♂ articulating subterminally .... *Arubolana*
   Marginal setae lacking only on endopod of pleopod 5; copulatory stylet of pleopod 2 endopod in ♂ articulating basally .... *Metacirolana*

5. Clypeus conical; uropodal endopod lacking notch in outer distal margin .... *Xylolana*
   Clypeus flattened; uropodal endopod with notch in outer distal margin .... *Excirolana*

Arubolana Botosaneanu and Stock, 1979

DIAGNOSIS Animal not able to conglobate. Blind, or with very small eyes. Anterior margin of frontal lamina broad. Antennular peduncle of three articles. Maxillipedal palp of four articles (articles 2 and 3 fused). Maxilla 2 reduced, endite unarmed, exopod with few marginal setae. Pereopods 1 and 2 and sometimes pereopod 3 prehensile; pereopods 4–7 ambulatory. Pleopods 1 and 2, rami undivided. Pleopod 2 ♂ with copulatory stylet articulating subterminally on endopod. Pleopods 3–5, exopods biarticulate; endopods lacking marginal setae, or with few setae on endopods of 3 and 4; pleopod 5 exopod with marginal setae interrupted.

Key to species of *Arubolana*

1. Eyes present, small .... *parvioculata*
   Eyes absent 2

2. Pleotelson posteriorly rounded; rostrum not distinct in dorsal view
   .... *aruboides*
   Pleotelson posteriorly subtruncate; rostrum distinct in dorsal view
   .... *imula*

Arubolana aruboides (Bowman and Iliffe, 1983)
Figure 65A–C

DIAGNOSIS ♂ 3.9 mm, ♀ 4.1 mm. Body about three times longer than wide. Eyes absent. Antennular peduncle article 3 longer than articles 1 and 2 together. Antenna reaching posteriorly to pereonite 6 or 7. Frontal lamina visible in dorsal view between antennal bases, anteriorly rounded and only slightly wider than proximally, with distally flared ridge on ventral (exposed) surface. Pleotelson as long as basal width, evenly narrowing to broadly rounded posterior margin, latter with few small serrations and setae. Uropodal exopod four times longer than wide; endopod length about twice basal width, distally obliquely truncate, with elongate setae on inner margin.

RECORDS Church Cave and Wonderland Cave, Bermuda.

Arubolana imula Botosaneanu and Stock, 1979
Figure 65D

DIAGNOSIS ♂ and ♀ 6.25 mm. Body 2.3 times longer than wide. Eyes absent. Antenna reaching posteriorly to pereonite 4 or 5. Rostrum distinct, anteriorly truncate, separating antennal bases, fused with rectangular frontal lamina ventrally. Pleotelson basally slightly wider than long, posterior margin broadly rounded to subtruncate, with irregular crenulations or faint teeth. Uropodal exopod apically acute, reaching to about midlength of endopod; latter distally broad, with slight tooth at distolateral angle.

RECORDS Mangel Cora Tunnel, Aruba.

Arubolana parvioculata Notenboom, 1984
Figure 65E,F

DIAGNOSIS ♂ 2.8 mm, ♀ 2.9 mm. Body 3.3 times longer than wide. Cephalon with tiny pigmented eyes. Antenna reaching posteriorly to pereonite 5. Frontal lamina projecting, dorsally visible. Pleotelson basally wider than long, tapering to broadly rounded/subtruncate posterior margin bearing about six low teeth. Uropodal exopod distally acute, almost three times longer than basal width; endopod distally serrate, apically acute.

RECORDS Interstitial water near Discovery Bay, Jamaica.

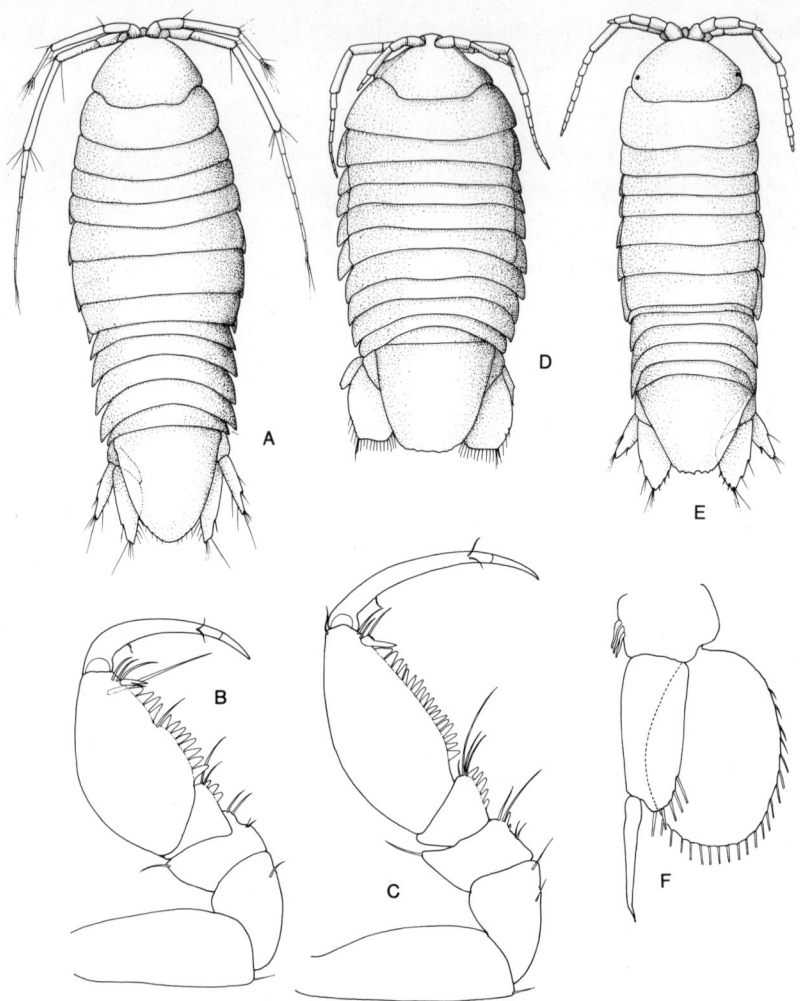

Figure 65. *Arubolana aruboides:* A, ♂; B, pereopod 1; C, pereopod 2. *Arubolana imula:* D, ♀. *Arubolana parvioculata:* E, ♂; F, pleopod 2, ♂.

Colopisthus Richardson, 1902

DIAGNOSIS Cephalon broader than long, becoming triangular between antennal bases. Pleon consisting of three short free pleonites (often obscured beneath pereonite 7), plus triangular pleotelson.

Colopisthus parvus Richardson, 1902
    Figure 66A

DIAGNOSIS ♀ 3.6 mm. Eyes large, well pigmented. Frontal lamina proximally narrow, anteriorly widened to truncate distal margin. Antennules and antennae short, latter reaching posteriorly to pereonite 1. Pleotelson with strong middorsal ridge.

RECORDS Bermuda; Puerto Rico, intertidal rocks and algae.

Eurydice Leach, 1815

DIAGNOSIS Antennular peduncle article 2 at right angle to article 1. Antennal peduncle of four articles. Frontal lamina usually slender; clypeus usually a ventrally directed triangular blade. Maxillipedal endite reduced, lacking coupling hooks. ♂ pleopod 2 with copulatory stylet articulating at midlength. Pleopod 5, endopod lacking marginal setae. Pleonite 5 with free lateral margins, not overlapped by pleonite 4. Uropodal sympod not produced along medial margin of endopod.

Key to species of *Eurydice*

1. Frontal lamina distally truncate to faintly bilobed 2
   Frontal lamina lanceolate, distally acute .... *personata*

2. Posterior margin of pleotelson between notches rounded, with four
      moderate spines .... *convexa*
   Posterior margin of pleotelson between notches almost straight, with
      four very short spines and several elongate setae .... *piperata*

Eurydice convexa Richardson, 1900
    Figures 66B–E, 67A,B

DIAGNOSIS ♂ 6.1 mm, ♀ 6.1 mm. Frontal lamina slender, anteriorly widening slightly and becoming truncate to faintly bilobed. Posterior margin of pleotelson between lateral notches convex, with four spines and few setae between serrations; spines between three and four times longer than wide. ♂: Plicate process on antennal flagellar articles about $1/5$ length of article. Pleopod 2 with copulatory stylet distally blunt, reaching well beyond rami.

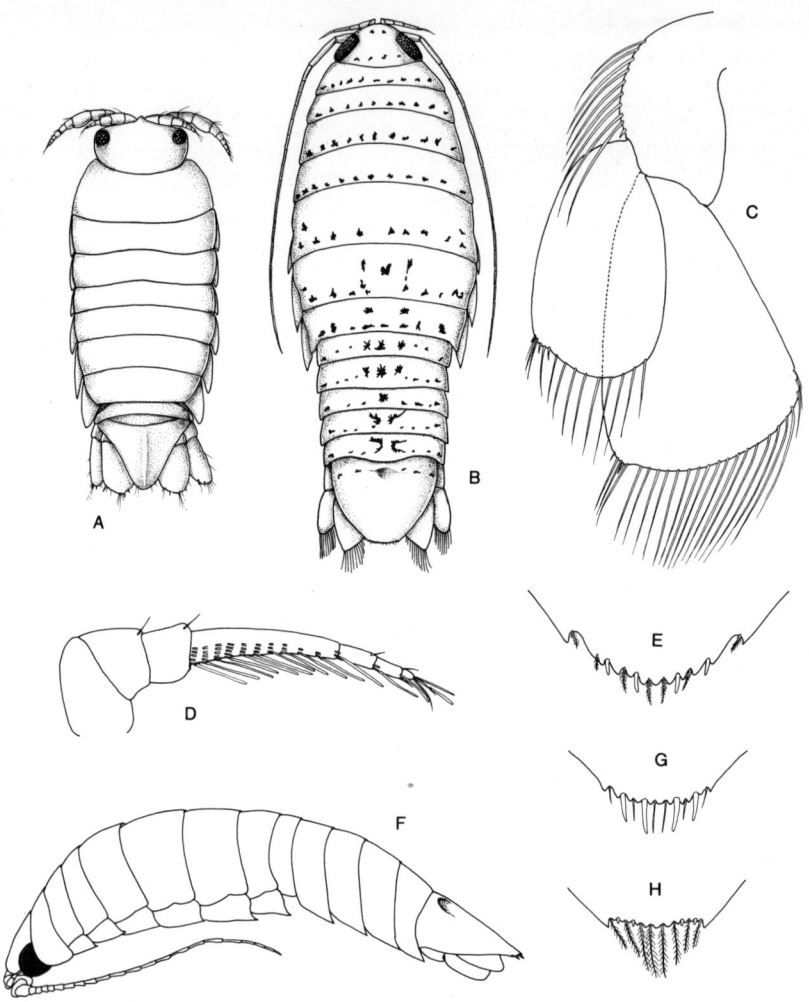

Figure 66. *Colopisthus parvus:* A, ♀. *Eurydice convexa:* B, ♂; C, uropod; D, antennule; E, pleotelsonic apex. *Eurydice personata:* F, ♂, lateral view; G, pleotelsonic apex. *Eurydice piperata:* H, pleotelsonic apex.

RECORDS South Carolina to Florida Keys; Bahamas; Gulf of Mexico and Caribbean.

REMARKS Whether *E. convexa* and *E. littoralis* are conspecific needs further investigation. Differences can be detected in the male plicate organs of the antennae, and in the mandibular palp spination, but range of variation in

these features is still unknown. It seems unlikely that the species recorded as
E. littoralis by Moreira (1972) from Brazil is the same species.

Eurydice personata Kensley, 1987a
Figures 66F,G; 67C

DIAGNOSIS ♂ 6.0 mm, ovigerous ♀ 5.1–6.4 mm. Frontal lamina slender, lanceolate, anteriorly acute. Posterior margin between notches faintly convex, with four relatively elongate spines (inner pair five or six times longer than wide) and few setae between dentitions. ♂: Plicate organ on antennal flagellar articles half length of article. Pleopod 2, copulatory stylet on endopod clavate, barely reaching beyond ramus.

RECORDS Bermuda; off Georgia, 18–27 m; off South Carolina, 34 m; off Miami, Florida; Puerto Rico, 13–17 m; Bahamas, 1–2 m and surface plankton tow; Haiti; Cuba; Venezuela.

REMARKS This recently discovered species has masqueraded under the names of *E. convexa* and *E. littoralis* for some time, which may explain some of the inconsistencies in the literature, especially in variation in the pleotelsonic apex.

Eurydice piperata Menzies and Frankenberg, 1966
Figure 66H

DIAGNOSIS ♂ 4.0 mm, ♀ 4.5 mm. Frontal lamina slender, widening anteriorly to become slightly bilobed. Posterior margin of pleotelson between notches straight to faintly convex, with four spines barely twice longer than wide, and several much longer setae between dentition. ♂: Plicate organ on antennal flagellar articles about ⅙ length of article but situated subdistally. Pleopod 2, copulatory stylet clavate, reaching well beyond ramus.

RECORDS Georgia to Florida, Gulf of Mexico, 37–150 m.

Excirolana Richardson, 1912a

DIAGNOSIS Cephalon with prominent rostrum separating antennular bases; fused with flattened frontal lamina. Clypeus with short, broadly triangular blade projecting anteroventrally. Antennal peduncle with four or five articles. Maxillipedal endite with single coupling hook. Pleopods 3–5, endopods

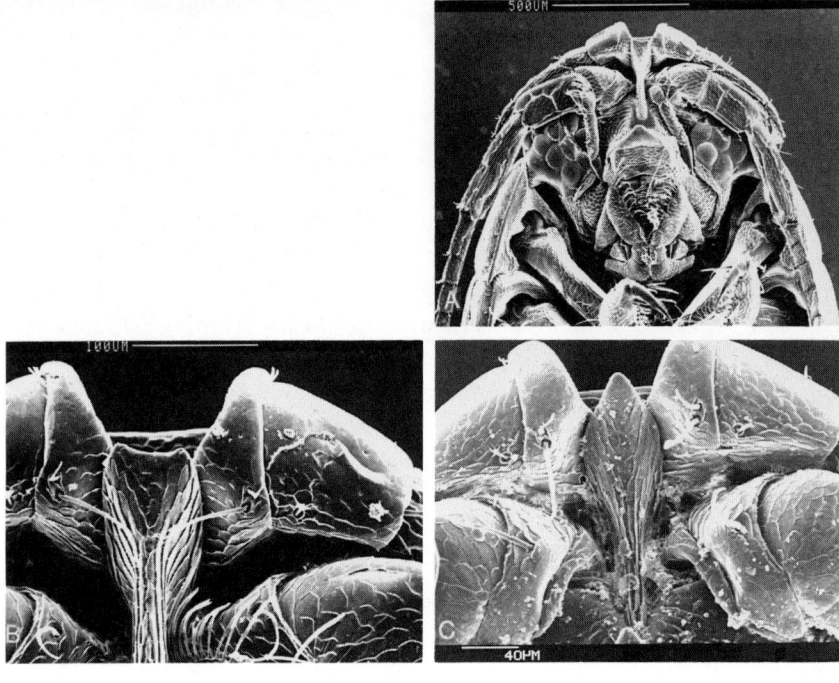

Figure 67. *Eurydice convexa:* A, anteroventral cephalon; B, frontal lamina. *Eurydice personata:* C, frontal lamina.

lacking marginal setae. Pleonite 5 with free lateral margins, not overlapped by pleonite 4. Uropodal sympod produced along mesial margin of endopod.

Key to species of *Excirolana*

1. Pleotelson with two anterior hollows clearly joined by impressed line; uropodal endopod about half length of exopod .... *braziliensis*
   Pleotelson with two anterior hollows not connected by impressed line; uropodal endopod about two-thirds length of exopod .... *mayana*

Excirolana braziliensis Richardson, 1912a
  Figures 68A–C, 69A–C

DIAGNOSIS ♂ 6.0 mm, ♀ 7.5 mm. Frontal lamina very slender between antennal bases, widening anteriorly into rounded structure between anten-

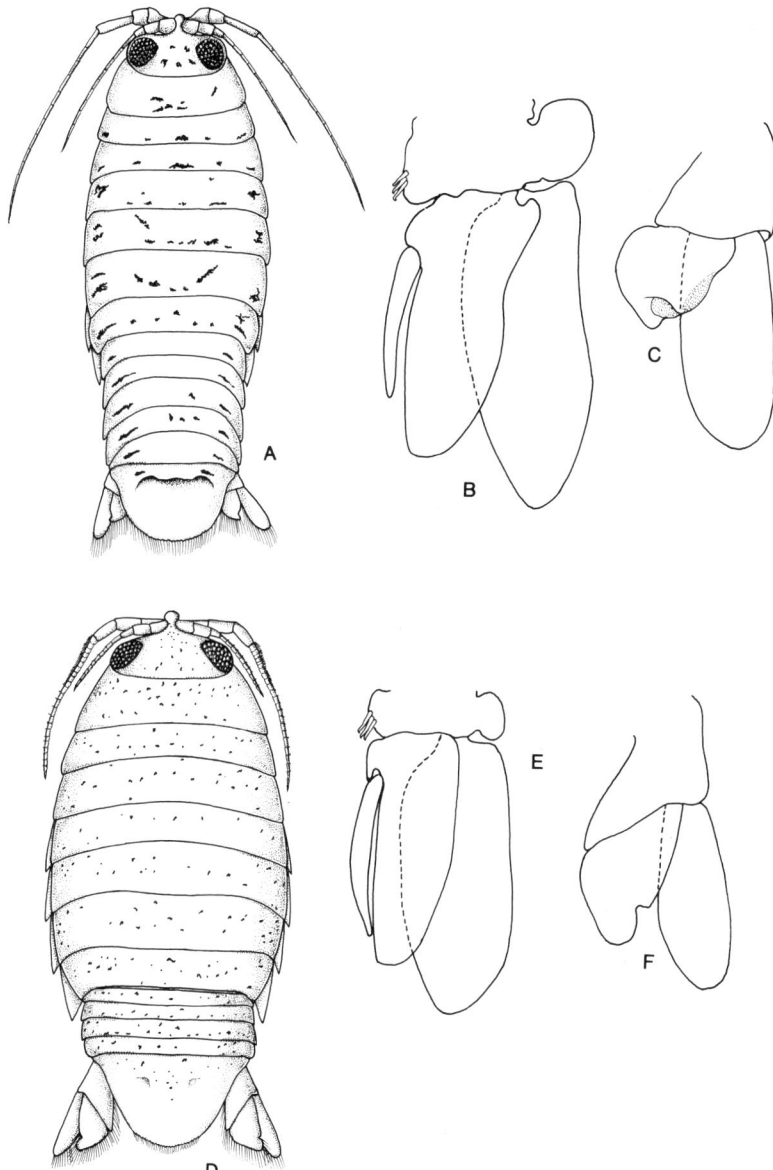

Figure 68. *Excirolana braziliensis:* A, ♂; B, pleopod 2 ♂; C, uropod, (setae and spines omitted). *Excirolana mayana:* D, ♀; E, pleopod 2 ♂; F, uropod, setae and spines omitted.

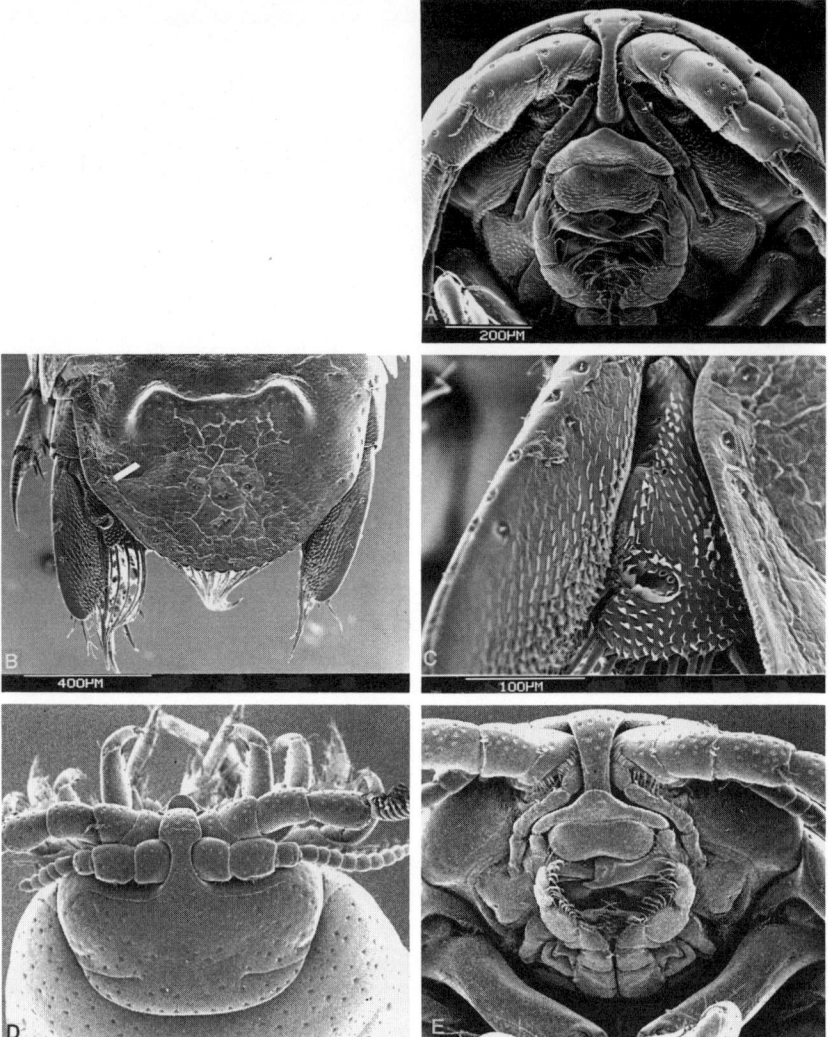

Figure 69. *Excirolana braziliensis:* A, anteroventral cephalon; B, pleotelson; C, uropodal endopod. *Excirolana mayana:* D, anterodorsal cephalon; E, anteroventral cephalon.

nular bases, joined to rostrum by very slim isthmus. Clypeus distally subacute. Uropodal endopod about half length of exopod. Pleotelson with two lateral hollows defined and connected by clear impressed line; posterior margin evenly convex, bearing numerous plumose setae.

RECORDS   Caribbean to Brazil; common in the intertidal of sandy beaches; Gulf of Mexico.
Gulf of California to Chile.

REMARKS   Glynn et al. (1975) produced a thorough study of the taxonomy, zonation, and distribution of this Pan-American species.

Excirolana mayana (Ives, 1891)
Figures 68D–F, 69D,E

DIAGNOSIS   ♂ 8.2 mm, ♀ 10.0 mm. Frontal lamina between antennal bases about half anterior width. Clypeus anteriorly rounded. Uropodal endopod about ⅔ length of exopod. Pleotelson with two faint lateral hollows in anterior half, not connected by impressed line.

RECORDS   Florida to Venezuela, intertidal.

Metacirolana Nierstrasz, 1931

DIAGNOSIS   Frontal lamina anteriorly dilated, free, projecting; clypeus triangular, projecting ventrally. Maxillipedal endite with one coupling hook. Pleon with five free segments, pleonite 5 not overlapped laterally by pleonite 4. Eyes often larger, and antennular flagellum of more articles in ♂ than in ♀.

Key to species of *Metacirolana*

1. Telson posteriorly truncate .... *halia*
   Telson posteriorly rounded or angulate 2

2. Posterior margin of telson an obtuse angle .... *agaricicola*
   Telson posteriorly rounded 3

3. Posterior margin of telson narrowly rounded; uropodal rami, margins strongly dentate .... *menziesi*
   Posterior margin of telson broadly rounded; uropodal rami, margins obscurely dentate .... *sphaeromiformis*

Metacirolana agaricicola Kensley, 1984
Figure 70A–C

DIAGNOSIS ♂ 2,6 mm, ovigerous ♀ 2.1 mm. Antennular flagellum of six or seven articles. Antennal flagellum of 10 articles. Posterior margin of telson finely dentate, with broadly obtuse median point. Uropodal exopod about half width of endopod, margins dentate, apically acute; endopod, margins dentate, distally angled, apically acute.

RECORDS Carrie Bow Cay, Belize, 1–20 m, in coral on reef slope, and spur and groove zone.

Metacirolana halia Kensley, 1984
Figure 70D–F

DIAGNOSIS ♂ 2.9 mm, ovigerous ♀ 2.7 mm. Antennular flagellum of 10 articles in ♀, 14 in ♂. Antennal flagellum of 10 articles in ♀, 11 in ♂. Posterior margin of telson truncate, bearing about eight sensory spines. Uropodal exopod distally broadly rounded, more than half distal width of endopod, outer margin dentate, with about 11 sensory spines; endopod distally broad, margin straight, bearing about 12 sensory spines.

RECORDS Carrie Bow Cay, Glover's Reef, Belize; intertidal to 23 m; Turks and Caicos Islands, 1 m; Bahamas; Jamaica; Cozumel, Mexico.

Metacirolana menziesi Kensley, 1984
Figure 71A,B

DIAGNOSIS ♂ 2.3 mm, ovigerous ♀ 2.4 mm. Antennular flagellum of six articles in ♀, eight in ♂. Antennal flagellum of nine articles in ♀, 10 in ♂. Posterior margin of telson broadly rounded, finely dentate. Uropodal exopod half width of endopod, margins dentate, apically acute; endopod, margins dentate, apically acute.

RECORDS Carrie Bow Cay, Belize, intertidal to 30 m, usually in coral rubble.

Metacirolana sphaeromiformis (Hansen, 1890)
Figure 71C,D

DIAGNOSIS ♂ 2.5 mm, ♀ 3.2 mm. Antennular flagellum of three articles. Antennal flagellum of eight articles. Posterior margin of telson narrowly

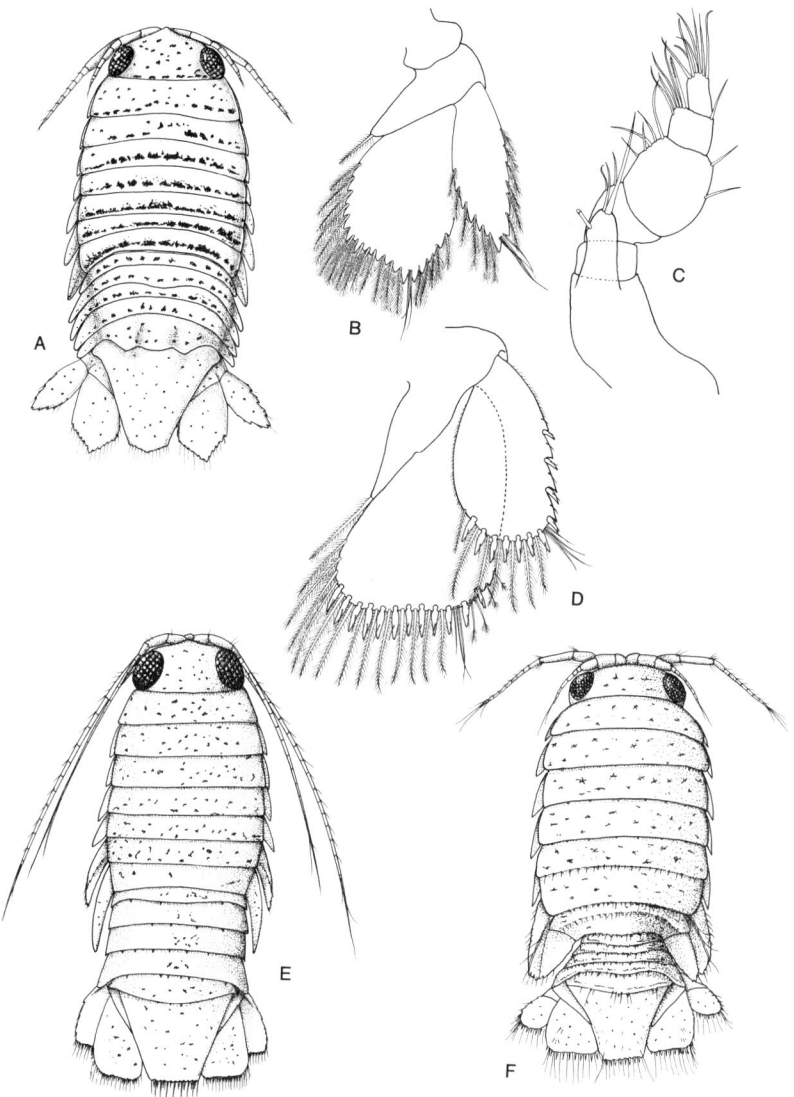

Figure 70. *Metacirolana agaricicola:* A, ♀; B, uropod; C, maxilliped. *Metacirolana halia:* D, uropod; E, ♂; F, ♀.

rounded, obscurely dentate. Telson with low rounded middorsal ridge and pair of lateral ridges. Uropodal exopod more than half width of endopod, margin dentate; exopod distally broadened, margin dentate, with few sensory spines.

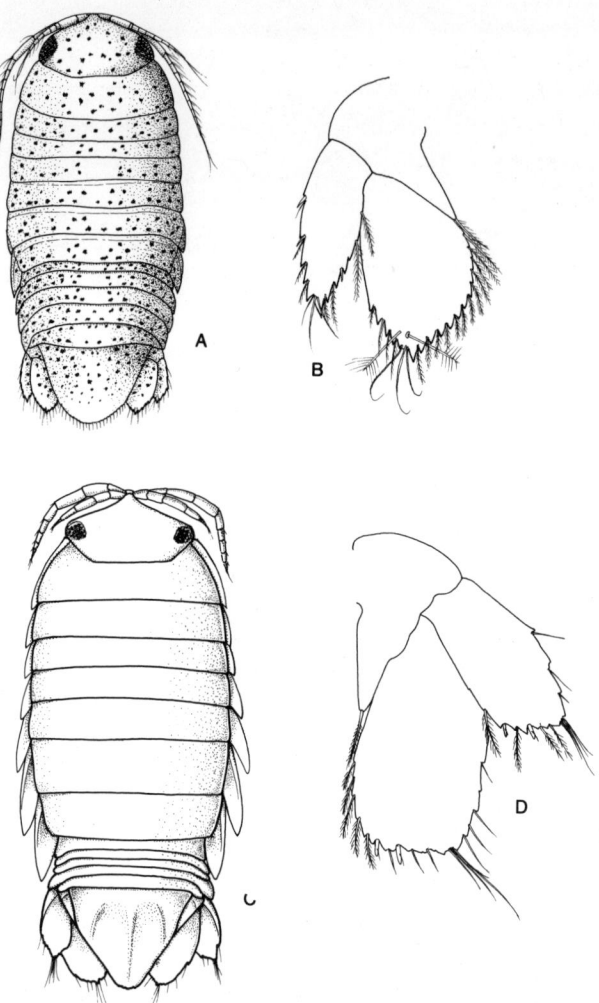

Figure 71. *Metacirolana menziesi:* A, ♀; B, uropod. *Metacirolana sphaeromiformis:* C, ♀; D, uropod.

RECORDS   Looe Key, Florida, intertidal reef crest; Turks and Caicos Islands, 1 m; St. Thomas, U.S. Virgin Islands.

Xylolana Kensley, 1987a

DIAGNOSIS   Frontal lamina and rostrum fused, broad, separating antennular bases. Clypeus conical, projecting. Maxillipedal endite reduced, lacking coupling hooks; palp of five articles. Copulatory stylet in ♂ articulating in distal half of mesial margin of pleopod 2 endopod. Pleopods 3–5, exopods biarticulate; endopods lacking marginal setae. Pleonite 5 lacking free lateral margins, overlapped by pleonite 4. Uropodal sympod produced along mesial margin of endopod.

Xylolana radicicola Kensley, 1987a
Figure 72

DIAGNOSIS ♂ 2.6 mm, ♀ 3.3 mm. Body about four times longer than greatest width. Uropodal exopod about ⅔ width of endopod, bearing single short subapical spine; both uropodal rami distally rounded. Pleotelson with lateral margins subparallel, with poorly defined middorsal longitudinal ridge.

RECORDS Twin Cays, Belize, in dead red mangrove roots, 1 m.

Family Corallanidae Hansen, 1890

DIAGNOSIS Outer ramus of maxilla 1 apex with single strong falcate spine, with single strong spine with one or more smaller hooked spines at base, or with two large recurved spines, occasionally with one to three smaller spines between them. Maxillipedal endite reduced or lacking.

Key to genera of Corallanidae

1. Maxilla 1 with single strong falcate spine 2
   Maxilla 1 with two large falcate spines .... *Alcirona*

2. Maxilla 2, distal article slender-elongate; article 2 of maxillipedal palp longest .... *Nalicora*
   Maxilla 2 distally bluntly bilobed; article 3 of maxillipedal palp longest .... *Excorallana*

Alcirona Hansen, 1890

DIAGNOSIS Antennular peduncle of two articles. Mandible lacking molar. Maxilla 1 having two large recurved spines, with one or more smaller spines between these. Maxilla 2 a simple rounded lobe. Maxilliped lacking endite. Posterior half of body hirsute.

Key to species of *Alcirona*

1. Golden-brown setae starting dorsally on pereonite 3; pereopod 1, dactylus having several elongate spines .... *insularis*
   Golden-brown setae starting dorsally on pereonites 5 or 6; pereopod 1, dactylus having accessory spine only .... *krebsi*

Alcirona insularis Hansen, 1890
Figure 73A

DIAGNOSIS ♂ and ♀, 5.0 mm. Posterior half of body, especially in ♂, bearing stiff golden-brown setae, these beginning as posterior row on pereonite 3, and becoming dense on posterior pereonites, pleonites and pleotelson. Pereopods 1–3, dactylus strongly falcate, having distal unguis equal in length to rest of article, and with three or four strong teeth on posterior margin. Apex of pleotelson rounded, bearing six short marginal spines in addition to setae.

RECORDS Looe Key, Florida, 0.5–6 m; St. Thomas, U.S. Virgin Islands, 40–46 m; Puerto Rico, from intertidal coral rubble and from gills of nurse shark *Ginglymostoma sirratum;* St. Lucia, from coral rubble.

Alcirona krebsii Hansen, 1890
Figure 73B–D

DIAGNOSIS ♂ 10 mm, ovigerous ♀ 15.5 mm. Posterior half of body, especially in ♂ bearing stiff golden-brown setae, these beginning in posterior row on pereonites 5 or 6, becoming dense on posterior pereonites, pleonites, uropods, and pleotelson. Pereopod 1, dactylus strongly falcate, with unguis equal in length to rest of article, and with one strong tooth and several low tubercles on posterior margin. Apex of pleotelson rounded, bearing six short spines in addition to setae.

RECORDS Bermuda, in sponges; Florida Keys; Quintana Roo, Yucatan Peninsula; St. Thomas, U.S. Virgin Islands; Venezuela.

REMARKS A single 8-mm ♂ specimen of *Alcirona* from Panama Bay (Pacific) has the characteristic pereopod 1 of *A. krebsii,* but has the rows of stiff setae beginning on about pereonite 3. The possibility that *A. krebsii* is another amphi-Panamic species needs to be investigated.

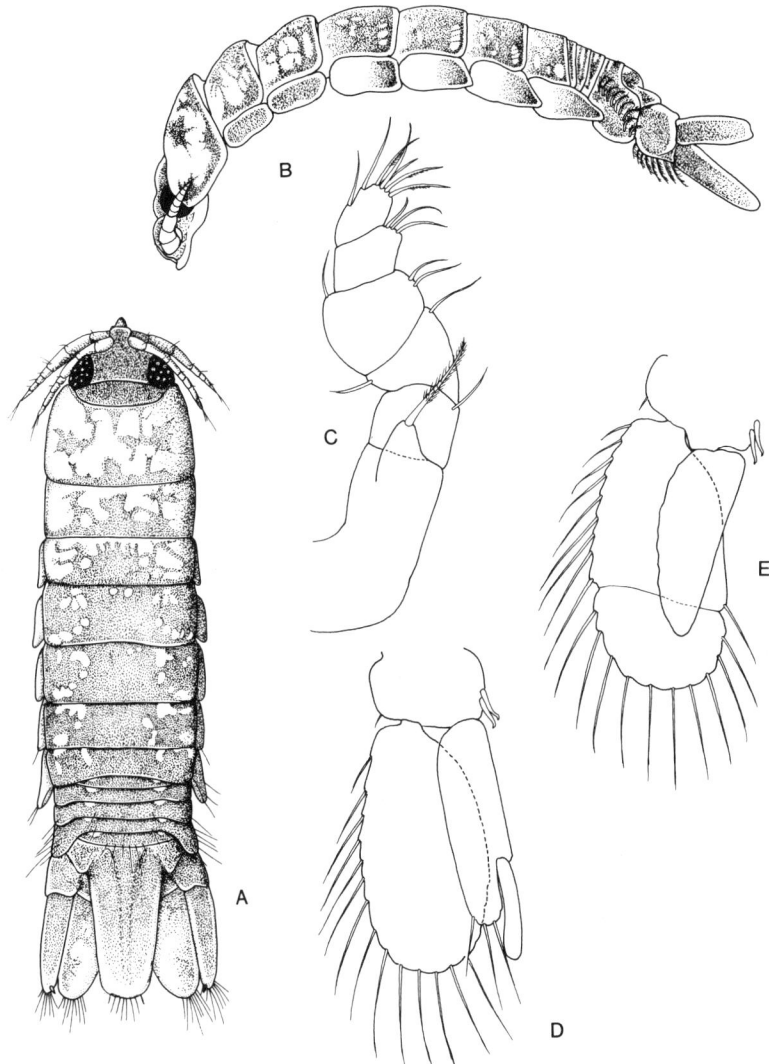

Figure 72. *Xylolana radicicola:* A, B, ♂; C, maxilliped; D, pleopod 2 ♂; E, pleopod 3.

Excorallana Stebbing, 1904

DIAGNOSIS Eyes well developed and pigmented, sometimes contiguous or nearly so. Maxilla 1, outer ramus a single falcate spine. Maxillipedal palp of five articles; endite reduced or absent. Pereopods 1–3 subprehensile or pre-

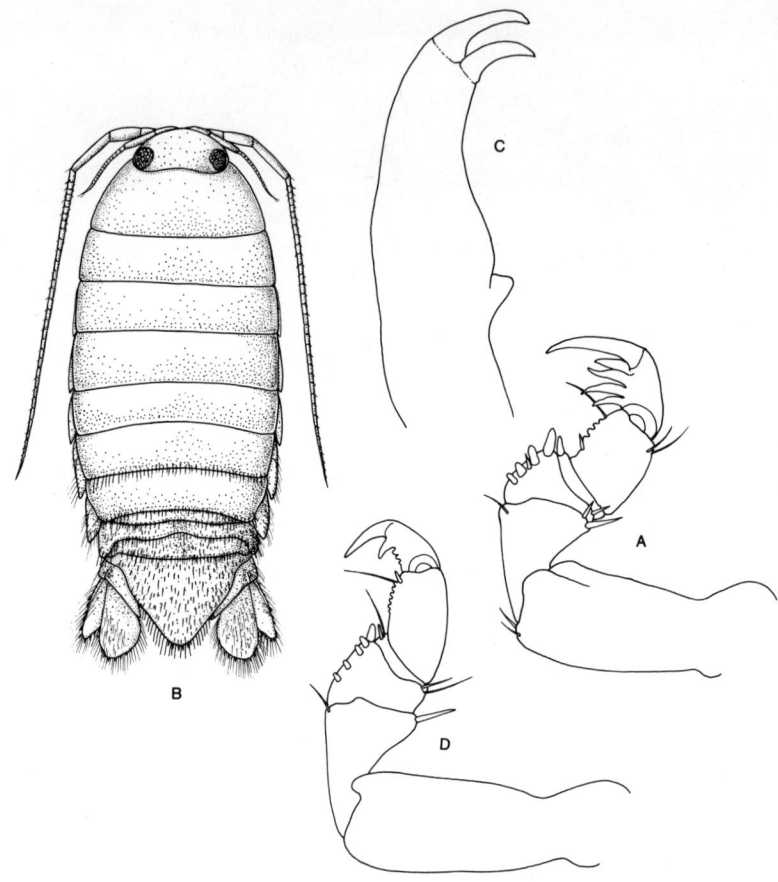

Figure 73. *Alcirona insularis:* A, pereopod 1. *Alcirona krebsi:* B, ♂; C, maxilla 1; D, pereopod 1.

hensile, pereopods 4–7 ambulatory. All rami of pleopods bearing marginal setae. Pleotelson often with characteristic spination and tuberculation; lateral margins often with incision.

REMARKS   *Excorallana subtilis* (Hansen, 1890) was described from St. Thomas, U.S. Virgin Islands, based on a specimen undergoing ecdysis; the true identity of this species remains uncertain.

In the key, two species have been included which are not illustrated. These are *E. mexicana* Richardson, 1905, from the Gulf of Mexico, and *E. delaneyi*

Stone and Heard, 1989, from the northeastern Gulf of Mexico. The latter species, particularly, could conceivably be encountered in the Florida Keys.

Delaney (1984) provides a useful review of the genus *Excorallana*, and of the distribution of the species.

Key to species of *Excorallana*

1. Eyes contiguous 2
   Eyes well separated 4
2. Apex of pleotelson with deep slit .... *fissicauda*
   Apex of pleotelson entire 3
3. Pleotelson with lateral incision .... *oculata*
   Pleotelson lacking lateral incision .... *warmingii*
4. Pleotelson with lateral incision 6
   Pleotelson lacking lateral incision 5
5. Frontal lamina linguiform, anteriorly rounded .... *berbicensis*
   Frontal lamina posteriorly with faintly concave margins, anteriorly
      subacute .... *delaneyi*
6. Frontal lamina strongly grooved for entire length 7
   Frontal lamina with ventral surface flat 8
7. Pleotelson with medial row of small tubercles flanked by row of larger
      tubercles in posterior half .... *mexicana*
   Pleotelson lacking rows of tubercles in posterior half .... *antillensis*
8. Frontal lamina distinctly bell shaped; ♂ cephalon with two pairs of
      tubercles and antennular bases not tuberculate .... *quadricornis*
   Frontal lamina, and ♂ cephalon and antennules not as above 9
9. Frontal lamina anteriorly broadly rounded, length 1.5 or less times
      basal width; ♂ cephalon with two pairs of tubercles and basal
      antennular article each with tubercle .... *sexticornis*
   Frontal lamina anteriorly narrowly rounded, length about twice basal
      width; ♂ cephalon with three tubercles, tubercles lacking on
      antennular bases .... *tricornis tricornis*

Excorallana antillensis (Hansen, 1890)
Figure 74A–D

DIAGNOSIS ♂ 11.0 mm, ovigerous ♀ 15.0 mm. Cephalon unornamented. Eyes well separated. Frontal lamina parallel sided, length twice basal width, anteriorly broadly rounded. Pleonites 2–4, posterolateral margins tuberculate, middorsal posterior margin excavate with strong middorsal tubercle; pleonite 5 with strongest tubercles submedian, posterior margin not excavate. Pleotelson with low middorsal ridge, strong basal tubercles; lateral incisions present; two submedian patches of spines; apex narrowly rounded.

RECORDS Florida Keys; St. Thomas, U.S. Virgin Islands; Puerto Rico; Quintana Roo, Mexico; Carrie Bow Cay, Belize, 5–18 m.

Excorallana berbicensis Boone, 1918
Figure 74E,F

DIAGNOSIS ♂ 9.9 mm, ♀ 12.0 mm. Eyes well separated. Frontal lamina about twice longer than wide, posteriorly parallel sided, widening anteriorly to broadly rounded apex. Subadult ♂ cephalon unornamented except for pair of very low tubercles mesial to eyes. Posterior margins of pleonites 3–5 very faintly tuberculate. Pleotelson with submedian pair of low tubercles basally; lacking lateral incisions; apex rounded.

RECORDS Guyana; French Guiana; Guadeloupe.

Excorallana fissicauda (Hansen, 1890)

DIAGNOSIS 11 mm. Cephalon unornamented. Eyes contiguous. Frontal lamina unknown. Pleonite 5 with three strong mesial, and several smaller lateral tubercles on posterior margin. Pleotelson with two strong submedian basal tubercles; lateral incision lacking; posterior margin with deep open incision.

RECORDS St. Thomas, U.S. Virgin Islands.

REMARKS This species was described from a single specimen, and has not been recorded since. After examining the holotype, Paul Delaney (in litt.) suspects that the terminal incision of the pleotelson may be the result of an injury.

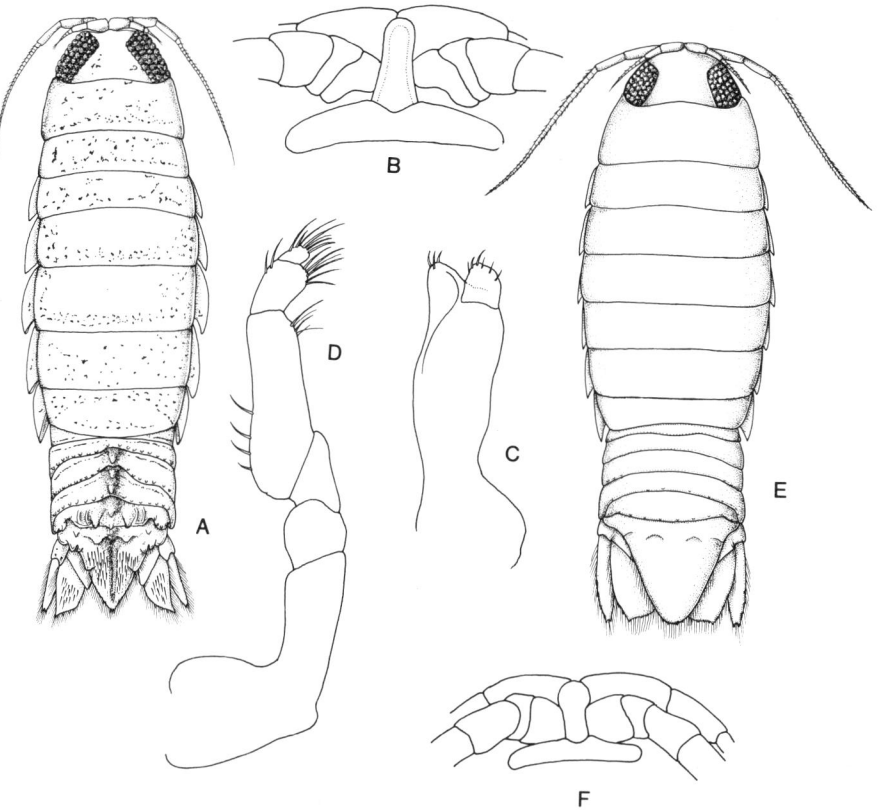

Figure 74. *Excorallana antillensis:* A, ♂; B, frontal lamina; C, maxilla 2; D, maxilliped. *Excorallana berbicensis:* E, ♂; F, frontal lamina.

Excorallana oculata (Hansen, 1890)
Figure 75A,B

DIAGNOSIS ♂ 6.9 mm, ♀ 8.5 mm. Eyes contiguous. Cephalon unornamented. Frontal lamina slender, linguiform, widest posteriorly. Pleonites 3–5 each with slightly hollowed middorsal area containing strong flattened tubercle; pleonite 5 with two strong flanking tubercles; pleotelson basally with low median ridge and two strong submedian tubercles; short strong spines in two roughly triangular submedian patches; lateral incisions present; apex narrowly rounded.

RECORDS Bahamas; Cuba, Puerto Rico, 40 m; Barbados. Brazil.

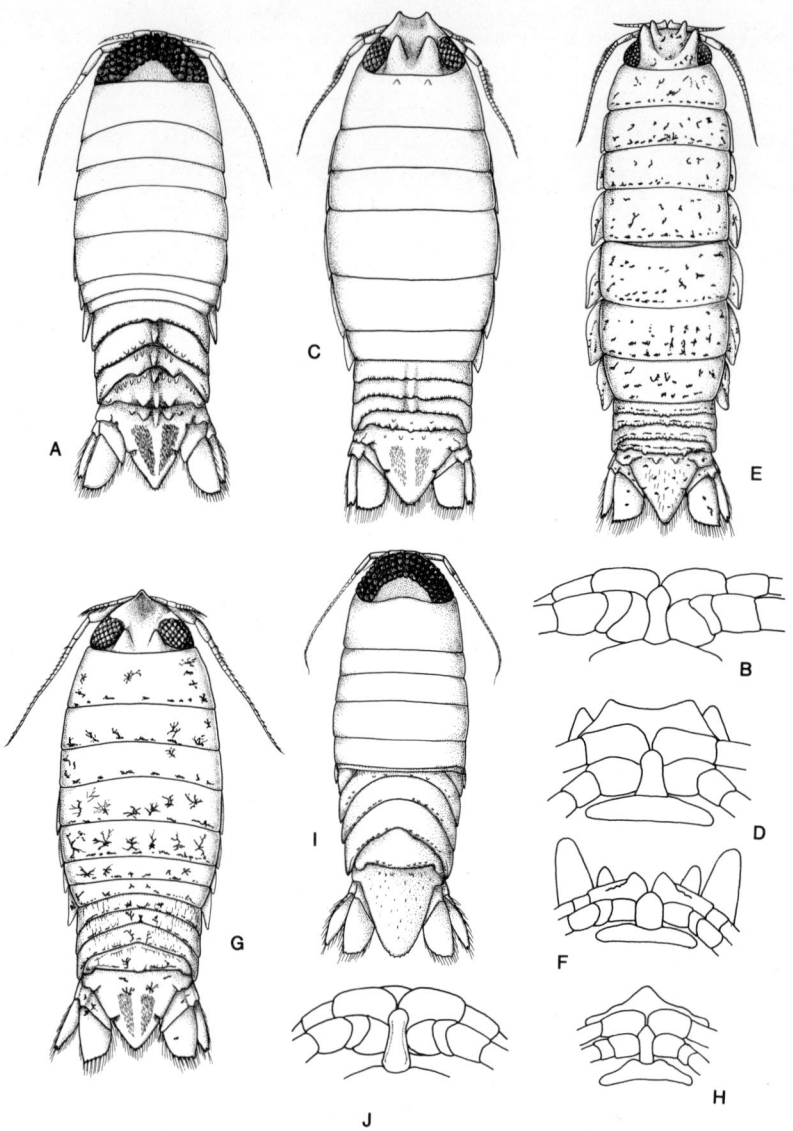

Figure 75. *Excorallana oculata:* A, ♂; B, frontal lamina. *Excorallana quadricornis:* C, ♂; D, frontal lamina. *Excorallana sexticornis:* E, ♂; F, frontal lamina. *Excorallana tricornis tricornis:* G, ♂; H, frontal lamina. *Excorallana warmingi:* I, ♂; J, frontal lamina.

Excorallana quadricornis (Hansen, 1890)
Figures 75C,D; 76A-C

DIAGNOSIS ♂ 13.2 mm, ♀ 12.1 mm. Eyes well separated. Cephalon in ♂ with two pairs of tubercles, anterior pair connected by low rounded ridge, posterior pair situated mesial to eyes. Frontal lamina bell shaped, broadest posteriorly. Pereonite 1 with submedian pair of low tubercles. Posterior margin of pereonite 7 and pleonites faintly tuberculate. Pleotelson with two submedian raised areas bearing short spines; lateral margins with incision; few low tubercles basally.

RECORDS Bermuda; St. Thomas, U.S. Virgin Islands; Jamaica, intertidal in grassflats and between mangrove roots; Martinique; Belize; Venezuela.

Excorallana sexticornis (Richardson, 1901)
Figures 75E,F; 76D-F

DIAGNOSIS ♂ 7.9 mm, ovigerous ♀ 6.9-8.3 mm. Eyes well separated. Basal antennular peduncular article in ♂ with short anterodorsally directed tubercle. Cephalon with two pairs of prominent tubercles, anterior pair shorter than posterior pair, latter situated mesial to eyes. Frontal lamina, length less than twice width, sides parallel to faintly converging anteriorly, apically broadly rounded. Posterior margins of pleonites 2-5 with low rounded tubercles, those near middorsal line largest. Pleotelson with two basal submedian tubercles, numerous scattered dorsal spines, lateral margins with incision, apex narrowly rounded.

RECORDS Key West, Florida; Cuba; Puerto Rico; Twin Cays, Belize, shallow infratidal from dead mangrove wood.

Excorallana tricornis tricornis (Hansen, 1890)
Figures 75G,H; 77

DIAGNOSIS ♂ 8.2 mm, ♀ 9.9 mm. Basal antennular article narrow, not dilated. Cephalon in ♂ with one median and two dorsal "horns." Eyes large, well separated. Frontal lamina between two and three times longer than basal width, sides subparallel, anteriorly rounded to subacute. Pereon smooth. Margins of lateral incision of pleotelson separated by gap; anterior margin of incision lined with short spines; scattered short spines on dorsum of telson, but especially concentrated in two submedian patches. Uropodal exopod, length 2.3-2.5 times width; apical notch nearly symmetrical.

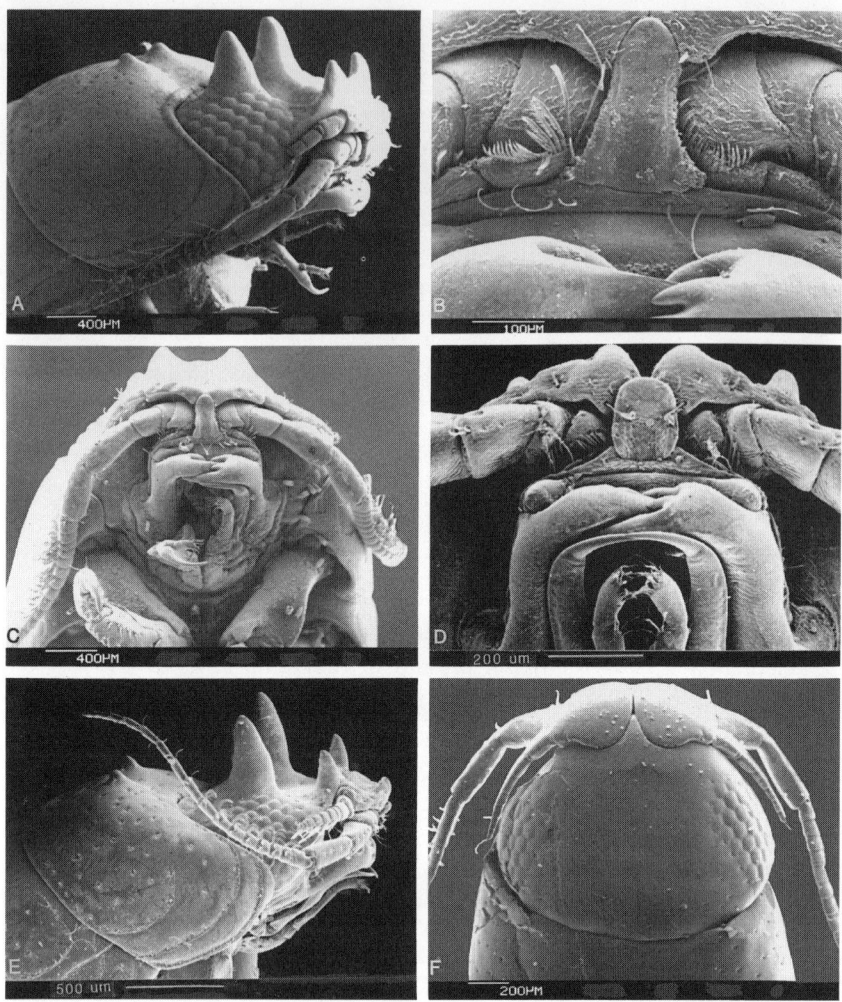

Figure 76. *Excorallana quadricornis:* A, cephalon and pereonite 1 ♂; B, ventral cephalon; C, frontal lamina enlarged. *Excorallana sexticornis:* D, cephalon and pereonite 1 ♂; E, cephalon ♀; F, ventral cephalon.

RECORDS Turks and Caicos Islands, 1 m; St. Thomas and St. Croix, U.S. Virgin Islands, 48–55 m; Cuba; Puerto Rico, on gills of rays *Aetobatus narinari* and *Dasyatis americana,* and on squirrel fish; Belize, intertidal to 15.2 m, in intertidal coral rubble, in coarse sediments in *Syringodium* and *Thalassia* seagrass beds, on brown alga *Turbinaria,* on *Madracis* sp. sponge, on *Agaricia* sp. coral; Gulf of Mexico.

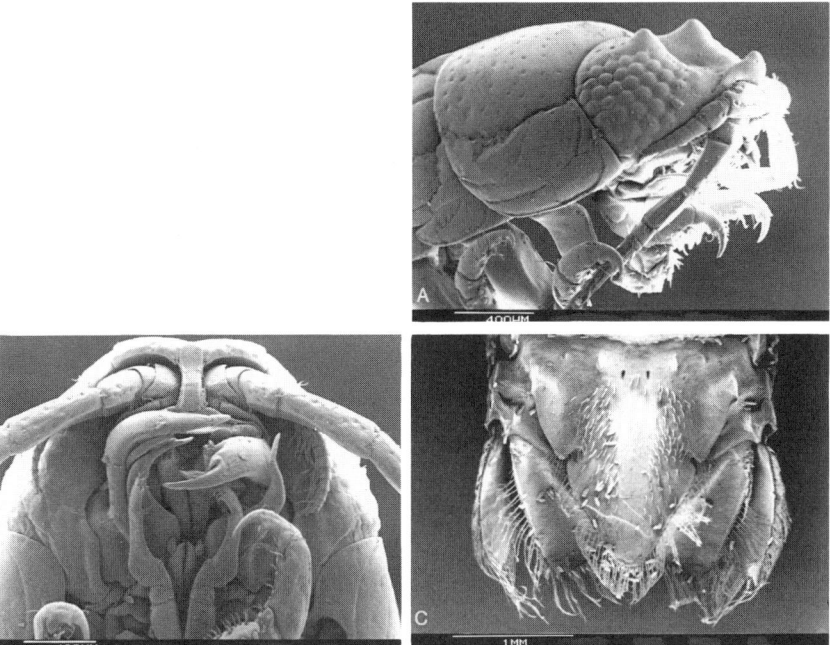

Figure 77. *Excorallana tricornis tricornis*: *A*, cephalon and pereonite 1 ♂; *B*, ventral cephalon; *C*, pleotelson and uropods.

REMARKS The subspecies *Excorallana tricornis occidentalis* Richardson, 1905a, from southern California, differs from the Gulf and Caribbean subspecies in lacking a gap between the margins of the pleotelsonic incision, and in having a relatively wider uropodal exopod which shows a distinctly asymmetrical apical notch.

Excorallana warmingii (Hansen, 1890)
Figure 75I,J

DIAGNOSIS ♂ 9.7 mm, ♀ 12.0 mm. Cephalon unornamented. Eyes contiguous, occupying most of dorsal surface of head. Posterior margins of pleonites very faintly tuberculate. Frontal lamina, length slightly more than twice basal width, tapering anteriorly to rounded apex. Pleotelson unornamented except for two faint submedian tubercles basally; lateral incisions lacking; apex broadly rounded, with five low but distinct marginal teeth.

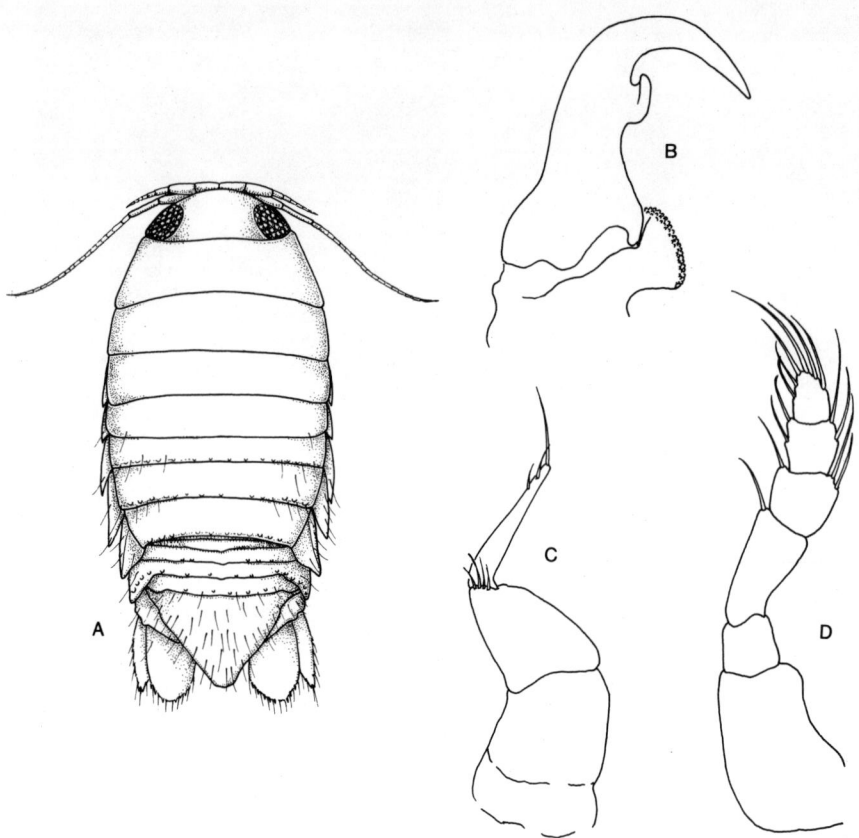

Figure 78. *Nalicora rapax:* A, ♀; B, maxilla 1; C, maxilla 2; D, maxilliped.

RECORDS   Bahamas; between Cuba and the Yucatan Pensinsula; Puerto Rico.
Off Brazil near Rio de Janeiro.

Nalicora Moore, 1901

DIAGNOSIS   Maxilla 1 exopod a single strongly falcate distal spine with knoblike mesial process, and basal caplike convex papilla-bearing structure. Maxilla 2 of four articles, distal article slender. Maxillipedal palp of five articles; endite lacking.

Nalicora rapax Moore, 1901
   Figure 78

DIAGNOSIS   ♂ 6.9 mm, ovigerous ♀ 10.0 mm. Eyes well developed. Frontal lamina basally slender, widening anteriorly, apex subacute. Posterior half of body bearing numerous scattered stiff setae. Pereonites 4–7 with row of low rounded tubercles near posterior margin. Posterior margins of pleonites 3–5 faintly tuberculate, more noticeable in ♂. Pleotelson wider than long; lateral margins faintly sinuous; apex rounded.

RECORDS   Florida Keys, 55 m; Puerto Rico, 50–150 m; Gulf of Mexico off Florida, 37–73 m.

Family Cymothoidae Leach, 1818

DIAGNOSIS   Antennules and antennae reduced, no clear distinction between peduncles and flagella. Mandibular palp of three articles. Maxilla 1 with four terminal spines. Maxilla 2 apically bilobed, armed with several spines. Maxillipedal palp of two articles, terminal article bearing hooks. All seven pairs of pereopods prehensile, ending in strongly hooked dactyli. Pleopods lacking marginal setae in adults.

REMARKS   The cymothoids are exclusively ectoparasites on marine, freshwater, and brackish-water fishes. Most cymothoids occur in shallow water, mainly in tropical and subtropical areas. The position of attachment on the host (externally, in the buccal cavity, or in the gill chamber) is usually genus- or species-specific. The body of gill parasites is often asymmetrical, being slightly twisted, perhaps an effect of the position on the host. The mouthparts are highly adapted for the parasitic mode of life, while all seven pairs of pereopods are strongly prehensile. The posterior pereopods of some genera have the basal article expanded and carinate, allowing for increased musculature. The secretion of anticoagulants in the juvenile stages further aids the blood-feeding habit. The surface area of the pleopods is often increased by the development of lobes on the bases or the lamellae, providing an increased respiratory ability.

The post-mancal juvenile stages (sometimes referred to as the aegathoid stage) have large eyes, and highly setose pleopods for active swimming. The juveniles will attach themselves indiscriminately to any convenient fish host, but eventually attach to the preferred host-species. The juvenile then develops into a functional male, losing the swimming setae of the pleopods. Both juveniles and males feed actively, drawing blood from the host fish. The

Key to genera of Cymothoidae

1. Antennule broader and usually longer than antenna; cephalon very weakly sunk into pereonite 1 2
   Antennule not broader or longer than antenna; cephalon distinctly immersed in, or not at all immersed in pereonite 1 4

2. Bases of antennules widely separated 3
   Bases of antennules contiguous .... *Glossobius*

3. Body curved to one side; pleonite 1 extended laterally more on one side than on other .... *Mothocya*
   Body rarely curved to one side; pleonite 1 extended equally on each side .... *Renocila*

4. Pereonites and coxal plates 4–7 strongly expanded on one side only .... *Agarna*
   No pereonites or coxal plates strongly expanded 5

5. Cephalon not immersed in pereonite 1; posterior margin of cephalon trisinuate 6
   Cephalon to some degree immersed in pereonite 1; posterior margin of cephalon not trisinuate 7

6. Posterolateral angles of pereonites 2–6 not produced; coxal plates short, rarely reaching posterior margin of their pereonites .... *Anilocra*
   Posterolateral angles of pereonites 2–6 posteriorly increasingly produced; coxal plates usually reaching to posterior margin of their pereonites .... *Nerocila*

7. Basal antennular articles expanded and contiguous .... *Ceratothoa*
   Basal antennular articles expanded but not contiguous, or basal antennular articles neither expanded nor contiguous 8

8. Basal antennular articles expanded but not contiguous .... *Kuna*
   Basal antennular articles neither expanded nor contiguous 9

9. Pleonal margins continuous with pereonal margins, pleon not abruptly narrowed, only weakly immersed in pereonite 7 .... *Lironeca*
   Pleon to some degree narrower than pereon; pleon usually deeply immersed in pereonite 7 .... *Cymothoa*

male eventually becomes a female (all cymothoids are protandrous) should a female not already be present. In some species, the female is nonfeeding. In those species which settle either in the mouth cavity or gill chamber of the host, integumental pigment is frequently lost, and the eyes become reduced.

Given the highly variable morphology of the cymothoids, in part imposed by the parasitic mode of life, and the existence of polymorphism and possible sibling species, the taxonomy of this family demands the examination of large numbers of specimens. As a further aid to identification, Table 3 is provided, giving host species, parasite, and site of attachment.

TABLE 3. CYMOTHOID PARASITES FROM THE CARIBBEAN AREA, LISTED BY FISH HOST SPECIES

Fish host	Cymothoid parasite	Site of attachment
Abudefduf saxatilis	*Anilocra abudefdufi*	beneath eye
	Kuna insularis	gill chamber
Acanthurus bahianus	*Anilocra acanthuri*	♀ at base of pectoral fin; immature on or near pectoral or pelvic fin
Acanthurus chirurgus	*Anilocra acanthuri*	♀ at base of pectoral fin; immature on or near pectoral or pelvic fin
Alutera schoepfi	*Nerocila acuminata*	on or at base of fin
Anchoa lamprotaenia	*Lironeca tenuistylis*	posterior to pectoral fin
Apogon lachneri	*Mothocya bohlkeorum*	in gill chamber
Apogon maculatus	*Renocila colini*	next to dorsal fin
Apogon townsendi	*Renocila colini*	next to dorsal fin
Arius felis	*Nerocila acuminata*	on or at base of fin
Astrapogon stellatus	*Mothocya bohlkeorum*	in gill chamber
Batrachoides surinamensis	*Nerocila acuminata*	on or at base of fin
Caranx hippos	*Cymothoa oestrum*	inside mouth
Caranx latus	*Cymothoa oestrum*	inside mouth
Caranx ruber	*Cymothoa oestrum*	inside mouth
Caranx sp.	*Cymothoa oestrum*	inside mouth
Chaetodipterus faber	*Nerocila acuminata*	on or at base of fin
Chaetodon capistratus	*Anilocra chaetodontis*	beneath eye
Chaetodon ocellatus	*Anilocra chaetodontis*	beneath eye
Chaetodon sedentarius	*Anilocra chaetodontis*	beneath eye
Chaetodon striatus	*Anilocra chaetodontis*	beneath eye

(continued)

TABLE 3. (*Continued*)

Fish host	Cymothoid parasite	Site of attachment
Chilomycterus schoepfi	*Nerocila acuminata*	on or at base of fin
Chromis cyaneus	*Anilocra chromis*	beneath eye
Chromis multilineatus	*Anilocra chromis*	beneath eye
Cynoscion nebulosus	*Cymothoa excisa*	inside mouth
Cynoscion sp.	*Cymothoa oestrum*	inside mouth
Epinephelus cruentatus	*Anilocra haemuli*	beneath eye
Epinephelus fulvus	*Anilocra haemuli*	beneath eye
Epinephelus guttatus	*Anilocra haemuli*	beneath eye
Epinephelus itajara	*Nerocila acuminata*	on or at base of fin
Epinephelus sp.	*Cymothoa oestrum*	on or at base of fin
Exocoetus spp.	*Glossobius impressus*	inside mouth
Gerres rhombeus	*Lironeca redmanni*	in gill chamber
Haemulon aurolineatum	*Anilocra haemuli*	beneath eye
Haemulon carbonarium	*Anilocra haemuli*	beneath eye
Haemulon chrysargyreum	*Anilocra haemuli*	beneath eye
Haemulon flavolineatum	*Anilocra haemuli*	beneath eye
Haemulon macrostomum	*Anilocra haemuli*	beneath eye
Haemulon plumieri	*Anilocra haemuli*	beneath eye
Haemulon sciurus	*Anilocra haemuli*	beneath eye
Hemirhamphus brasiliensis	*Glossobius hemirhamphi*	inside mouth
Hirundichthys speculifer	*Glossobius impressus*	inside mouth
Holacanthus tricolor	*Anilocra holacanthi*	beneath eye
Holocentrus ascensionis	*Anilocra holocentri*	♀ between eyes, ♂ and immature beneath eye
Hyporhamphus unifasciatus	*Mothocya nana*	in gill chamber
Leiostomus xanthurus	*Nerocila acuminata*	on or at base of fin
	Cymothoa excisa	inside mouth
	Lironeca redmanni	in gill chamber
Lepiosteus spatula	*Nerocila acuminata*	on at base of fin
Lutjanus analis	*Cymothoa excisa*	inside mouth
Lutjanus mahogoni	*Cymothoa excisa*	inside mouth
Lutjanus synagris	*Cymothoa excisa*	inside mouth
Megalops atlanticus	*Cymothoa oestrum*	inside mouth
Monacanthus ciliatus	*Nerocila acuminata*	on or at base of fin
Mugil cephalus	*Nerocila acuminata*	on or at base of fin
Myripristis jacobus	*Anilocra myripristi*	♀ between eyes, immature beneath eye
Ocyurus chrysurus	*Cymothoa excisa*	inside mouth
Orthopristis chrysoptera	*Cymothoa excisa*	inside mouth

Fish host	Cymothoid parasite	Site of attachment
Orthopristis ruber	*Anilocra haemuli*	beneath eye
Paranthias furcifer	*Anilocra haemuli*	beneath eye
Phaeoptyx conklini	*Mothocya bohlkeorum*	in gill chamber
Phaeoptyx pigmentaria	*Mothocya bohlkeorum*	in gill chamber
Pogonias cromis	*Nerocila acuminata*	on or at base of fin
Pomacentrus partitus	*Anilocra partiti*	beneath eye
Priacanthus arenatus	*Cymothoa oestrum*	inside nouth
Scomberomorus cavalla	*Lironeca redmanni*	in gill chamber
Scomberomorus maculatus	*Lironeca redmanni*	in gill chamber
Scomberomorus regalis	*Lironeca redmanni*	in gill chamber
Selar crumenophthalmus	*Cymothoa oestrum*	inside mouth
Serranus tigrinus	*Renocila bowmani*	next to dorsal fin
	Renocila waldneri	next to dorsal fin
Sphoeroides maculatus	*Nerocila acuminata*	on or at base of fin
Synodus foetens	*Cymothoa excisa*	inside mouth

Agarna Schioedte and Meinert, 1883

DIAGNOSIS Cephalon with posterior margin not trilobed; immersed in pereonite 1. Antennular bases contiguous. Pereonites 4–7 on one side flattened and expanded; coxal plates of pereopods 4–7 also expanded and flattened but generally hidden by lateral expansion of pereonites. Bases of posterior three pereopods with well-formed carinae. Pleonites 1 and 2 immersed in pereonite 7; pleonites 2–5 with free fingerlike lateral margins.

Agarna cumulus (Haller, 1880)
Figure 79

DIAGNOSIS ♀ 18 mm. Eyes present, indistinct. Pereon strongly "humped" dorsally. Uropod about $1/3$ length of pleotelson; uropodal exopod slightly longer, and twice width of endopod. Pleotelson triangular, length $3/4$ basal width, apex rounded.

RECORDS No host recorded: Key West, Florida.

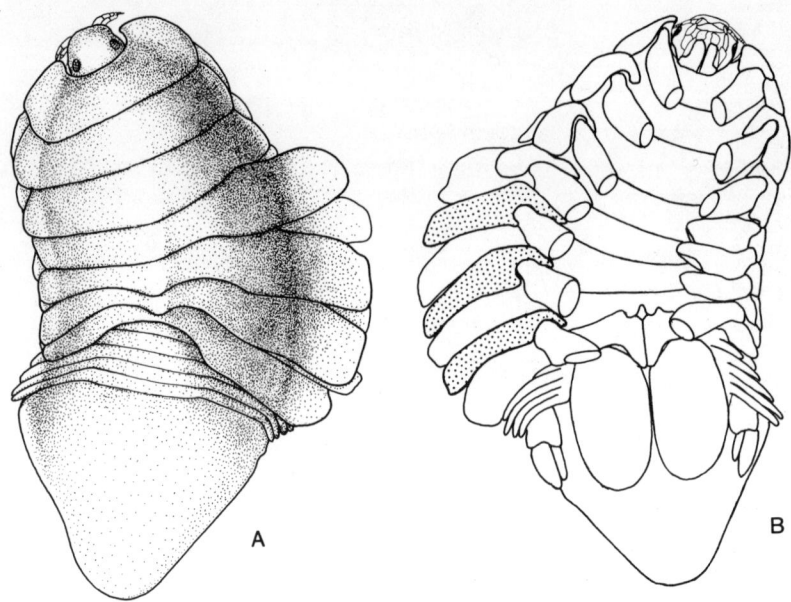

Figure 79. *Agarna cumulus:* A, ♀, dorsal view; B, ♀, ventral view, coxal plates stippled.

Anilocra Leach, 1818

DIAGNOSIS Cephalon usually narrowed anteriorly to triangular apex folded ventrally between bases of antennules; posterior margin trilobed; not immersed, or only weakly immersed in pereonite 1. Coxal plates small, compact, not reaching level of posterior margin of their respective pereonites. Pereopods increasing in length posteriorly, pereopod 7 often markedly longer than 6. Pleon not immersed or only slightly immersed in pereonite 7. Pleopods 3–5 often formed into deep pockets or pleats. Uropods often extending beyond pleotelsonic apex.

REMARKS Williams and Williams (1981) have provided a comprehensive treatment of this genus and nine of its species in the West Indies. Table 1 in this latter paper provides characters for separating these nine species. This table also indicates that for each species, the site of attachment of the adult to the host fish is specific, with six species attaching under the eye of the host.

Key to species of *Anilocra*

1. Pereopods 2–4 with swelling on outer margin of dactylus 2
   Pereopods 2–4 lacking swelling on outer margin of dactylus 5

2. Body axis distorted by more than 10° .... *holacanthi*
   Body axis distorted by less than 5° 3

3. Dactylus of pereopod 7 longer than propodus .... *partiti*
   Dactylus of pereopod 7 shorter than propodus 4

4. Posteroventral angle of pereonite 7 overlapping pleonite 1 only
   .... *abudefdufi*
   Posteroventral angle of pereonite 7 overlapping pleonites 1 and 2
   .... *chaetodontis*

5. Posteroventral angle of pereonite 7 produced 6
   Posteroventral angle of pereonite 7 not produced 7

6. Uropod reaching posterior margin of pleotelson .... *myripristis*
   Uropod not reaching posterior margin of pleotelson .... *haemuli*

7. Posteroventral angle of pereonite 7 overlapping pleonite 1 .... *holocentri*
   Posteroventral angle of pereonite 7 not overlapping pleonite 1 8

8. Uropod reaching posterior margin of pleotelson .... *acanthuri*
   Uropod not reaching posterior margin of pleotelson .... *chromis*

Anilocra abudefdufi Williams and Williams, 1981
   Figure 80A–C

DIAGNOSIS   Ovigerous ♀ 19.0–31.0 mm, ♂ 7.0–8.5 mm. Pereopods 2–4 with swelling on outer margin of dactylus. Posteroventral angle of pereonite 6 slightly produced, of pereonite 7 more produced, overlapping pleonite 1. Uropodal endopod variable, not reaching, to extending well beyond, apex of exopod. Color: upper lateral half to three-fourths of dorsal surface of ♀ when attached to host is dark brown; rest of dorsal surface light brown to yellow. Attaching beneath eye of host.

RECORDS   Sergeant major *Abudefduf saxatilis:* Panama; Colombia.

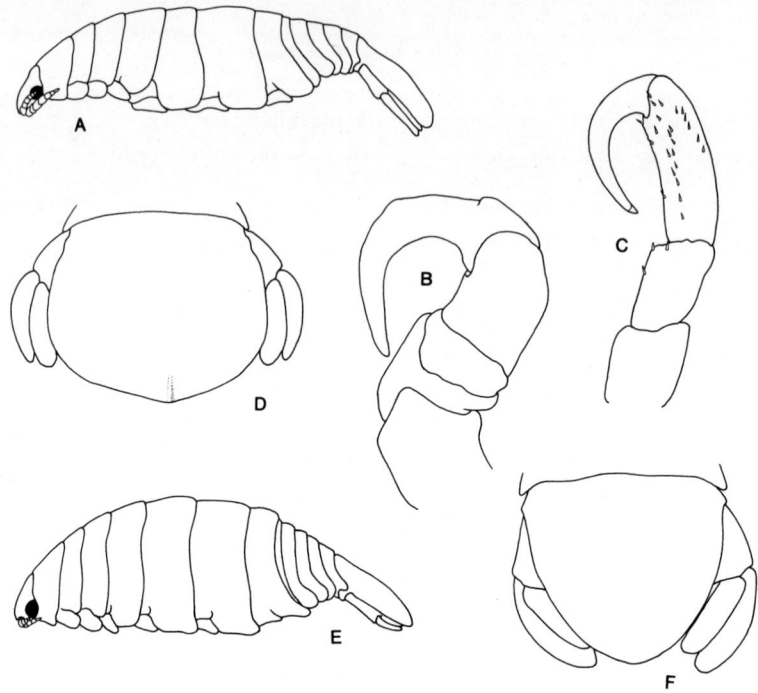

Figure 80. *Anilocra abudefdufi*: A, ♀, lateral view; B, pereopod 3; C, pereopod 7. *Anilocra acanthuri*: D, pleotelson and uropods. *Anilocra chaetodontis*: E, ♀, lateral view. *Anilocra chromis*: F, pleotelson and uropods.

Anilocra acanthuri Williams and Williams, 1981
Figure 80D

DIAGNOSIS Ovigerous ♀ 29.0–40.0 mm, ♂ 4.0–8.0 mm. Pereopods 2–4 without swelling on outer margin of dactylus. Posteroventral angles of pereonites not produced. Uropod not reaching posterior margin of pleotelson. Endopod of uropod variable, not reaching, to extending well beyond, apex of exopod. Color: dorsal surface of ♀ black to lead gray, ventral surface gray. Attaching under pectoral fin of host.

RECORDS Doctorfish *Acanthurus chirurgus*: Florida Keys; Bahamas; Puerto Rico; U.S. Virgin Islands. Ocean surgeon *Acanthurus bahianus*: Florida Keys; Bahamas; Cuba; Jamaica; Dominican Republic; Puerto Rico; U.S. Virgin Islands.

Anilocra chaetodontis Williams and Williams, 1981
Figure 80E

DIAGNOSIS   Ovigerous ♀ 18–28 mm, ♂ 4–5 mm. Pereopods 2–4 with swelling on outer margin of dactylus. Posteroventral angles of pereonites 4–7 becoming progressively produced, that of pereonite 7 overlapping pleonite 2. Uropod not reaching posterior margin of pleotelson; uropodal endopod extending beyond apex of exopod. Pleotelson as wide as long to slightly wider than long. Color: dorsal surface of ♀ black to lead gray, ventral surface gray. Attaching beneath eye of host.

RECORDS   Foureye butterflyfish *Chaetodon capistratus:* Bahamas; Puerto Rico; British and U.S. Virgin Islands. Banded butterflyfish *Chaetodon striatus:* Bahamas; Puerto Rico; British Virgin Islands. Spotfin butterflyfish *Chaetodon ocellatus:* Bahamas; Puerto Rico; U.S. Virgin Islands. Reef butterflyfish *Chaetodon sedentarius:* Puerto Rico.

Anilocra chromis Williams and Williams, 1981
Figure 80F

DIAGNOSIS   Ovigerous ♀ 16–28 mm, ♂ 4–9 mm. Pereopods 2–4 lacking swelling on outer margin of dactylus. Posteroventral angles of pereonites not produced. Uropod extending beyond posterior margin of pleotelson; uropodal endopod not reaching beyond exopod. Color: upper lateral one-fourth to two-thirds of dorsal surface of ♀ when attached is dark gray, shading to off-white lower lateral area. Attaching beneath eye of host.

RECORDS   Brown chromis *Chromis multilineatus:* Puerto Rico; British and U.S. Virgin Islands. Blue chromis *Chromis cyaneus:* Bahamas; Dominican Republic. No host recorded: Anguilla.

Anilocra haemuli Williams and Williams, 1981
Figure 81A,B

DIAGNOSIS   Ovigerous ♀ 21–40 mm, ♂ 7 mm. Body axis distorted less than 5°. Pereopods 2–4 lacking swelling on outer margin of dactylus. Posteroventral angle of pereonites 6 and 7 produced, latter overlapping pleonite 1. Uropod not reaching posterior margin of pleotelson; uropodal endopod reaching beyond apex of exopod. Color: dorsal surface of ♀ yellow to light brown. Attaching beneath eye of host.

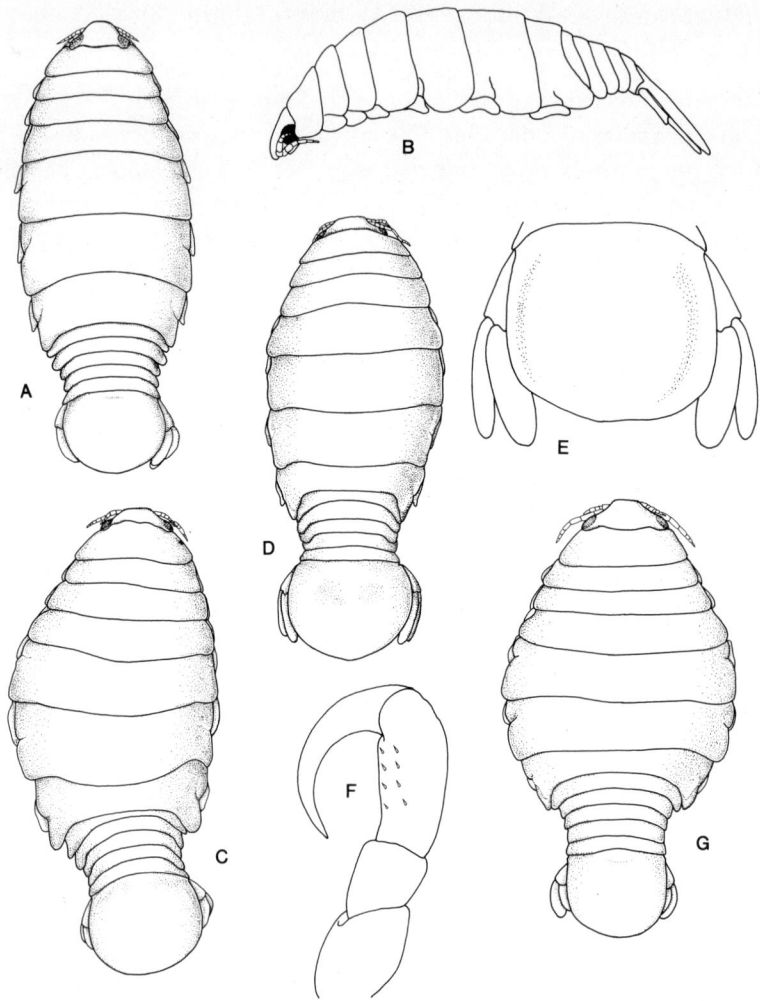

Figure 81. *Anilocra haemuli: A*, ♀, dorsal view; *B*, ♀, lateral view. *Anilocra holacanthi: C*, ♀. *Anilocra holocentri: D*, ♀. *Anilocra myripristis: E*, pleotelson and uropods. *Anilocra partiti: F*, ♀; *G*, pereopod 7.

RECORDS French grunt *Haemulon flavolineatum:* Florida Keys; Puerto Rico; British and U.S. Virgin Islands. Tomtate *Haemulon aurolineatum:* Jamaica; Puerto Rico. Smallmouth grunt *Haemulon chrysargyreum:* Puerto Rico; U.S. Virgin Islands. Caesar grunt *Haemulon carbonarium:* Puerto Rico; U.S. Virgin

Islands. Spanish grunt *Haemulon macrostomum:* Puerto Rico. White grunt *Haemulon plumieri:* Florida Keys; Yucatan Peninsula. Bluestriped grunt *Haemulon sciurus:* Florida Keys. Cora cora *Orthopristis ruber:* Margarita Island, Venezuela. Coney *Epinephelus fulvus:* Bahamas; Dominican Republic; Puerto Rico; U.S. Virgin Islands; Guadeloupe. Red hind *Epinephelus guttatus:* Puerto Rico; British and U.S. Virgin Islands. Graysby *Epinephelus cruentatus:* Bahamas; Dominican Republic; U.S. Virgin Islands. Creole-fish *Paranthias furcifer:* Dominican Republic; Puerto Rico; Colombia. No host recorded: Cuba; Jamaica; Dominica; Barbados; Venezuela; Brazil.

Anilocra holacanthi Williams and Williams, 1981
Figure 81C

DIAGNOSIS Ovigerous ♀ 21–33 mm, ♂ 4–7 mm. Body axis distorted by more than 10°. Pereopods 2–4 with swelling on outer margin of dactylus. Posteroventral angles of pereonites 5–7 progressively more produced, that of pereonite 7 overlapping pleonite 1. Uropod not reaching posterior margin of pleotelson; uropodal endopod reaching beyond apex of exopod. Color: dorsal surface of ♀ black to lead gray. Attaching beneath eye of host.

RECORDS Rock beauty *Holacanthus tricolor:* Bahamas; Jamaica; Dominican Republic; Puerto Rico; British and U.S. Virgin Islands.

Anilocra holocentri Williams and Williams, 1981
Figure 81D

DIAGNOSIS Ovigerous ♀ 32–46 mm, ♂ 5–9 mm. Body axis distorted less than 5°. Pereopods 2–4 lacking swelling on outer margin of dactylus. Posteroventral angle of pereonite 7 produced, overlapping pleonite 1. Uropod not reaching posterior margin of pleotelson; uropodal endopod reaching beyond apex of exopod. Color: dorsal surface of ♀ dark brown, ventral surface light brown. ♀ attaching between eyes of host; ♂ or transitional stage beneath eye.

RECORDS Squirrelfish *Holocentrus ascensionis:* Puerto Rico; U.S. Virgin Islands.
No host recorded: Patagonia, Straits of Magellan.

Anilocra myripristis Williams and Williams, 1981
   Figure 81E

DIAGNOSIS   Ovigerous ♀ 29–40 mm, ♂ 6–7 mm. Body axis distorted less than 5°. Pereopods 2–4 lacking swellings on outer margin of dactylus. Posteroventral angle of pereonites 6 and 7 produced, latter overlapping pleonite 1. Uropod reaching beyond posterior margin of pleotelson; uropodal endopod reaching beyond apex of exopod. Color: dorsal surface of ♀ light reddish brown, ventral surface yellow. ♀ attaching between eyes of host; immature or transitional forms sometimes beneath eye.

RECORDS   Blackbar soldierfish *Myripristis jacobus:* Bahamas; Dominican Republic; Puerto Rico.

Anilocra partiti Williams and Williams, 1981
   Figure 81F,G

DIAGNOSIS   Ovigerous ♀ 12–16 mm, transitional 7.6–9.0 mm. Body axis distorted less than 5°. Pereopods 2–4 with swelling on outer margin of dactylus. Pereopod 7 with dactylus longer than propodus. Posteroventral angle of pereonite 7 produced, overlapping pleonite 1. Uropod not reaching posterior margin of pleotelson; uropodal endopod not reaching apex of exopod. Color: dorsal surface black to slate gray. Attaching beneath eye of host.

RECORDS   Bicolor damselfish *Pomacentrus partitus:* Jamaica.

Ceratothoa Dana, 1852

DIAGNOSIS   Cephalon more or less immersed in pereonite 1, posterior margin not trisinuate. Bases of antennules expanded, contiguous. Coxal plates compact; anterior plates not extending beyond posterior margins of their respective pereonites; posterior coxal plates may or may not be produced beyond the posterior margins of the pereonites. Anterior pleonites narrowed, immersed in pereonite 7. Copulatory stylet lacking on pleopod 2 of ♂ of some species.

Ceratothoa deplanata Bovallius, 1885
   Figure 82A

DIAGNOSIS   ♀ 18 mm. Cephalon subtriangular, anterior margin rounded. Pereopods 4–7 with strongly carinate bases. Uropod reaching or extending

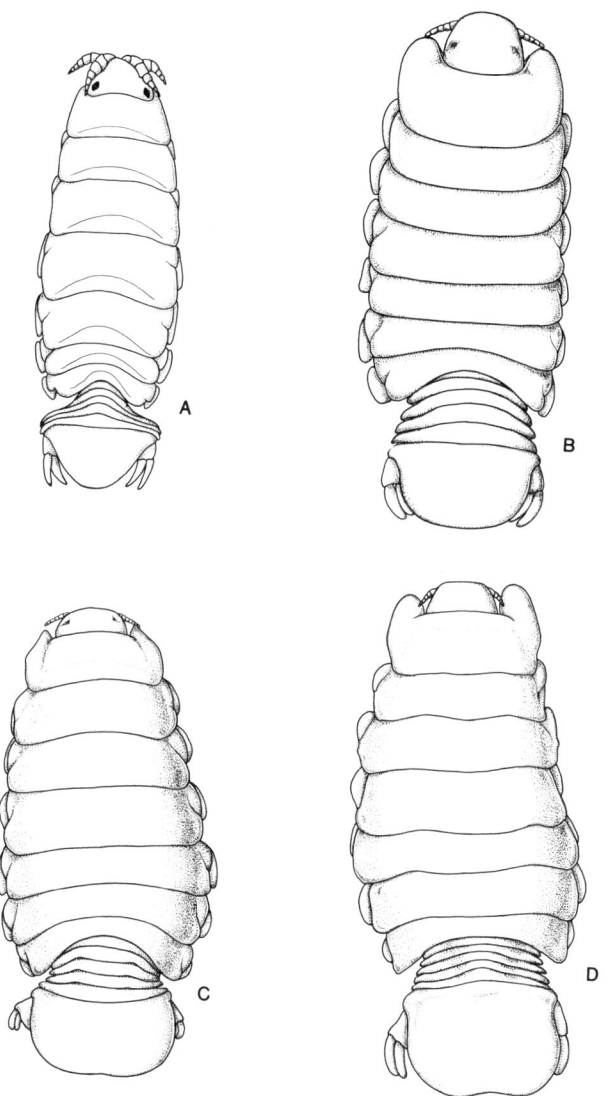

Figure 82. *A, Ceratothoa deplanata* (from Bovallius, 1885); *B, Cymothoa caraibica*; *C, Cymothoa excisa*; *D, Cymothoa oestrum*.

slightly beyond posterior margin of pleotelson; rami subequal in length and width. Pleotelson basally wider than long, posterior margin broadly rounded. Color: bright yellow.

RECORDS Haiti, host not recorded.

Cymothoa Fabricius, 1793

DIAGNOSIS Body usually not distorted. Cephalon with posterior margin not trilobed; more or less immersed in pereonite 1; latter with anterolateral corners produced to embrace cephalon. Bases of antennules not expanded, well separated. Anterior coxal plates not reaching posterior borders of their respective pereonites, posterior coxal plates nearly reaching or extending beyond posterior borders of pereonites. Pleon narrower than, and immersed in pereonite 7. Pleonites increasing in length and width posteriorly.

Key to species of *Cymothoa*

1. Anterolateral angles of pereonite 1 reaching to half length of cephalon or less; eyes or traces of eyes present 2
   Anterolateral angles of pereonite 1 broad, reaching to anterior margin of cephalon; eyes absent .... *oestrum*

2. Anterolateral angles of pereonite 1 narrow, subacute .... *excisa*
   Anterolateral angles of pereonite 1 broad, rounded .... *caraibica*

Cymothoa caraibica Bovallius, 1885
Figure 82B

DIAGNOSIS ♀ 17 mm, ♂ 12–16 mm. Anterior margin of cephalon broadly rounded. Eyes large, distinct. Broadly rounded anterolateral angles of pereonite 1 reaching to about midlength of cephalon. Bases of pereopods 4–7 with strong, rounded carina. Uropodal rami subequal in length, equal to peduncle in length. Pleotelson width about twice length, posterolateral margin broadly rounded.

RECORDS Puerto Rico; Gulf of Mexico.

Cymothoa excisa Perty, 1833
Figure 82C

DIAGNOSIS Ovigerous ♀ 20–24 mm. Anterior margin of cephalon in dorsal view truncate to slightly excavate; eyes small, indistinct. Anterolateral angles of pereonite 1 narrowly rounded to subacute, reaching anteriorly to about midlength of cephalon. Pereopods 4–7 with high rounded carina on basis.

Uropods hardly reaching halfway along lateral margin of pleotelson; exopod slightly longer than endopod. Pleotelson about twice wider than long; broadly rounded and somewhat bilobed.

RECORDS Yellowtail snapper *Ocyurus chrysurus:* Yucatan Peninsula, Mexico; Carrie Bow Cay, Belize; Margarita Island, Venezuela; Panama. Mutton snapper *Lutjanus analis:* Yucatan Peninsula, Mexico; Panama. Lane snapper *Lutjanus synagris:* Panama. Mahogany snapper *Lutjanus mahogoni:* Panama. Pigfish *Orthopristis chrysoptera:* Florida, Gulf of Mexico. Spot *Leiostomus xanthurus:* Texas, Gulf of Mexico. Spotted seatrout *Cynoscion nebulosus:* Texas, Gulf of Mexico. Inshore lizardfish *Synodus foetens:* Texas, Gulf of Mexico. No host recorded: Massachusetts; South Carolina; Georgia; Florida Keys; Bahamas; Cuba; Trinidad; Brazil.

Cymothoa oestrum (Linnaeus, 1793)
Figure 82D

DIAGNOSIS Ovigerous ♀ 38 mm. Cephalon in dorsal view with anterolateral angles rounded, anterior margin slightly excavate; eyes absent. Anterolateral angles of pereonite 1 expanded, broadly rounded, reaching to level of anterior margin of cephalon. Pereonites 4–7 with high rounded carina on basis. Uropod reaching posteriorly beyond midlength of pleotelson; exopod slightly longer than endopod. Pleotelson length slightly more than half basal width.

RECORDS Bigeye scad *Selar crumenophthalmus:* Bermuda; U.S. Virgin Islands. Bigeye *Priacanthus arenatus:* Bermuda. Bar jack *Caranx ruber:* Florida Keys; Carrie Bow Cay, Belize. Horse-eye jack *Caranx latus:* Bahamas; Barbados. Crevalle jack *Caranx hippos:* Venezuela. Jack *Caranx* sp.: Jamaica; Curaçao. Hind *Epinephelus* sp.: Grenada. Parrotfish: Jamaica. Seatrout *Cynoscion* sp.: Panama. Tarpon *Megalops atlantica:* Texas, Gulf of Mexico. No host recorded: Honduras; Haiti.

Glossobius Schioedte and Meinert, 1883

DIAGNOSIS Cephalon not immersed in pereonite 1; excavate on either side in anterior half, forming broad and anteriorly rounded median area; antennae fitting into excavate areas. Bases of antennules contiguous, expanded. Antennules broader and longer than antennae. Bases of pereopods 4–7 with posterior margin expanded and flattened. Pleonites 1–3 immersed in pereonite 7.

Key to species of *Glossobius*

1. Coxal plates of pereonites 1 and 2 anteroventrally protruding   *impressus*
   Coxal plates of pereonites 1 and 2 close to body, not protruding
   .... *hemirhamphi*

Glossobius hemiramphi Williams and Williams, 1985a
   Figure 83A

DIAGNOSIS   Ovigerous ♀ 27 mm. Eyes small but distinct. Cephalon pointed anteriorly. Fused coxa of pereonite 1 and free coxa of pereonite 2 carinate but not protruding. Coxa of pereonite 7 semicircular in dorsal view. Pleotelson with middorsal length more than half basal width; lateral margins somewhat tapered; posterior margin variable, sinuate or excavate. Uropods reaching to or slightly beyond posterior pleotelsonic margin; rami subequal in length, exopod slightly broader than endopod.

RECORDS   Ballyhoo *Hemiramphus brasiliensis:* Puerto Rico.

Glossobius impressus (Say, 1818)
   Figure 83B

DIAGNOSIS   Ovigerous ♀ 33 mm. Eyes small but distinct. Cephalon rounded anteriorly. Fused coxal plate of pereonite 1 and distinct coxal plate of pereonite 2 protruding strongly in oblique anteroventral direction. Uropod reaching to posterior half of pleotelson; exopod shorter and narrower than endopod. Pleotelson basal width twice length, posteriorly broadly bilobed. Attaching inside mouth of host.

RECORDS   Flyingfish *Exocoetus* spp.: Rio de Janeiro, Brazil; North Atlantic, especially in the Gulf Stream.
   Mirrorwing flyingfish *Hirundichthys speculifer:* North Atlantic. No host record: Senegal, West Africa.

Kuna Williams and Williams, 1986

DIAGNOSIS   Cephalon somewhat immersed in pereonite 1. Anterior margin of pereonite 1 not trisinuate. Number of articles in antennules and antennae

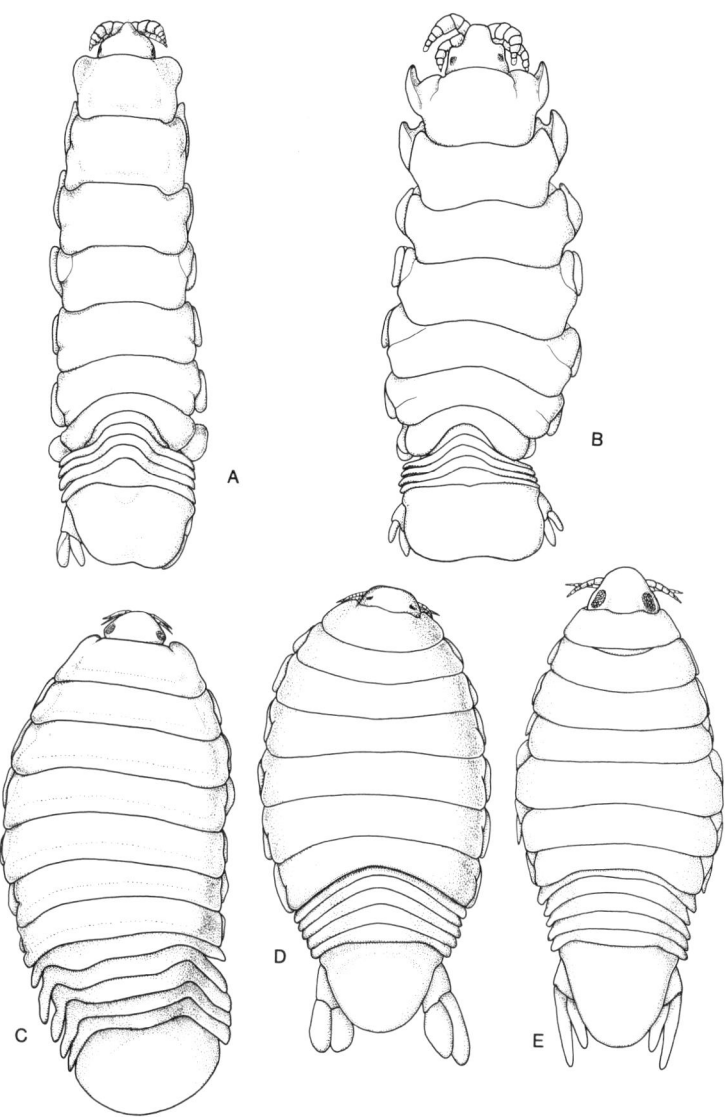

Figure 83. A, *Glossobius hemiramphi*; B, *Glossobius impressus*; C, *Kuna insularis*; D, *Lironeca redmani*; E, *Lironeca tenuistylis*.

reduced. Antennule somewhat expanded; basal article expanded but not contiguous. Copulatory stylet present on pleopods 1–3 in ♂. Pleonites dorsally strongly convex, not immersed in pereonite 7.

Kuna insularis (Williams and Williams, 1985b)
Figure 83C

DIAGNOSIS Ovigerous ♀ 11.1–17.2 mm, ♂ 4.2–8.7 mm, transitional 9.6–9.8 mm. Antennules and antennae consisting of four articles each. Uropods short, not reaching posterior margin of pleotelson. Clavate copulatory stylet present on pleopods 1–3 in ♂. Pleotelson basally broader than long, posterior margin broadly rounded.

RECORDS Sergeant major *Abudefduf saxatilis:* Carrie Bow Cay, Belize; Curaçao; Panama.

Lironeca Leach, 1818

DIAGNOSIS Cephalon weakly to deeply immersed in pereonite 1; posterior border rarely trisinuate. Bases of antennules not expanded, well separated. Posterior pereopods with carinae on bases in ♂, carinae present or absent in ♀. Pleonites subequal in width; pleonites 1 and 2 rarely narrowed and weakly to moderately immersed in pereonite 7. Pleopods highly folded, and with lamellar or digitiform accessory gills in some species.

Key to species of *Lironeca*

1. Uropodal endopod about twice longer than wide; pleon somewhat immersed in pereon .... *redmanni*
   Uropodal endopod about three times longer than wide; pleon barely immersed in pereon .... *tenuistylis*

Lironeca redmanni Leach, 1818
Figure 83D

DIAGNOSIS Ovigerous ♀ 19.5–25.0 mm. Cephalon barely immersed in pereonite 1. Pleon somewhat immersed in pereon, but lateral margins of pleonite 1 free. Pleotelson basally wider than long. Uropodal rami reaching well beyond posterior margin of pleotelson; exopod longer than endopod, both rami somewhat broad, endopod about twice longer than wide. Attaching to gills of host.

RECORDS New Jersey to Florida; gills of kingfish, Jamaica; Cuba; St. Christopher; Spanish mackerel *Scomberomorus maculatus* and cero *Scomberomorus regalis,* Puerto Rico; king mackerel *Scomberomorus cavalla,* Colombia; *Gerres rhombeus,* Panama; spot *Leiostomus xanthurus,* Gulf of Mexico.
Brazil.

Lironeca tenuistylis (Richardson, 1912b)
Figure 83E

DIAGNOSIS ♀ 13 mm. Cephalon barely immersed in pereonite 1. Uropodal rami reaching beyond rounded posterior margin of pleotelson; exopod longer than endopod; endopod slender, about three times longer than wide. Pleonite 1 barely immersed in pereonite 7. Pleotelson basally wider than long. Attaching to host between pectoral and anal fin.

RECORDS Longnose anchovy *Anchoa lamprotaenia:* Panama.

Mothocya Costa, 1851

DIAGNOSIS Cephalon more or less immersed in pereonite 1. Bases of antennules widely separated; antennules longer and more robust than antennae. Coxae nearly reaching or extending beyond posterior margin of respective pereonites. Pleon somewhat immersed in pereonite 7. Uropodal exopod longer than endopod.

REMARKS Bruce (1986b) revised the genus *Mothocya.* The species of *Mothocya* are almost entirely gill parasites on the fish families Hemiramphidae, Apogonidae, Belonidae, and Atherinidae.

Key to species of *Mothocya*

1. Cephalon anteriorly narrowed, slightly immersed in pereonite 1; pleotelson subrectangular .... *bohlkeorum*
   Cephalon anteriorly broad, deeply immersed in pereonite 1; pleotelson subtriangular .... *nana*

Mothocya bohlkeorum Williams and Williams, 1982
Figure 84B

DIAGNOSIS   Ovigerous ♀ 7.6–8.5 mm, ♂ 3.7 mm. Cephalon anteriorly narrowed in dorsal view, ventrally flexed, broadly rounded; slightly immersed in pereonite 1. Pleotelson subrectangular. Uropods extending slightly beyond posterior margin of pleotelson; exopod only slightly longer than endopod. ♀ lateral lobes of pleopodal peduncles not developed. Endopods of pleopods 3–5 with small proximomedial lobe.

RECORDS   Whitestar cardinalfish *Apogon lachneri:* Puerto Rico. Dusky cardinalfish *Phaeoptyx pigmentaria:* Bahamas. Freckled cardinalfish *Phaeoptyx conklini:* Florida Keys; Bahamas. Conchfish *Astrapogon stellatus:* Leeward Islands.

Mothocya nana (Schioedte and Meinert, 1884)
Figure 84A

DIAGNOSIS   Ovigerous ♀ 11.0–17.0 mm, ♂ 7.9–8.3 mm. Cephalon deeply immersed in pereonite 1; rostrum anteroventrally narrowly rounded. Uropodal exopod markedly longer than endopod. Pleotelson broad, with posterior margin rounded sufficiently to give appearance of being subtriangular.

RECORDS   Halfbeak *Hyporhamphus unifasciatus:* Chesapeake Bay, Maryland; Georgia; Florida; Colon, Panama. Halfbeak *Hemiramphus bermudensis:* Bermuda.

Nerocila Leach, 1818

DIAGNOSIS   Body generally more depressed than in most cymothoid genera, rarely curved. Cephalon with anterior margin convex, narrowly rounded, or concave; not, or only slightly, immersed in pereonite 1. Pereonite 1 anterior margin trisinuate. Posterolateral angles of pereonites weakly to strongly produced, increasing in length posteriorly. Coxal plates prominent, usually almost reaching or extending to posterior margin of their respective pereonites. Juveniles and ♂ usually with spines on posterior pereopods; ♀ lacking these spines. Pleon not immersed in pereonite 7. Pleonites subequal in length; pleonites 1 and 2 usually produced posterolaterally. Pleopods typically with small lamellar accessory gills; pleopods 3–5 folded into deep pockets or pleats. Uropods usually extending beyond pleotelsonic apex.

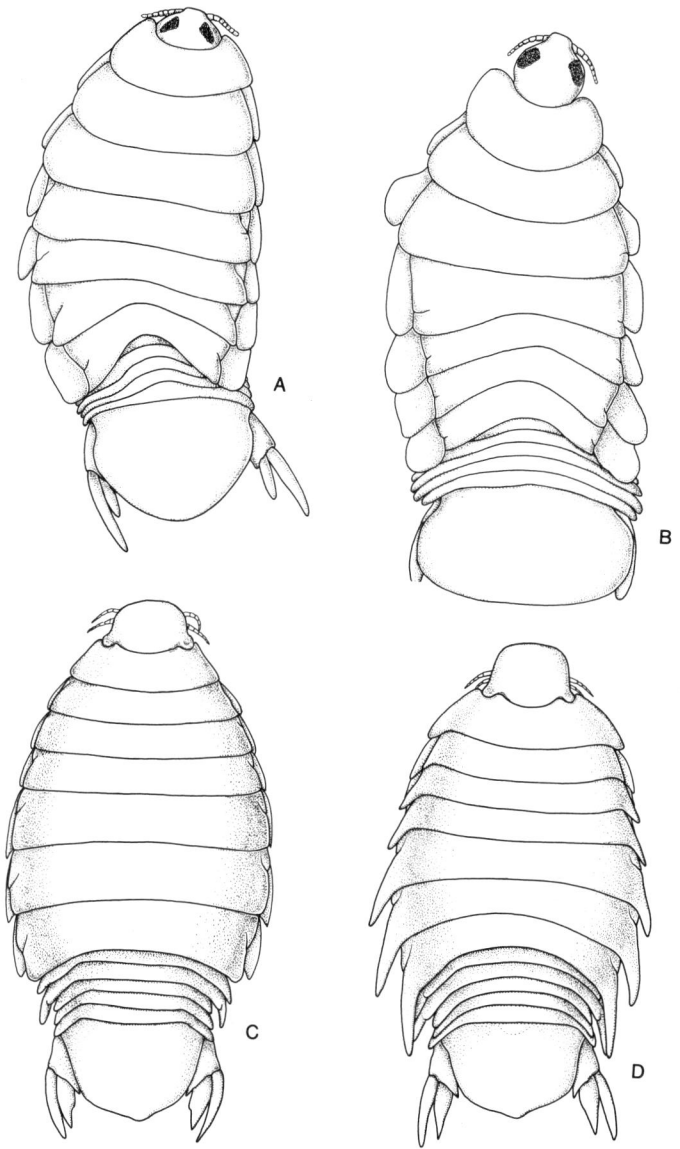

Figure 84. *A, Mothocya nana; B, Mothocya bohlkeorum; C, Nerocila acuminata* f. *acuminata; D, Nerocila acuminata* f. *aster.*

Nerocila acuminata Schioedte and Meinert, 1881

DIAGNOSIS Ovigerous ♀ 16.2–19.0 mm. Cephalon with anterior margin convex. Posterolateral angles of all, or of posterior pereonites only, produced into acute or subacute angles.

RECORDS Striped burrfish *Chilomycterus schoepfi:* Texas, Gulf of Mexico. Northern puffer *Sphoeroides maculatus:* New York. Striped mullet *Mugil cephalus:* Texas, Gulf of Mexico. Jewfish *Epinephelus itajara:* Texas, Gulf of Mexico. Hogfish: Bermuda. Alligator gar *Lepisosteus spatula:* Louisiana, Gulf of Mexico. Hardhead catfish *Arius felis:* Texas, Gulf of Mexico. Sawfish: Florida (Atlantic). Black drum *Pogonias cromis:* Texas, Gulf of Mexico. Orange filefish *Alutera schoepfi:* Texas, Gulf of Mexico. Toadfish *Batrachoides surinamensis:* Colon, Panama. Spot *Leiostomus xanthurus:* Florida, Gulf of Mexico. Spadefish *Chaetodipterus faber:* Florida, Gulf of Mexico; Virginia. Fringed filefish *Monacanthus ciliatus:* Florida, Gulf of Mexico. No host recorded: Massachusetts; Florida Keys; Florida, Gulf of Mexico. Louisiana, Gulf of Mexico. Texas, Gulf of Mexico.

REMARKS Brusca (1981) has shown that this highly variable species occurs on both sides of the Isthmus of Panama, in two relatively distinct forms. Intergrades between the two forms do occur but are uncommon. Brusca (1981:159) also lists all the host-records for this species in the eastern Pacific.

Nerocila acuminata Schioedte and Meinert, 1881, forma *acuminata*
Figure 84C

DIAGNOSIS Cephalon width equal to or greater than length; frontal margin narrowly rounded. Posterolateral angles of anterior pereonites weakly produced, rounded to subacute; of posterior pereonites more strongly produced, subacute to acute. Coxal plates 3–7, 4–7, or 5–7 with acute posterolateral angles; coxae rarely reaching beyond posterior margins of their respective pereonites.

Nerocila acuminata Schioedte and Meinert, 1881, forma *aster*
Figure 84D

DIAGNOSIS Cephalon always wider than long; anterior margin broadly rounded. Posterolateral angles of all pereonites strongly produced, acute, all reaching well beyond posterior margins of their respective pereonites. Coxal plates 2–7 strongly produced with acute posterior angles.

Renocila Miers, 1880

DIAGNOSIS Body rarely curved. Cephalon anteriorly weakly to distinctly truncate. Antennular bases well separated. Antennules and antennae somewhat flattened, antennules usually broader and longer than antennae. Pereonites 5–7 with posterolateral corners more or less strongly produced. Pleonites not laterally incised.

REMARKS Williams and Williams (1980) provide a key to nine species of *Renocila*.

Key to species of *Renocila*

1. Posteroventral angle of pereonite 7 reaching pleonite 1 .... *colini*
   Posteroventral angle of pereonite 7 reaching beyond pleonite 1 2

2. Dorsal surface of body brown; posteroventral angle of pereonite 7
   reaching pleonite 2 .... *waldneri*
   Dorsal surface of body black; posteroventral angle of pereonite 7
   reaching pleonite 3 .... *bowmani*

Renocila bowmani Williams and Williams, 1980
Figure 85A

DIAGNOSIS ♀ 18.0 mm, ♂ 11.5 mm. Posteroventral angles of pereonites 5–7 produced, that of pereonite 7 overlapping pleonites 1–3. Pereopods 1–3 lacking swelling on dactylus. Pereopods 6–7 subequal in length. Uropodal exopod longer than endopod. Pleotelson length ¾ basal width. Color: dorsal surface of body and appendages uniform black. Attached to dorsum of body close to dorsal fin.

RECORDS Harlequin bass *Serranus tigrinus:* Dominican Republic.

Renocila colini Williams and Williams, 1980
Figure 85B,C

DIAGNOSIS Ovigerous ♀ 12.0–17.5 mm, ♂ 7.5–13.0 mm. Pereonites 5–7 with posteroventral angle produced, that of pereonite 7 overlapping pleonite 1 only. Pereopods 1–3 lacking swelling on dactyli; pereopods 6–7 subequal in length. Uropod reaching beyond pleotelson, endopod more than half length

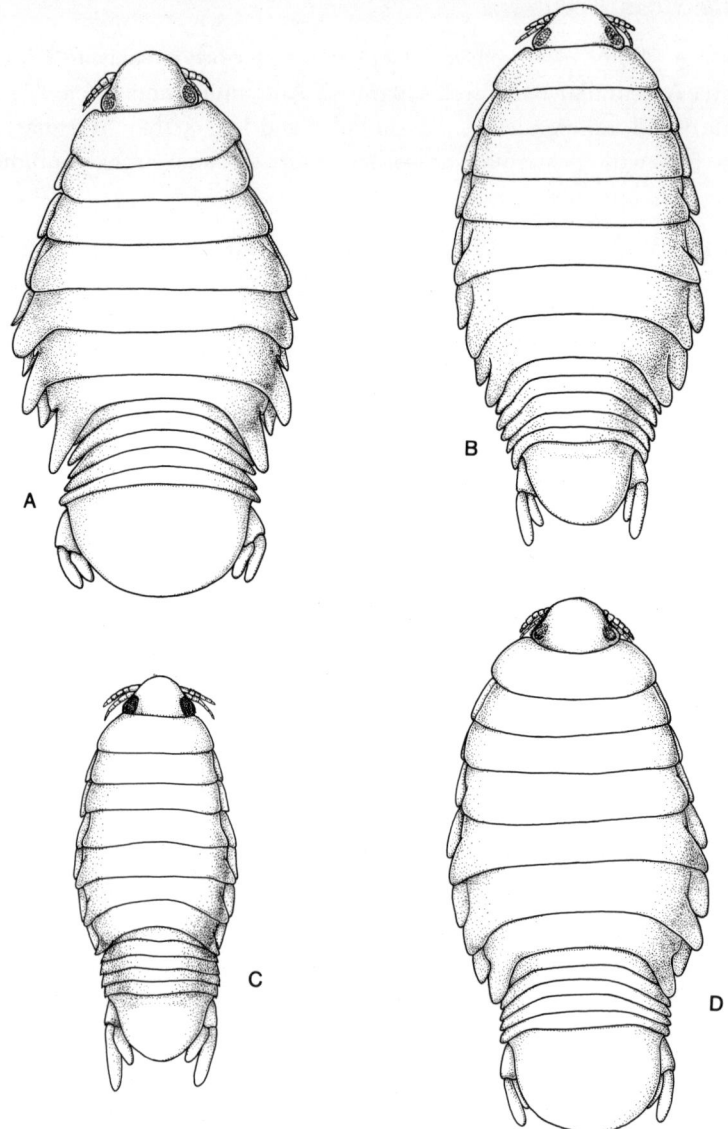

Figure 85. *A, Renocila bowmani. Renocila colini: B,* ♀; *C,* ♂. *D, Renocila waldneri.*

of exopod. Pleotelson $1/7$ to $1/2$ wider than long, with slight rounded apex. Color: dorsal surface of body and appendages uniformly yellowish brown. Attached to dorsum of body, close to dorsal fin.

RECORDS Flamefish *Apogon maculatus:* Puerto Rico. Belted cardinalfish *Apogon townsendi:* Puerto Rico.

Renocila waldneri Williams and Williams, 1980
    Figure 85D

DIAGNOSIS Ovigerous ♀ 15.3–19.3 mm, ♂ 5.0–10.8 mm. Posteroventral angle of pereonite 5 moderately produced, of pereonites 6–7 more strongly produced, that of pereonite 7 overlapping pleonites 1 and 2. Pereopods 1–3 without swelling on dactyli. Pereopods 6 and 7 subequal in length. Uropodal exopod slightly longer than endopod. Pleotelson basally wider than long; posterior margin broadly and evenly rounded. Color: dorsal surface of body uniform brown; appendages yellowish brown. Attached to dorsum of body close to dorsal fin.

RECORDS Harlequin bass *Serranus tigrinus:* Dominican Republic.

Family Limnoriidae Harger, 1879

DIAGNOSIS Body ovate in cross section, often becoming more setose posteriorly. Cephalon subspherical, freely articulating with pereonite 1; eyes lateral. Antennules and antennae well separated at bases. Mandible with strong incisor; lacking molar and well-defined lacinia mobilis, but with species-distinctive lacinioid bristle or seta; palp usually of three articles. Maxillipedal palp of five articles; endite well developed. Coxae present on pereonites 2–7. Pleon consisting of five free pleonites plus pleotelson; latter subcircular, set obliquely to axis of body, usually with anterolateral crests. Uropod with strong protopod inserted ventrolaterally.

Key to genera of Limnoriidae

1. Uropodal rami very unequal 2
   Uropodal rami subequal .... *Paralimnoria*
2. Mandibular incisors possessing rasp and file .... *Limnoria*
   Mandibular incisors lacking rasp and file .... *Phycolimnoria*

REMARKS   This family includes a number of species that are of considerable economic importance. Given that species of *Limnoria* are wood borers, wooden structures such as wharf pilings that are immersed in sea water and even in water of reduced salinity are vulnerable to attack by these gribbles. Prolonged exposure can lead to weakening and eventual collapse of these structures (see Ray, 1959). Even creosote-treated wood is not fully protected; *Limnoria tuberculata* will bore into such wood to where the creosote has not penetrated.

The isopods rasp at the wood fibres with the rasp and file structures of the mandibles, usually following the grain of the wood. With this boring activity, saprophytic fungi and bacteria invade the wood and assist in the breakdown process. *Limnoria* lack cellulase-secreting microflora in their gut, but probably secrete a cellulase themselves (Boyle and Mitchell, 1978). It is also probable that the fungi and bacteria, the latter often densely aggregated on the setae of the isopod, form part of the animals' diet. In the natural environment, *Limnoria* perform an important role in the breakdown of dead wood, especially in mangrove areas.

Sexual dimorphism of the pleotelson does occur in some species. This aspect of the morphology, however, has hardly been investigated.

Limnoria Leach, 1814

DIAGNOSIS   Antennular flagellum of four articles. Antennal flagellum of three to five articles. Incisor of right mandible equipped with filelike structure on upper surface; incisor of left mandible with rasplike structure. Rami of pleopod 5 lacking marginal setae. Uropodal exopod much shorter than endopod, bearing terminal claw. Pleotelson smooth, or variously ornamented with tubercles and ridges.

Limnoria indica Becker and Kampf, 1958
Figure 86A,B

DIAGNOSIS   ♂ 3.0 mm, ovigerous ♀ 3.0 mm. Pleonite 5 with submedian pair of strong rounded ridges, converging slightly posteriorly. Pleotelson basally with two pairs of submedian tubercles and pair of lateral tubercles.

RECORDS   Cozumel, Mexico; Man o'War Cay, Belize.
India; Hong Kong; Philippines; east coast of Australia.

Key to species of *Limnoria*

1. Dorsal surface of pleotelson lacking prominent tubercles, ridges, or carinae (*L. simulata* may appear to lack ornamentation; in this species the tubercles are very small) 2
Dorsal surface of pleotelson bearing tubercles, ridges, or carinae 3

2. Pleotelson flat; pleonite 5 with broadly rounded middorsal ridge
.... *platycauda*
Pleotelson cup shaped; pleonite 5 with strong narrowly rounded middorsal ridge .... *insulae*

3. Pleotelson with basal tubercles but lacking ridges 4
Pleotelson with ridges but lacking freestanding tubercles 7

4. ♂ pleotelson with single strong middorsal tubercle .... *unicornis*
Pleotelson with more than one basal tubercle 5

5. Pleotelson with three basal tubercles .... *tuberculata*
Pleotelson with more than three basal tubercles 6

6. Pleotelson with four basal tubercles in line (difficult to detect) *simulata*
♂ pleotelson with six basal tubercles .... *indica*

7. Pleotelson with single middorsal longitudinal ridge ...... *multipunctata*
Pleotelson with two rounded basal ridges 8

8. Pleonite 5 with strong Y-shaped ridge .... *pfefferi*
Pleonite 5 with two posteriorly converging ridges .... *saseboensis*

Limnoria insulae Menzies, 1957
Figure 86C

DIAGNOSIS ♂ 3.0 mm, ovigerous ♀ 3.4 mm. Pleonite 5 with strong middorsal ridge. Pleotelson cup shaped, lateral crests extended anteromesially, separated basally by distinct gap; posterior margin and lateral crests not tuberculate.

RECORDS Twin Cays, Belize.
Fiji; Guam; Palmyra Island; Caroline Islands.

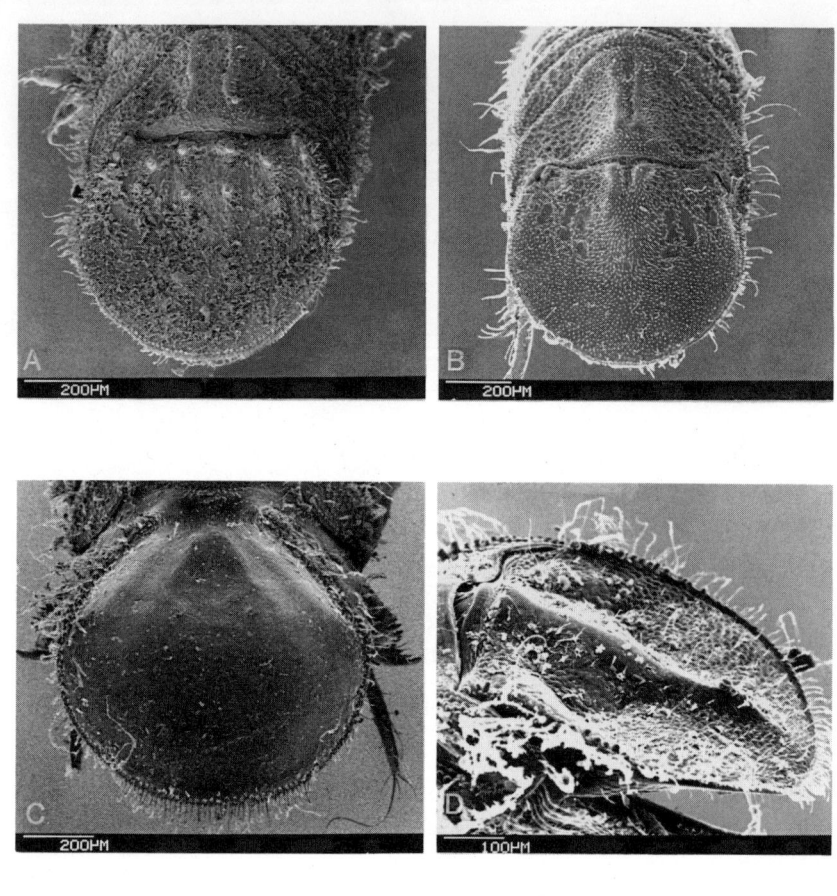

Figure 86. *Limnoria indica:* A, pleotelson, ♂; B, pleotelson, ♀. *Limnoria insulae:* C, pleotelson. *Limnoria multipunctata:* D, pleotelson in oblique-lateral view.

Limnoria multipunctata Menzies, 1957
  Figures 86D; 87A

DIAGNOSIS ♂ 2.8 mm, ovigerous ♀ 3.0 mm. Pleonite 5 dorsally smooth. Pleotelson with middorsal longitudinal rounded ridge bearing several

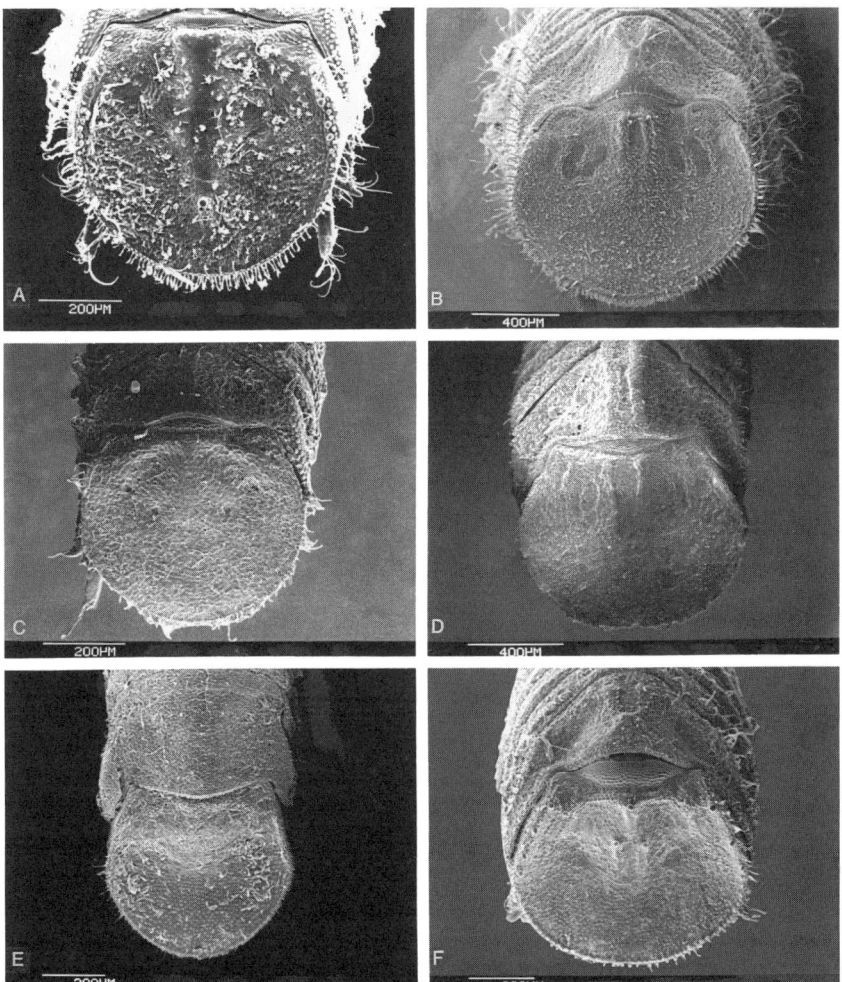

Figure 87. *Limnoria multipunctata:* A, pleotelson; *Limnoria pfefferi:* B, pleotelson; *Limnoria platycauda:* C, pleotelson; *Limnoria saseboensis:* D, pleotelson; *Limnoria simulata:* E, pleotelson; *Limnoria tuberculata:* F, pleotelson.

button-shaped tubercles in posterior half; posterior margin and lateral crests tuberculate.

RECORDS   Puerto Rico; Jamaica; Twin Cays, Belize. Japan; Kai Islands, South Pacific.

Limnoria pfefferi Stebbing, 1904
Figure 87B

DIAGNOSIS ♂ 3.8 mm, ovigerous ♀ 4.0 mm. Pleonite 5 with conspicuous middorsal Y-shaped carina. Pleotelson basally with pair of submedian rounded ridges; lateral crests lacking tubercles.

RECORDS Florida Keys; Bahamas; Puerto Rico; U.S. Virgin Islands; Twin Cays and Man o'War Cay, Belize; Yucatan Peninsula, Mexico.
Minikoi Atoll and Aldabra Atoll, Indian Ocean; Philippines; New Guinea; Panama.

Limnoria platycauda Menzies, 1957
Figure 87C

DIAGNOSIS ♂ 2.5 mm, ovigerous ♀ 2.6 mm. Pleonite 5 with broad middorsal longitudinal rounded ridge. Pleotelson lacking dorsal ornamentation; posterior margin and lateral crests bearing tubercles.

RECORDS Cuba; Puerto Rico to Curaçao; Cozumel, Mexico; Twin Cays and Man o'War Cay, Belize.
Aldabra Atoll, Indian Ocean.

Limnoria saseboensis Menzies, 1957
Figure 87D

DIAGNOSIS ♂ 3.5 mm. Pleonite 5 with submedian pair of ridges, converging slightly posteriorly. Pleotelson basally with submedian pair of anteriorly tuberculate ridges; posterior margin and lateral crests tuberculate.

RECORDS Miami, Florida.
Japan; Fiji.

Limnoria simulata Menzies, 1957
Figure 87E

DIAGNOSIS ♂ 3.8 mm, ovigerous ♀ 4.0 mm. Pleonite 5 with obscure median longitudinal groove. Pleotelson basally with submedian pair of tubercles and small lateral tubercles, latter often difficult to detect; lateral crests tuberculate.

RECORDS Florida Keys; U.S. Virgin Islands; Gulf of Mexico.

Limnoria tuberculata Sowinsky, 1884
Figure 87F

DIAGNOSIS ♂ 2.8 mm, ovigerous ♀ 3.0 mm. Pleonite 5 with two anterior tubercles, one middorsal posterior tubercle, area between tubercles depressed. Pleotelson basally with middorsal tubercle, followed by pair of submedian tubercles, all three tubercles having short obscure carina; posterior margin and lateral crests tuberculate.

RECORDS Rhode Island to Venezuela; Cuba; Man o'War Cay, Belize; Gulf of Mexico.
Uruguay; West Africa; Mediterranean; Black Sea; India; Hong Kong; Hawaii; Australia; California.

REMARKS This species has frequently been recorded under the name *Limnoria tripunctata* Menzies, 1951a.

Limnoria unicornis Menzies, 1957
Figure 88A,B

DIAGNOSIS ♂ 2.6 mm, ovigerous ♀ 2.6 mm. Mandibular palp of one article. Pleonite 5 with somewhat obscure Y-shaped ridge middorsally. Pleotelson in ♂ with strong basal slightly curved middorsal tubercle; lateral crests lacking tubercles.

RECORDS Bahamas; Man o'War Cay and Twin Cays, Belize.
Caroline Islands; Palau; Society Islands.

Paralimnoria Menzies, 1957

DIAGNOSIS Antennular flagellum of five articles. Antennal flagellum of five or six articles. Mandibular incisor with rasp and file. Pleopod 5, rami bearing marginal setae. Uropodal rami subequal in length, each with clawlike apex.

Paralimnoria andrewsi (Calman, 1910)
Figure 88C,D

DIAGNOSIS ♂ 2.6 mm, ♀ 2.6 mm. Pleonite 5 with or without triangular middorsal depressed area. Pleotelson with basal submedian pair of tubercles either obscurely or strongly carinate; lateral crest tubercles of variable strength.

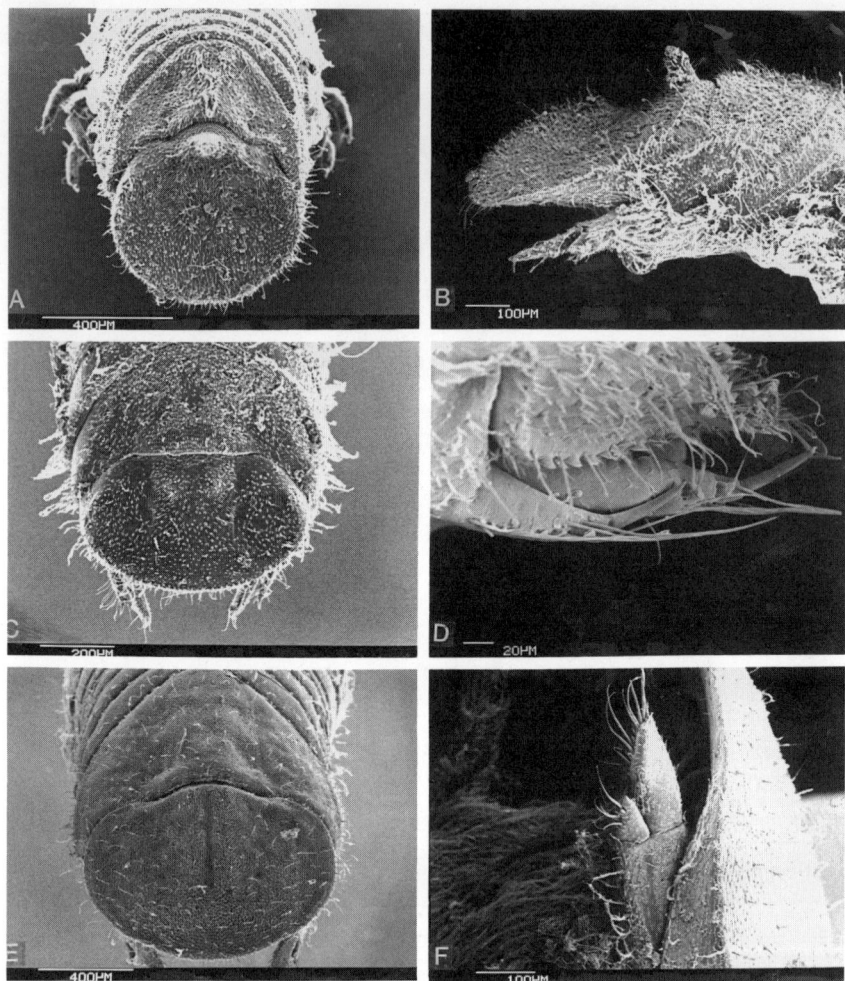

Figure 88. *Limnoria unicornis:* A, pleotelson, ♂; B, pleon, ♂, in lateral view. *Paralimnoria andrewsi:* C, pleonite 5 and pleotelson; D, uropod. *Phycolimnoria clarkae:* E, pleonite 5 and pleotelson; F, uropod and pleotelson in lateral view.

RECORDS   Florida Keys; Puerto Rico; Twin Cays, Belize; Curaçao. Christmas Islands, Indian Ocean; Samoa; Hawaii; Japan.

REMARKS   Menzies (1957) discusses three forms of this species: Forma *typica*, which lacks a central depressed area dorsally on pleonite 5 and has a pair of submedian obscurely carinate tubercles on the pleotelson; Forma A, which has a triangular depressed area dorsally on pleonite 5 and a pair of submedian tubercles supported by strong carinae on the pleotelson; Forma B, having a triangular depressed area dorsally on pleonite 5 and an obscurely carinate pair of tubercles on the pleotelson. Given that at least two of these forms have been recorded occurring together, it would seem that this is merely a highly variable species.

Phycolimnoria Menzies, 1957

DIAGNOSIS   Mandibular incisor lacking rasp and file. Uropodal rami unequal, exopod longer than endopod, latter usually with clawlike apex.

REMARKS   Most species of *Phycolimnoria* are algal borers, frequently encountered in the holdfasts of brown algae such as *Macrocystis, Laminaria,* and *Sargassum.* The one species recorded from the Caribbean, *P. clarkae,* however, has only been taken from decaying wood.

Phycolimnoria clarkae Kensley and Schotte, 1987
Figure 88E,F

DIAGNOSIS   ♂ 4.3 mm, ovigerous ♀ 3.3–4.4 mm. Uropodal exopod less than half length of endopod, straight, tipped with short squat claw. Pleonite 5 with broad raised middorsal region having irregular bumps. Pleotelson wider than long, with two rounded submedian ridges basally, becoming obsolete posteriorly.

RECORDS   Bahamas; Twin Cays, Belize. Aldabra Atoll, Indian Ocean.

Family Serolidae Dana, 1852

DIAGNOSIS   Body dorsoventrally depressed. Eyes present or absent. Cephalon fused with pereonite 1 dorsally. Mandible bearing palp. Maxillipedal palp of one to four articles. Pereonites 2–4 with coxae demarked; pereonites 5 and 6 with coxae not demarked; pereonite 7 narrow, lacking free lateral margins. Pereopod 1 in ♂ and ♀ subchelate, pereopod 2 subchelate or ambulatory in ♂, ambulatory in ♀. Pleonites 1 and 2 free, articulated, remainder of pleonites fused with telson. Pleopods 1–3 small, natatory; pleopods 4 and 5 large, operculate. Uropods lateral, biramous.

REMARKS   The serolids reach their greatest diversity (and their greatest size of up to 80 mm in length) in the southern oceans, with few species extending into the subtropics and tropics. The deep- and abyssal-dwelling species usually lack eyes. The animals are epibenthic, living in the upper few centimeters of the bottom sediment, where they are scavengers and carnivores.

Serolis Leach, 1818

DIAGNOSIS Body markedly dorsoventrally flattened. Coxal plates produced laterally. Mandible having lacinia mobilis and single spine. Maxillipedal palp of three articles (rarely two to four). Pereopod 2 exhibiting sexual dimorphism, subchelate in ♂, ambulatory in ♀. Pleopods 1–3, peduncles elongate, rami subelliptical. Pleopod 3, exopod uniarticulate.

Serolis mgrayi Menzies and Frankenberg, 1966
Figure 89

DIAGNOSIS ♂ 4.5 mm, ovigerous ♀ 4.7 mm. Eyes present. Cephalon with two middorsal tubercles. Pereonites 2–4 each with faint rounded tubercle just mesial to coxal suture. Pereon and pleon with faint middorsal longitudinal carina bearing small blunt tubercle on posterior margin of each segment. Pleonites 1 and 2 with lateral margins not contributing to body outline, overlapped by pereonite 6. Pleotelson broadly triangular, with lateral carina in anterior half; apex truncate. Uropodal rami reaching to or slightly beyond pleotelsonic apex.

RECORDS Off North Carolina, 18–34 m; off South Carolina, 22 m; off Georgia, 18–47 m; Florida Keys, 18–88 m; Trinidad; Venezuela, 95 m; Florida, Gulf of Mexico, 11–88 m.

Family Sphaeromatidae H. Milne Edwards, 1840

DIAGNOSIS Antennular peduncle of three articles, antennal peduncle of five articles. Mandible stout, lacinia mobilis and molar usually well developed, palp of three articles. Maxillipedal palp of five articles. Mouthparts in some genera metamorphosed and somewhat reduced in ovigerous ♀. Pleon of five partially or completely fused pleonites, often indicated by lateral sutures, plus dorsally convex and sometimes inflated pleotelson. Uropods lateral, exopod free if present, endopod fused with sympod. Sexual dimorphism often marked, especially in pleotelsonal structure. Animal often capable of conglobating or folding over. Young brooded in internal pouches or anterior or posterior pockets; oostegites variable in number, if present.

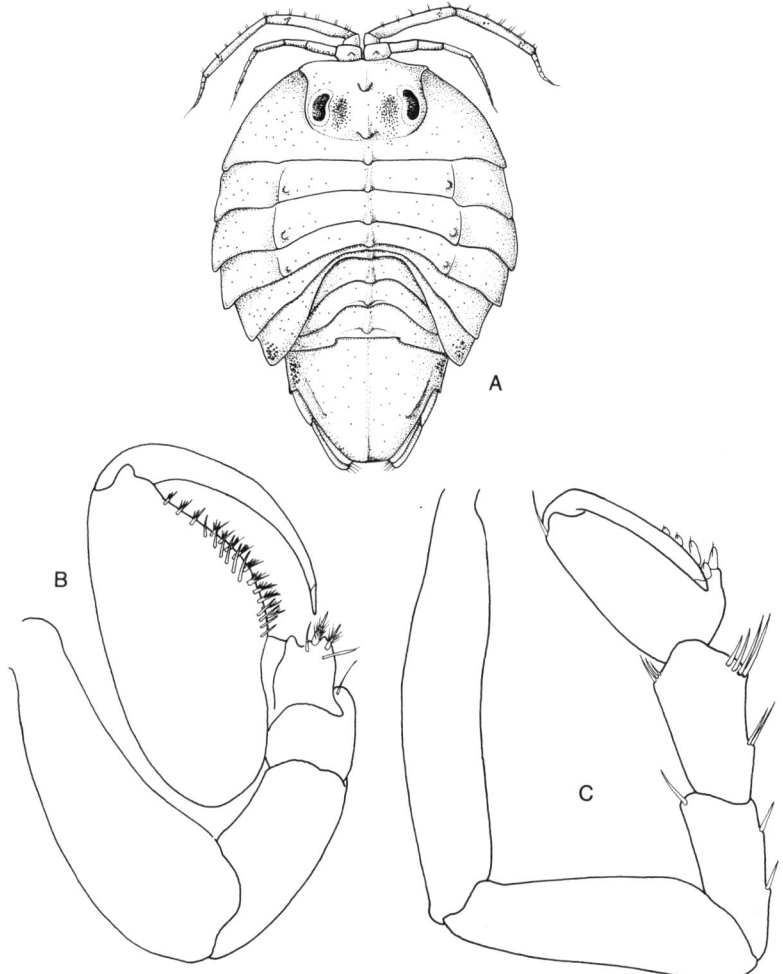

Figure 89. *Serolis mgrayi:* A, ♂; B, pereopod 1; C, pereopod 2, ♂.

REMARKS Right into the 1980s this family was routinely divided into three groups, based on the structure of the two posterior pairs of pleopods: Platybranchiatae—pleopods 4 and 5 with both rami membranous and lacking branchial pleats; Hemibranchiatae—pleopods 4 and 5 with branchial pleats on endopods only; Eubranchiatae—pleopods 4 and 5 with branchial pleats on both rami. These three "groups" were recognized formally as subfamilies by Hurley and Jansen (1977) but the names were not based on contained genera and were replaced with current subfamily names by Bowman (1981) and Iverson (1982), the latter providing diagnoses for all five subfamilies. Four of these are represented in the Caribbean area; the fifth, the Tecticipitinae, contains only the single primarily Pacific genus *Tecticeps*.

While the subfamilial status now appears to be resolved, many of the genera still require unambiguous diagnoses. The work of Harrison (1984) on the structure of the female broodpouch, with its various components of oostegites, internal pouches, and anterior and posterior pockets (Figure 90), along with the metamorphosis of the female mouthparts (see Figure 96) has helped enormously to standardize the genera. Nevertheless, these features of the female remain unknown in several genera. Further, with this stabilization based on females, many problems of incorrect generic designation have been uncovered. In this work, Harrison's generic diagnoses are followed as far as possible. Where uncertainty exists, this is indicated. In some cases, we may still be unaware of existing problems: future work will without doubt result in the shifting of species to different genera, as well as in the creation of new genera.

Key to subfamilies of Sphaeromatidae

1. Pereopod 1 prehensile in both sexes; pereopod 2 prehensile only in ♂ .... **Ancininae**
   Pereopods 1 and 2 ambulatory 2

2. Pleopods 4 and 5 lacking branchial pleats .... **Cassidininae**
   Pleopods 4 and 5 with branchial pleats on endopods 3

3. Pleopods 4 and 5 with branchial pleats on both rami ... **Dynameninae**
   Pleopods 4 and 5 with branchial pleats on endopods only .... **Sphaeromatinae**

Subfamily Ancininae Tattersall, 1905

DIAGNOSIS Body markedly dorsoventrally depressed. Cephalon fused medially with pereonite 1. Pereopod 1 prehensile in ♂ and ♀. Pereopod 2 prehensile in ♂ only. Pleopods 4 and 5 similar, lacking branchial pleats. Uropods uniramous.

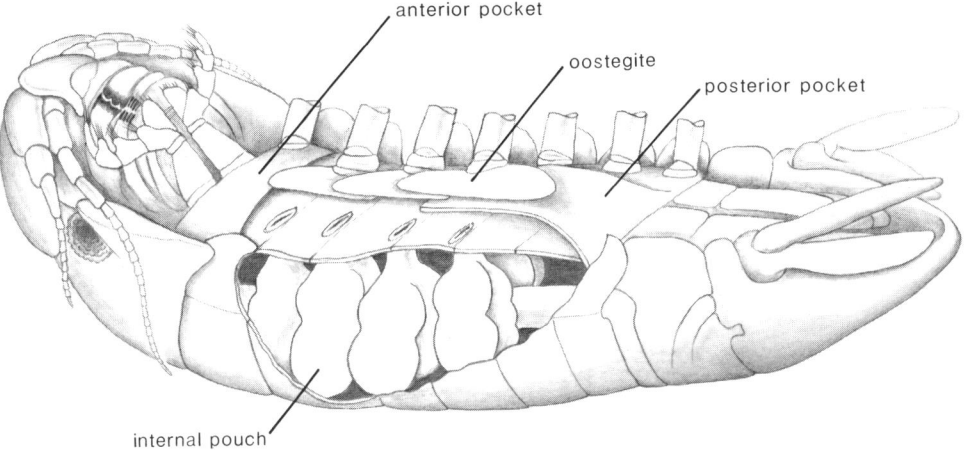

Figure 90. Diagrammatic representation of ♀ sphaeromatid, showing marsupial structures (adapted from Harrison, 1984).

Ancinus H. Milne Edwards, 1840

DIAGNOSIS Eyes dorsal. ♀ mouthparts not metamorphosed. Mandibular molar absent; palp of three articles. Maxilla 1 of single ramus, endite rudimentary. Maxilla 2 of two rami. ♀ with oostegites absent; brood held in two opposing pockets, opening as narrow ventral slit between pereopods 4. Pleon consisting of short anterior pleonite with free lateral margin, plus broadly triangular pleotelson. Pleopod 1 uniramous, endopod absent. Pleopod 2 operculiform. Pleopod 3, exopod of single article. Uropod lacking exopod, sympod not laterally expanded.

Key to species of *Ancinus*

1. Pleotelson as long as basal width, apex narrowly rounded ... *brasiliensis*
   Pleotelson with basal width greater than length, apex subtruncate
   .... *belizensis*

Ancinus belizensis Kensley and Schotte, 1987
Figure 91A–C

DIAGNOSIS ♂ 4.1 mm, ♀ 2.8 mm. Body oval, about twice longer than wide. Dorsal integument strongly pitted. Antennular flagellum of 12 articles; an-

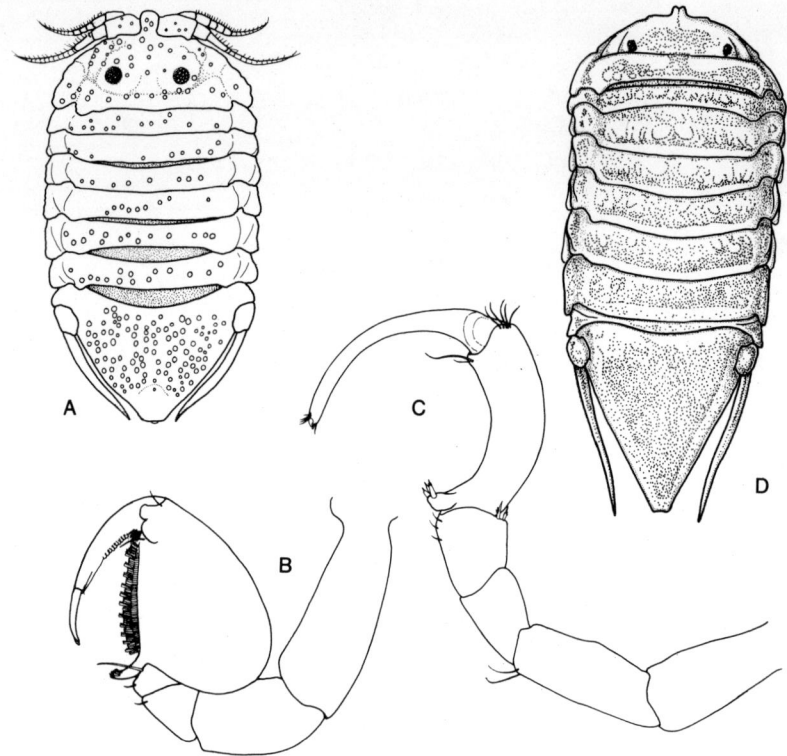

Figure 91. *Ancinus belizensis: A,* ♀; *B,* pereopod 1 ♂; *C,* pereopod 2 ♂. *Ancinus braziliensis: D,* adult (from Glynn and Glynn, 1974).

tennal flagellum of 10 articles. ♂ pereopod 2, dactylus strongly curved, reaching to proximal lobe of propodus. Pleopod 2 about 2.5 times longer than basal width.

RECORDS   Carlson Point, Belize, in seagrass flats, 0.5 m.

Ancinus brasiliensis Lemos de Castro, 1959
   Figure 91D

DIAGNOSIS   ♂ 7.0 mm, ♀ 6.0 mm. Body about twice longer than wide. Dorsal integument smooth. Antennular flagellum of 17 articles; antennal flagellum of 10 articles. ♂ pereopod 2, dactylus strongly curved, reaching to midlength of posterior margin of carpus. Pleopod 2 almost three times longer than basal width.

RECORDS   Brazilian coast from Rio de Janeiro northward, 1.5 m; Costa Rica, Panama; shallow infratidal below sandy beaches.

REMARKS   Glynn and Glynn (1974) discussed color polymorphism in this species.

Subfamily Cassidininae Iverson, 1982

DIAGNOSIS   Cephalon not medially fused with pereonite 1. Pereopod 1 ambulatory. Pleopods 4 and 5, both rami lacking transverse pleats, outer rami unsegmented. Pleopod 5, outer ramus with low subapical squamiferous protuberances. Pleotelsonic apex entire. Uropods with exopods reduced.

REMARKS   The genus *Dies* has twice been recorded from the Caribbean: *D. arndti* Ortiz and Lalana, 1980, from Cuba, and *D. barnardi* Carvacho, 1977, from Guadeloupe. This genus is distinguished from *Cassidinidea* solely on the basis of the penial structure: biramous in *Cassidinidea*, uniramous in *Dies*. Harrison (1984) has pointed out that the separation of these two genera has not been satisfactorily resolved. The penis of neither the Cuban nor the Guadeloupan species has been illustrated, but the whole-animal illustrations of both look suspiciously like *Cassidinidea ovalis*. Examination of material of *D. barnardi* from the Paris Museum supports the view that this species was based on immature material of *C. ovalis*. Neither of the so-called species of *Dies* are dealt with in this work, both being regarded as junior synonyms of *C. ovalis*.

Key to genera of Cassidininae

1. Frontal lamina visible dorsally between antennular bases; two basal articles of antennular peduncle not expanded .... *Cassinidinea*
   Frontal lamina not visible between antennular bases; two basal articles of antennular peduncle broadly expanded .... *Paraleptosphaeroma*

Cassidinidea Hansen, 1905b

DIAGNOSIS   Body strongly dorsoventrally depressed. Eyes dorsal, situated at posterolateral corners of cephalon. Latter somewhat sunken into pereonite 1. Frontal lamina expanded, visible dorsally between antennular bases. Antenna directed laterally. Pleon consisting of one free pleonite having short free lateral margin, plus broadly triangular pleotelson. Uropodal endopod well developed, fused with sympod; exopod markedly reduced. Penial rami elongate, separate. ♀ mouthparts not metamorphosed. Oostegites absent. Brood housed in pouch formed by opposing pockets overhanging ventrum, opening by slit between fourth pereopods.

Key to species of *Cassidinidea*

1. Posterior margin of pleotelson truncate .... *ovalis*
   Posterior margin of pleotelson rounded .... *mosaica*

Cassidinidea mosaica Kensley and Schotte, 1987
   Figure 92A

DIAGNOSIS   ♂ 1.8 mm, ovigerous ♀ 1.6 mm. Body twice longer than wide. Dorsal integument bearing close-packed flattened tubercles. Pleotelson triangular, with posterior margin narrowly rounded, dorsally convex, basally inflated.

RECORDS   Carrie Bow Cay, Belize, 1.5–10 m; in silty sand and rubble between patch reefs and coral buttresses.

Cassidinidea ovalis (Say, 1818)
   Figure 92B–E

DIAGNOSIS   ♂ and ♀ 3.6 mm. Body width slightly less than half length. Dorsal integument smooth. Pleotelson with raised anteromesial area, but lacking sculpture; posterior margin truncate.

RECORDS   New Jersey to Florida, in marsh mud and among dead leaves, 0–1 m; Trinidad; Belize; Panama; Dominica; Louisiana and Vera Cruz, Gulf of Mexico. Known from waters of less than 1‰ to 35‰.

Paraleptosphaeroma Buss and Iverson, 1981

DIAGNOSIS   Body oval in outline, entire circumference with transparent flange of fused setae on two expanded basal articles of antennule, on pereonites, pleonite 1, and uropods. Expanded basal articles of antennules con-

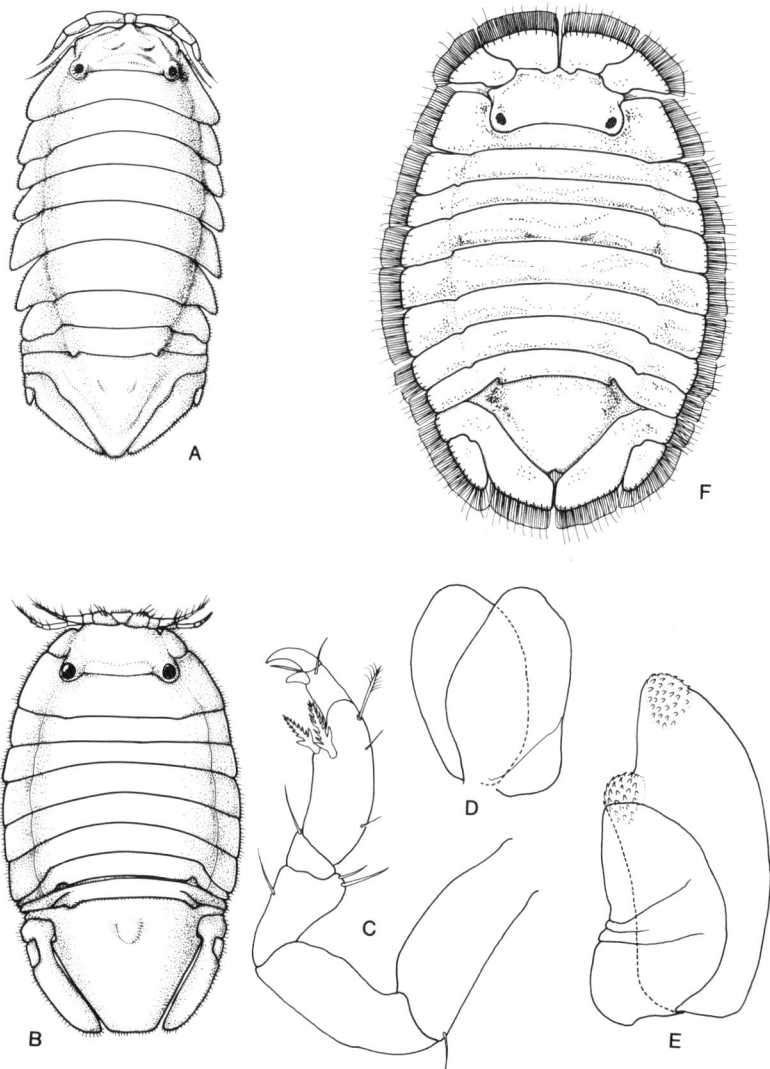

Figure 92. *Cassidinidea mosaica*: A, ♂. *Cassidinidea ovalis*: B, ♂; C, pereopod 1; D, pleopod 4; E, pleopod 5. *Paraleptosphaeroma glynni*: F, ♂.

tiguous in midline. Single articulated pleonite with short free lateral margin. Uropodal sympod and endopod fused; exopod articulated, much shorter than fused endopod.

Paraleptosphaeroma glynni Buss and Iverson, 1981
 Figure 92F

DIAGNOSIS ♂ 2.58 mm, ovigerous ♀ 2.38 mm. Pleotelson basally broad, tapering to notched posterior margin. Fused uropodal endopod and sympod of each side almost touching posterior to pleotelsonic apex.

RECORDS Portsmouth, Dominica, intertidal rock pools. Punta Paitilla, Pacific Panama.

REMARKS Buss and Iverson (1981) demonstrated that this species displays sequential protogynous hermaphroditism, and that the change from female to male seems to be mediated by social conditions, especially the proportion of males to females. The principal food source for this species was shown to be abascan bryozoans.

Key to genera of Dynameninae

1. Pleotelson very similar in both sexes 2
   Pleotelson showing marked sexual dimorphism 3

2. Cephalon and pleotelson smooth, lacking ridges .... *Ischyromene*
   Pleotelson and cephalon with ridges .... *Cerceis*

3. Uropods lamellar in both sexes 4
   Uropods lamellar in ♀, endopod reduced, exopod elongate-cylindrical in ♂ 5

4. Ovigerous ♀ lacking oostegites; ♂, pleopod 2 copulatory stylet basally broad, distally tapering, extending to or beyond ramus .. *Dynamenella*
   Ovigerous ♀ with one pair of oostegites on pereonite 4; ♂, pleopod 2 copulatory stylet narrow, extending well beyond ramus .... *Paradella*

5. ♂, strong median lobe in pleotelsonic notch reaching well beyond margin; ovigerous ♀ with three pairs of oostegites .... *Discerceis*
   ♂, pleotelsonic notch with short median lobe, if present; ovigerous ♀ with three or four pairs of oostegites 6

6. Ovigerous ♀ with three pairs of oostegites; ♂ pleotelsonic notch lacking marginal teeth or median lobe .... *Geocerceis*
   Ovigerous ♀ with four pairs of oostegites; ♂ pleotelsonic notch with marginal teeth and/or median lobe .... *Paracerceis*

Subfamily Dynameninae Bowman, 1981

DIAGNOSIS Cephalon not fused with pereonite 1. Pereopods 1 and 2 ambulatory. Pleopods 4 and 5, both rami having branchial pleats. Pleopod 4, exopod unjointed, usually lacking setae, endopod with few setae at most. Pleotelsonic apex often with terminal notch or foramen, especially in ♂. Uropods biramous.

Cerceis H. Milne Edwards, 1840

DIAGNOSIS Mouthparts metamorphosed in ♀. Broodpouch of four pairs of oostegites on pereonites 1–4, overlapping in midline. Brood held in four pairs of internal pouches. Pockets absent.

"*Cerceis*" *carinata* Glynn, 1970
Figure 93A

DIAGNOSIS ♂ 3.8 mm, ♀ 3.9 mm. Dorsal integument, especially posteriorly, finely pustulose. Cephalon with three pairs of rounded dorsolateral carinae, one middorsal carina, not reaching posterior margin. Pereonite 1 with two ventrolateral carinae. Pleopod 3, exopod biarticulate. Pleotelson similar in ♂ and ♀. Pleotelson with basal middorsal inflated area flanked by two smaller lateral swellings, with carina in midline almost reaching posterior margin; posterolateral margins converging to narrow, slightly concave posterior margin. ♂: Penes elongate, basally fused. Pleopod 2, copulatory stylet basally broad, tapering distally, articulating mediodistally on endopod; exopod with three enlarged distal plumose setae.

RECORDS Venezuela, 5–7 m.

REMARKS Several differences (in the male penial structure, copulatory stylet, antennular peduncle, and pleonal sutures) between *Cerceis carinata* and the definition of the genus (Harrison and Holdich, 1982; Harrison, 1984) indicate that this species has not been placed in the correct genus. Until fresh ovigerous females and mature males are available, the generic position must remain uncertain.

Discerceis Richardson, 1905

DIAGNOSIS Mouthparts in ♀ metamorphosed. Broodpouch formed by four pairs of oostegites on pereonites 1–4, overlapping in midline. Brood held in

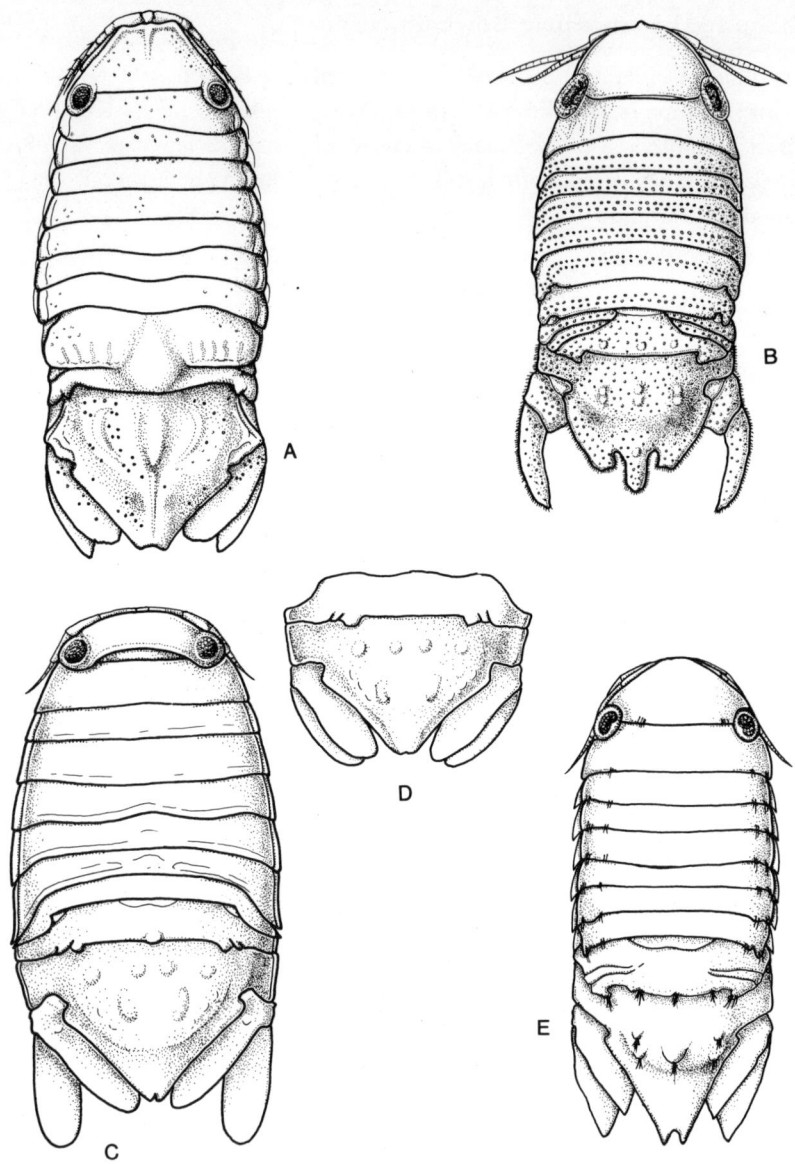

Figure 93. *"Cerceis" carinata:* A, ♂. *Discerceis linguicauda:* B, ♂. *Dynamenella acutitelson:* C, ♀; D, pleon ♂. *Dynamenella angulata:* E, ♀.

internal pouches (number unknown). Pockets absent. ♂ with uropodal endopod and sympod fused, very short; exopod elongate, cylindrical.

Discerceis linguicauda (Richardson, 1901)
Figure 93B

DIAGNOSIS ♂ 7.2 mm. Dorsal integument, especially of posterior half, with numerous scattered granular tubercles. Uropodal endopod and sympod fused, very short, exopod elongate, subcylindrical and slightly bowed. Anterior half of pleotelson inflated, with three elongate rounded ridges (each composed of two contiguous tubercles) ending posteriorly in subacute tubercle; posterior margin trilobed, median lobe broadly rounded, with subacute tubercle at base, lateral lobe truncate, well separated from median lobe. Head and pereonite 1 not fused. Frontal lamina visible dorsally between antennal bases. Penes short, separate. Copulatory stylet basally relatively broad, distally broadly rounded.

RECORDS Cape Catoche, Yucatan, Mexico, 48–50 m.

REMARKS This species is known only from the four male syntypes.

Dynamenella Hansen, 1905b

DIAGNOSIS Species exhibiting obvious sexual dimorphism. Both sexes lacking processes on pereon and pleon. Uropodal rami lamellar. Exopod of pleopod 3 with or without articulation. ♀: Mouthparts not metamorphosed. Broodpouch lacking oostegites, but formed by two opposing ventral pockets opening in midline between fourth pereopods. Apex of pleotelson with notch,

Key to species of *Dynamenella*

1. ♂ with pleotelsonic foramen 2
   ♂ lacking foramen but with notch, or appearing entire; ♀ pleotelson with faint notch visible .... *acutitelson*

2. ♂ with four strong pleotelsonic ridges; ♀ with subcircular pleotelsonic foramen .... *quadrilirata*
   ♂ lacking pleotelsonic ridges; ♀ with posterior margin of pleotelson entire .... *perforata*

groove, or foramen. ♂: Penes basally fused, rami long, tapering. Copulatory stylet proximally broad, tapering to acute tip, reaching to or just beyond apex of endopod. Uropods broader than in ♀. Posterior pleotelson with dorsally directed foramen connected to apex by narrow slit.

REMARKS   The species described by Richardson (1901) as *Dynamene angulata* from No Name Key, Florida, and referred to by some authors as a *Dynamenella*, while figured here (Figure 93E), is not included in the present key. The species is known only from immature females; correct generic placement is thus not possible.

Dynamenella acutitelson Menzies and Glynn, 1968
Figure 93C,D

DIAGNOSIS   ♂ 3.5 mm, ♀ 2.3 mm. ♂: Pereonites 4–6 with transverse ridge over dorsum, ridge interrupted to form short median section. Pleotelson with two submedian and two lateral rounded tubercles basally, two submedian, poorly defined ridges in central area; posterior margin tapering in dorsal view, with slit either just visible or appearing entire. In lateral view, posterior pleotelson seen to be laterally compressed, forming narrow groove.

RECORDS   Puerto Rico, intertidal rocks and algae.

REMARKS   Menzies and Glynn (1968) described this species with two varieties, the holotype as *D. acutitelson* var. *typica*, and 11 paratypes as *D. acutitelson* var. *glabrothorax*. The major difference between these varieties lay in the presence of transverse ridges on pereonites 4–6 in *typica* and their absence in *glabrothorax*. The holotype, however, at 3.5 mm, would seem to be a mature male, while all the paratypes are smaller. The differences described by Menzies and Glynn (1968) may thus be due to immaturity. As further comparative material is lacking, these varieties (or whatever their true status) are not recognized here.

Menzies and Glynn (1968, fig. 30a) illustrate *D. acutitelson* var. *glabrothorax* as having scattered tiny granules over the dorsal integument. These were not seen when the type material was reexamined.

Harrison and Holdich (1982) placed this species in *Paradella*, based on the literature. However, the penes for both varieties are shown as short and separate, as in *Ischyromene*. Again, until further mature males and ovigerous females are seen, the generic placement of this species must remain in doubt.

Dynamenella perforata (Moore, 1901)
Figure 94A,B

DIAGNOSIS ♂ 3.2 mm, ♀ 2.6 mm. ♂: Pleon bearing two low rounded submedian "mounds." Pleotelson with strongly convex anterior two-thirds, with T-shaped foramen. Pleon and pleotelson with numerous scattered small tubercles. Uropodal rami broadly ovate, outer margins crenulate. ♀: Pleotelson broadly rounded in dorsal view, posterior margin entire. Inner uropodal ramus distally subacute.

RECORDS Bermuda to Puerto Rico, intertidal coral rubble and algae, and under chiton *Acanthopleura granulata;* Dominican Republic; Cuba.

Dynamenella quadrilirata Kensley, 1984
Figure 94C–H

DIAGNOSIS ♂ 2.6 mm, ♀ 2.5 mm. ♂: Two low rounded submedian tubercles on last pleonite. Anterior half of pleotelson inflated, with four rounded longitudinal ridges; posterior half tapered, somewhat dorsally flexed, with cordate foramen. Uropodal rami distally rounded, outer margins crenulate to dentate. ♀: Lacking pleonal tubercles. Pleotelson inflated, unornamented, posterior margin forming subcircular foramen.

RECORDS Carrie Bow Cay, and Twin Cays, Belize; intertidal to 3 m.

Geocerceis Menzies and Glynn, 1968

DIAGNOSIS Ovigerous ♀ with mouthparts metamorphosed. Broodpouch with three pairs of oostegites, on pereonites 2–4, just overlapping in midline. Brood held in internal pouches (number unknown). Pockets absent. Uropodal rami lamellar, shorter than pleotelson. ♂ uropodal endopod fused with sympod, very short; exopod elongate, club shaped. Pleopod 2 with copulatory stylet articulating distally on endopod.

Geocerceis barbarae Menzies and Glynn, 1968
Figure 95A–C

DIAGNOSIS ♂ 3.3 mm, ♀ 2.5 mm. Pleopod 3 exopod of single article. Pleon with two elongate sutures reaching lateral pleon margin. ♂: Frontal lamina expanded into ventrally directed beaklike process. Penes separate, relatively

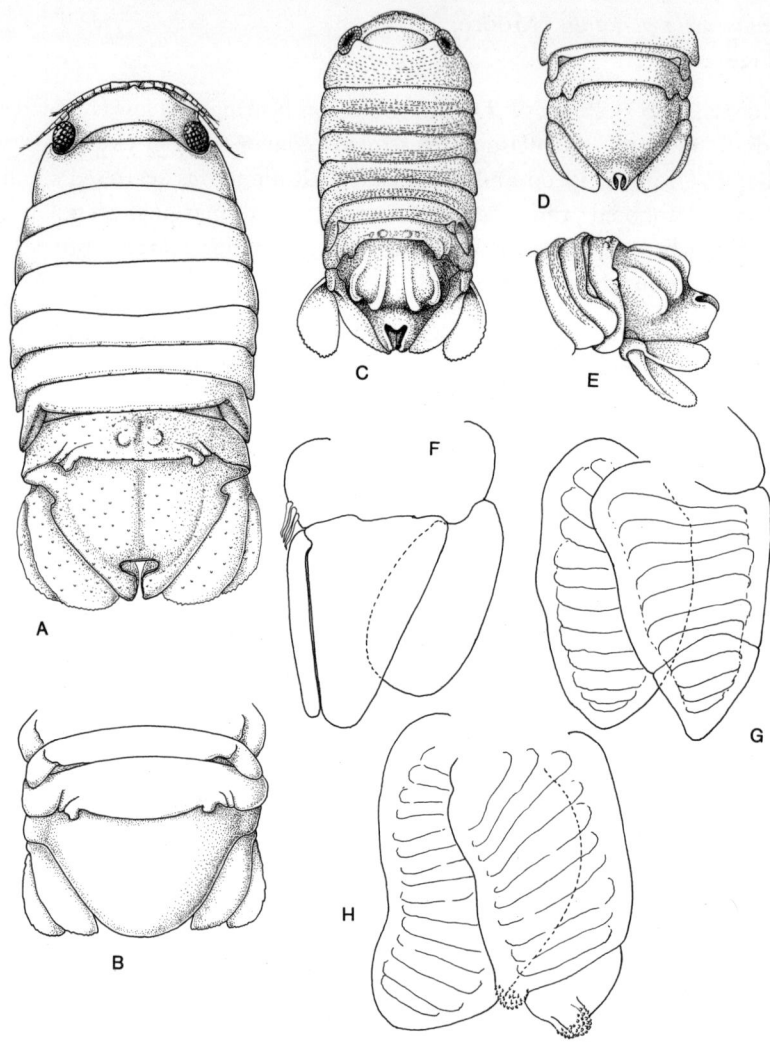

Figure 94. *Dynamenella perforata:* A, ♂; B, pleon ♀. *Dynamenella quadrilirata:* C, ♂; D, pleon ♀; E, pleon ♂, lateral view; F, pleopod 2 ♂; G, pleopod 4; H, pleopod 5.

elongate. Pleonite 5 with three dorsal tubercles near posterior margin. Pleotelson with raised anterocentral area having two lateral longitudinal rounded ridges; apex notched. ♀: Uropodal exopod and endopod subequal, lamellar. Pleotelson as in ♂. Frontal lamina not produced as in ♂.

RECORDS   Puerto Rico, intertidal to 3 m, in coral rubble.

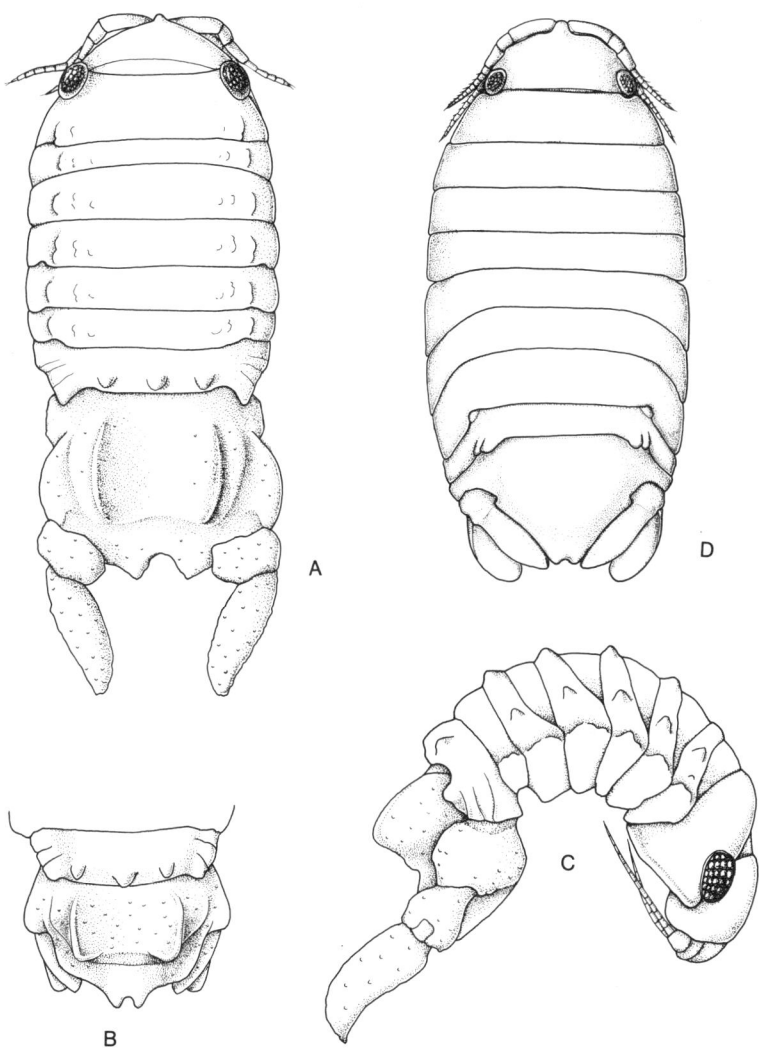

Figure 95. *Geocerceis barbarae:* A, ♂; B, pleon ♀; C, ♂, lateral view. *Ischyromene barnardi:* D, ♂.

Ischyromene Racovitza, 1908

DIAGNOSIS ♀ mouthparts not metamorphosed. Broodpouch of three pairs of oostegites on pereonites 2–4, overlapping in midline. Large posterior pocket covering posterior ventrum, opening anteriorly between fourth pereopods. Brood housed in ventral body wall. Sexual dimorphism not pronounced. Uropodal rami lamellar. ♂ pleopod 2 with copulatory stylet basally narrow, reaching to or just beyond distal margin of endopod.

Ischyromene barnardi (Menzies and Glynn, 1968)
Figure 95D

DIAGNOSIS   ♂ 4.5 mm, ♀ 3.7 mm. Both sexes lacking processes on pereon and pleon. Accessory unguis of pereopods often bifid. Pleopod 3, exopod of single article. Uropodal rami lamellar. ♂: Pereonite 7, posterior margin bilobed. Penes short, separate to base.

RECORDS   Puerto Rico, intertidal.

Paracerceis Hansen, 1905b

DIAGNOSIS   Pleopod 3 exopod with transverse suture in distal half. Pleon with two long sutures reaching to posterolateral margin. ♂: Penial rami short, separate. Pleotelson with basal area strongly vaulted; deep posterior notch sometimes having denticles on inner margins, and/or median tooth at base of notch. Uropodal endopod short, fused with sympod; exopod elongate, club shaped. ♀: Mouthparts metamorphosed. Mandible fused with cephalon. Broodpouch of four pairs of oostegites, three posterior pairs overlapping. Brood retained in internal pouches. Uropodal rami subequal, lamellar. Pleon usually less ornamented than in ♂, and with shallower median notch lacking teeth.

Key to species of *Paracerceis* (*P. nuttingi* not included)

1. ♂, pleotelsonic notch narrow, with median basal tooth; ♀, pleotelson dorsally unornamented 2

   ♂, pleotelsonic notch wide, with lateral teeth; ♀, pleotelson dorsally with tubercles 3

2. ♂, median tooth of pleotelsonic notch almost as long as notch; subacute median tubercle on anterior pleotelson; ♀, pleotelson with posterior margin faintly concave, not notched .... *edithae*

   ♂, median tooth less than half length of notch; pleotelson with blunt rounded median tubercle; ♀, pleotelson with distinct posterior notch .... *glynni*

3. ♂, pleotelsonic notch deep, margins usually with two teeth on each side; strong median tubercle on anterior pleotelson bluntly bifid; ♀, pleotelson with one or two rounded median tubercles and 2 smaller tubercles on each side .... *caudata*
   ♂, pleotelsonic notch shallow, with tiny lateral denticles; median tubercle of pleotelson conical, acute; ♀, pleotelson with three large conical acute tubercles and several smaller scattered tubercles in anterior half .... *cohenae*

Paracerceis caudata (Say, 1818)
   Figure 96

DIAGNOSIS ♂ 8.1 mm, ♀ 6.4 mm. ♂: Pleotelson with blunt median bifid tubercle, with two smaller tubercles on each side. Pleotelsonic notch usually with two strong denticles on each margin, basal median tooth lacking. Uropodal exopod reaching well beyond pleotelson, slightly bowed, with 2–4 setose bumps on outer margin. ♀: Pleonite 5 with three low tubercles. Pleotelsonic apex broadly rounded in dorsal view, with two rounded median tubercles and two smaller tubercles on each side. Uropodal rami subequal, lamellar, outer distal angle of each acute.

RECORDS Bermuda; New Jersey to Florida Keys; Yucatan to Venezuela; Turks and Caicos Islands; Cuba; Puerto Rico; Bahamas; Jamaica; Haiti; St. Maartens, 0.2–127 m; St. Lucia; Gulf of Mexico. Found in the following algae: *Caulerpa, Halimeda, Turbinaria, Amphiroa, Laurencia, Dictyota;* between sponges and tunicates on red mangrove roots; in coral rubble; in spur and groove zone of reefs, lagoon, back reef, seagrass flats, and fringing mangroves.

REMARKS Menzies and Glynn (1968:55, fig. 22f) named and figured *P. caudata* var. *brevipes* from Puerto Rico. This variant was characterized as having the margins of the pleotelsonic notch lacking denticles. Given the considerable variation in ornamentation in this species, we feel that no validity can be given to the name "brevipes."
   This is the commonest sphaeromatid in the Caribbean, and it has very broad ecological requirements, being found in a wide range of habitats and depths.

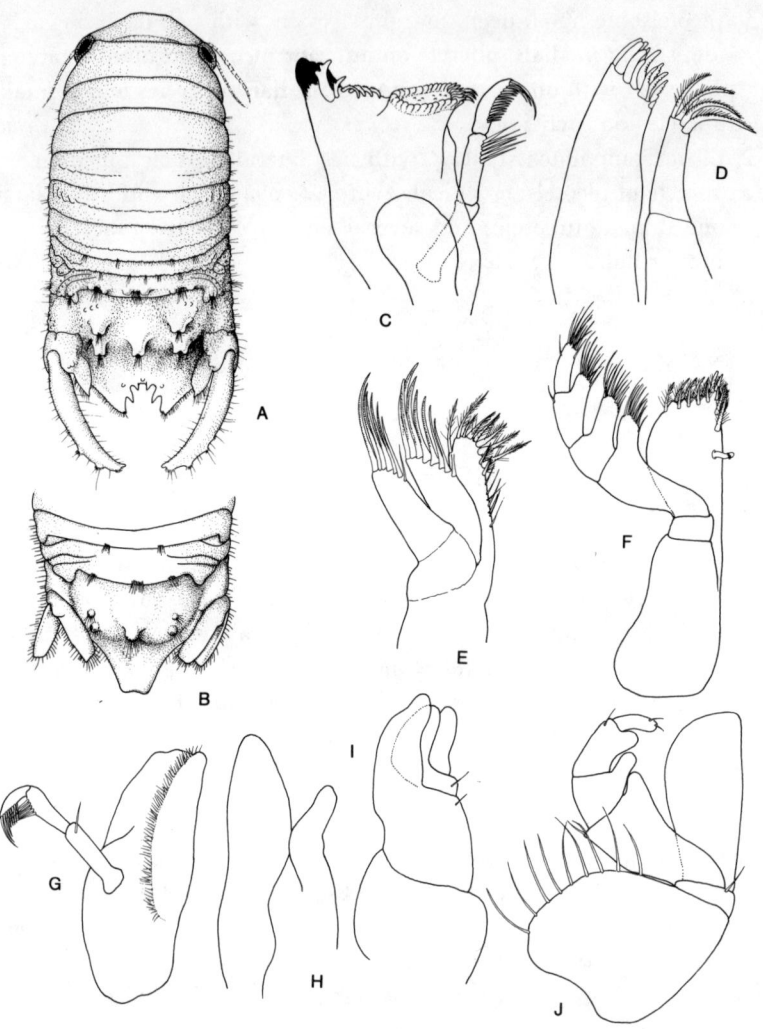

Figure 96. *Paracerceis caudata*: A, ♂; B, pleon ♀; C, mandible ♂; D, maxilla 1 ♂; E, maxilla 2 ♂; F, maxilliped ♂; G, mandible ♀; H, maxilla 1 ♀; I, maxilla 1 ♀; J, maxilliped ♀.

Paracerceis cohenae Kensley, 1984
Figure 97A,B

DIAGNOSIS ♂ 10.0 mm, ♀ 7.9 mm. ♂: Pereonites each with median tubercle and several smaller lateral tubercles near posterior margin of somite. Pleonite 5 with large median conical tubercle. Anterior two-thirds of pleotelson inflated, faintly tripartite, with strong median conical tubercle; notch in posterior margin shallow, with low median tooth and tiny lateral denticles; posterolateral margins finely dentate. Uropodal exopod cylindrical, distally denticulate, six to seven times longer than basal width. ♀: Pereon and pleon much as in ♂, but pleotelsonic notch shallower and posterolateral margins not denticulate. Uropodal rami subequal, lamellar, exopod with distolateral angle acute.

RECORDS   Carrie Bow Cay, Belize, 15–16 m. Only known from sponge *Callispongia plicifera* growing on outer reef slope.

Paracerceis edithae Boone, 1930
Figure 97C–E

DIAGNOSIS   ♂ 4.0 mm, ♀ 3.1 mm. ♂: Posterior three pereonites and pleonites each with irregular row of small tubercles near posterior margin, densely setulose tubercles becoming spinose more posteriorly. Pleotelson with strong median conical tooth in anterior half, flanked by convex spinose mound. Pleotelsonic notch deep, with elongate median basal tooth bearing strong acute tooth at its base. Lobes of posterior pleotelsonic margin broad, flattened, margins denticulate. Uropodal exopod tuberculate, tapering, apically acute. ♀: Integument much less tuberculate-spinose than in ♂. Immature ♀, posterior margin of pleotelson with faintly rounded median lobe. In mature ♀, posterior margin distinctly trilobed. Uropodal rami subequal, lamellar, distally rounded, with tiny distolateral spine on exopod.

RECORDS   Bahamas, 60–66 m, in vase sponge; Haiti; Puerto Rico, 20–25 m.

Paracerceis glynni Kensley, 1984
Figure 97F,G

DIAGNOSIS   ♂ 6.4 mm, ♀ 5.2 mm. ♂: Integument becoming strongly setose and tuberculate posteriorly from about pereonite 5. Posterior margin of inflated anterior area of pleotelson bearing strong median conical tubercle and smaller acute lateral tubercle, with low swelling beneath each lateral tubercle. Posterior notch deep, narrow, with small basal median tooth, lobes forming notch tricuspid, outer cusps recurved dorsally. Uropodal exopod fairly straight, cylindrical, apically acute. ♀: Body far less setose and tuberculate than ♂. Pleotelson with strongly inflated anterior area having very faint middorsal tubercle; notch well marked, formed by triangular lobes of

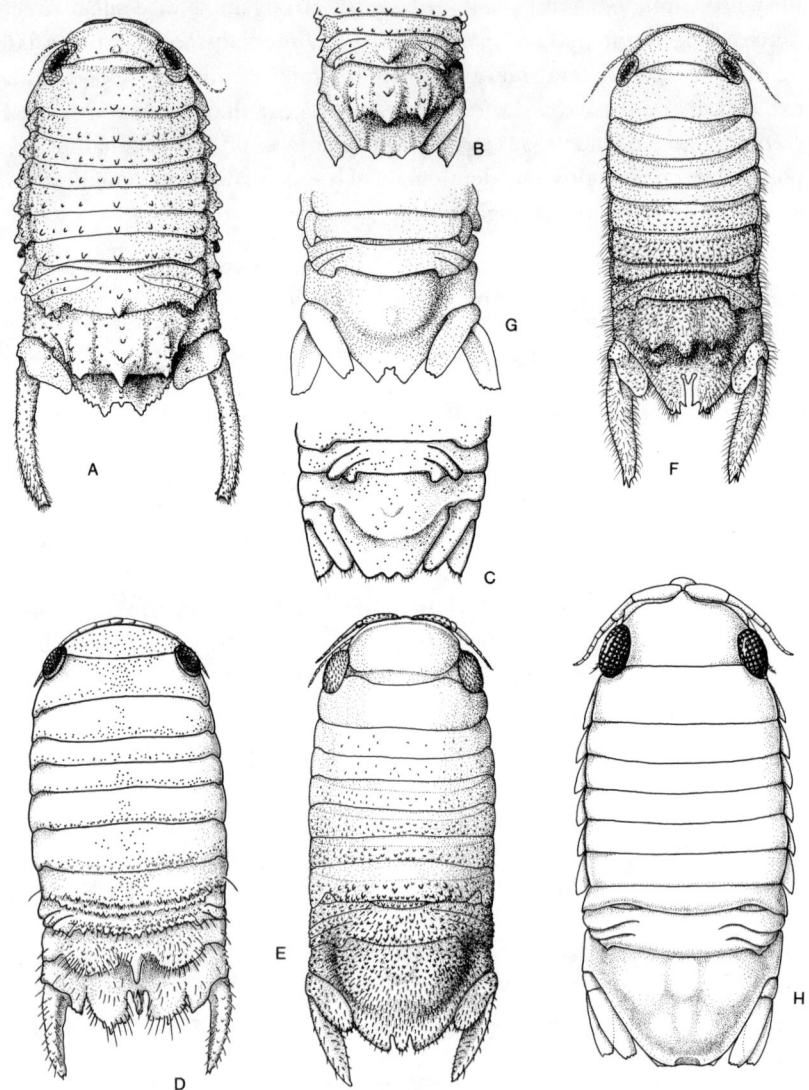

Figure 97. *Paracerceis cohenae:* A, ♂; B, pleotelson, ♀. *Paracerceis edithae:* C, pleotelson, ♀; D, mature ♂; E, immature ♂. *Paracerceis glynni:* F, ♂; G, pleotelson, ♀. *Paracerceis nuttingi:* H, ♀.

pleotelsonic margin. Uropodal rami subequal, flattened, endopod with distal margin faintly trituberculate; exopod with few distal tubercles.

RECORDS   Alligator Light, Florida, 11 m; Carrie Bow Cay, Belize, 11–15.2

m, from green alga *Halimeda* sp. on forereef, and from sponge *Aphysina fistularis*.

Paracerceis nuttingi (Boone, 1921)
   Figure 97H

RECORDS   Barbados; Puerto Rico, 1.5 m, from *Cymodocea* seagrass, and coral rubble and sponges.

REMARKS   The types of this species from Barbados consist only of females (total length 4.1 mm). Menzies and Glynn (1968) record an immature male from Puerto Rico with an incipient pleotelsonic notch. This specimen, however, still has the subequal lamellar uropodal rami. The mature male, with the characteristically reduced uropodal endopod and cylindrical exopod, is unknown. The possibility exists that this is not a true *Paracerceis*.

Paradella Harrison and Holdich, 1982

DIAGNOSIS   Both sexes lacking processes on pereon and pleon. Marked sexual dimorphism. Accessory unguis of pereopods simple, not bifid. Pleopod 3

Key to species of *Paradella*

1. Pereonite 7 with projecting bilobed flange; pleon and pleotelson finely but distinctly granulate 2
   Pleon and pleotelson smooth 3

2. ♂ with pleotelsonic foramen distinctly heart shaped, with median point; four submedian tubercles of pleotelson in ♂ and ♀ somewhat elongate; ♀ pleotelson posteriorly narrowed, slit visible dorsally
   .... *dianae*
   ♂ with pleotelsonic foramen wider than long, but lacking median point; four submedian tubercles of pleotelson in ♂ small, rounded, obscure in ♀; ♀ pleotelson posteriorly truncate, slit not visible dorsally .... *plicatura*

3. Tubercles on pleotelson in ♂ and ♀ small to obscure; ♂ with pleotelsonic foramen subcircular .... *quadripunctata*
   Tubercles on pleotelson broadly rounded mounds in ♂ and ♀; ♂ with pleotelsonic foramen wider than long .... *tumidicauda*

exopod with articulation. Uropodal rami lamellar. ♀: Mouthparts not metamorphosed. One pair of oostegites arising from pereonite 4, short, not reaching midline. Brood held in pouch formed by two opposing pockets covering entire ventrum, opening by transverse slit between 4th pereopods. ♂: Penes long, basally fused. Copulatory stylet basally narrow, extending further beyond endopod than in *Dynamenella*. Uropods broader than in ♀.

Paradella dianae (Menzies, 1962b)
Figure 98A–C

DIAGNOSIS ♂ 3.4 mm, ovigerous ♀ 3.7 mm. ♂: Pereonite 7, posterior margin broadly bilobed. Pleonite 5 with two rounded submedian tubercles. Pleotelsonic foramen distinctly heart shaped with median point. Pleotelson with four submedian and two lateral tubercles in ♂ and ♀, plus median tubercle at base of foramen in ♂; tubercles tending to be elongate and subcarinate. Uropodal rami lamellar, margins finely crenulate, relatively broader than in ♀. ♀: Pleotelsonic slit wide, dorsally visible.

RECORDS Key West, Florida; Puerto Rico, intertidal.
Baja California, intertidal.

Paradella plicatura (Glynn, 1970)
Figure 98D,E

DIAGNOSIS ♂ 4.1 mm, ♀ 3.6 mm. Pleon and pleotelson with tiny scattered tubercles. ♂: Pereonites 5–7 with posteriorly directed flanges, that of pereonite 7 largest, bilobed. Pleotelson with four submedian and two lateral discreet rounded tubercles. Pleotelsonic foramen wider than long, with basal bulge but lacking median point. ♀: Pleotelson with wide posterior slit not visible dorsally, posterior margin appearing truncate. Pleotelsonic tubercles less marked than in ♂.

RECORDS Jamaica, under red mangroves; Margarita Island, Venezuela, shallow infratidal.

Paradella quadripunctata (Menzies and Glynn, 1968)
Figure 98F,G

DIAGNOSIS ♂ 2.5 mm, ♀ 2.5 mm. Pleonite 5 with two low rounded submedian tubercles. Pleotelson with four low submedian rounded tubercles and

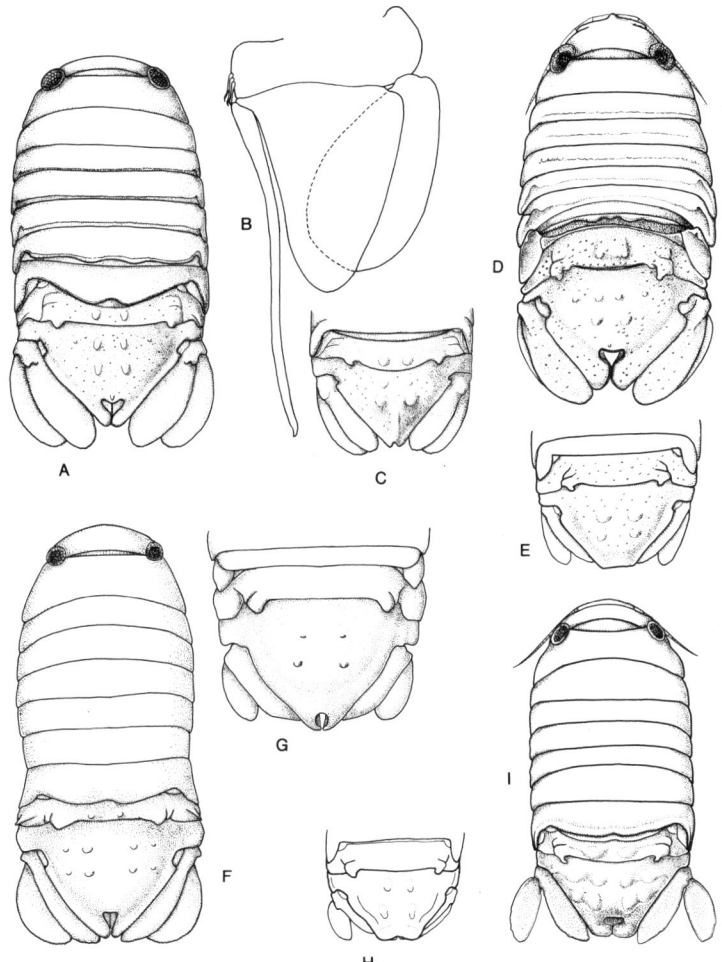

Figure 98. *Paradella dianae:* A, ♂; B, pleopod 2 ♂; C, pleon ♀. *Paradella plicatura:* D, ♂; E, pleon ♀. *Paradella quadripunctata:* F, ♂; G, pleon ♀. *Paradella tumidicauda:* H, pleon ♀ (from Glynn, 1970); I, ♂.

four smaller lateral tubercles. ♂: Pleotelsonic foramen subcircular, ventral margins of foramenal tube barely touching. ♀: Pleotelson posteriorly narrowly tapered, slit becoming tubelike, dorsally visible; four submedian tubercles less marked than in ♂.

RECORDS   Bermuda; Florida; Dominican Republic; Puerto Rico, intertidal; U.S. Virgin Islands, intertidal to 1 m.

Paradella tumidicauda (Glynn, 1970)
  Figure 98H,I

DIAGNOSIS ♂ 6.7 mm, ♀ 6.5 mm. ♂: Last pleonite with two submedian swellings. Pleotelson with four submedian broadly rounded swellinglike tubercles and two pairs of lateral tubercles. Foramen wider than long, posterior contiguous borders of foramen each bearing rounded swelling. ♀: Pleotelson with four submedian swellinglike tubercles, sometimes with two obscure lateral tubercles; posterior slit not visible dorsally, area surrounding slit swollen, horseshoe shaped.

RECORDS Margarita Island, Venezuela, from among intertidal barnacles.

Subfamily Sphaeromatinae H. Milne Edwards, 1840

DIAGNOSIS Cephalon not fused with pereonite 1. Pereopods 1 and 2 ambulatory. Pleopods 4 and 5, endopods having branchial pleats, exopods unpleated, membranous, of two articles. Uropods biramous.

Key to genera of Sphaeromatinae

1. Uropodal exopod with outer margin serrate .... *Sphaeroma*
   Uropodal exopod with outer margin entire or faintly crenulate 2
2. ♂, pleotelsonic notch with median lobe; ♀, pleotelsonic apex barely notched with rounded median lobe; ♀, mouthparts metamorphosed
   3
   Pleotelson entire to very faintly notched in ♂ and ♀; ♀, mouthparts not metamorphosed; with three pairs of oostegites ..... *Exosphaeroma*
3. Mature ♂, uropodal exopod about twice length of endopod; ♀ with three pairs of oostegites .... *Harrieta*
   ♂, uropodal rami subequal or exopod shorter than endopod; ♀ with four pairs of oostegites .... *Cymodoce*

Cymodoce Leach, 1814

DIAGNOSIS Pleon with two elongate straight parallel incomplete sutures on each side. Pleotelsonic apex with marked notch bearing median tooth.

Pleopod 5, exopod of two articles, distal article with apex and internal margin covered with fine teeth, anterior surface with long distally toothed boss; proximal article with two small toothed bosses at internodistal angle. ♂: Maxillipedal palp articles 2–4 bearing setigerous lobes. Penial rami elongate, separate. Pleon usually more tuberculate than in ♀. Uropodal exopod lamellar, shorter than endopod. ♀: Mouthparts metamorphosed. Broodpouch formed by four pairs of oostegites arising from pereonites 1–4, overlapping in midline. Brood housed in five pairs of internal pouches.

"*Cymodoce*" *barrerae* (Boone, 1918)
Figure 99A,B

DIAGNOSIS ♀ 7.5 mm. ♀: Body dorsally strongly vaulted, unornamented. Frontal lamina distally broadly rounded, lateral shoulders rounded. Mouthparts not metamorphosed. Pleotelson anteriorly strongly inflated with barest indication of two submedian swellings; posterior margin trilobed, with median lobe strong, narrowly rounded, outer lobes much smaller and ventral to median lobe. Uropodal endopod distally obliquely truncate; exopod distally acute.

RECORDS Cabanas, Cuba.

REMARKS This species is known only from the nonovigerous female holotype. Loyola e Silva (1960) placed the species in *Cymodoce*, based on a female specimen from Brazil. As the mouthparts are not metamorphosed, this does not agree with the present concept of *Cymodoce*, but with neither ovigerous females nor males available, the correct generic placement cannot be determined.

Cymodoce ruetzleri Kensley, 1984
Figure 99C–G

DIAGNOSIS ♂ 5.0 mm, ♀ 4.2 mm. ♂: Integument with numerous small tubercles, becoming densely setose posteriorly. Pleonite 4, posterior margin broadly bilobed. Pleotelson bearing pair of strong conical tubercles with acute tips, each tubercle flanked by low rounded tubercle; apex trilobed, outer lobes triangular, acute, sharp spine at base of incision, median lobe apically blunt. Uropodal exopod apically acute, oval in cross section, endopod and sympod fused, somewhat flattened, apex triangular with strong tooth. ♀: Pleotelson with two conical apically acute tubercles, apex barely notched, with short rounded lobe slightly offset from posterior margin. Both

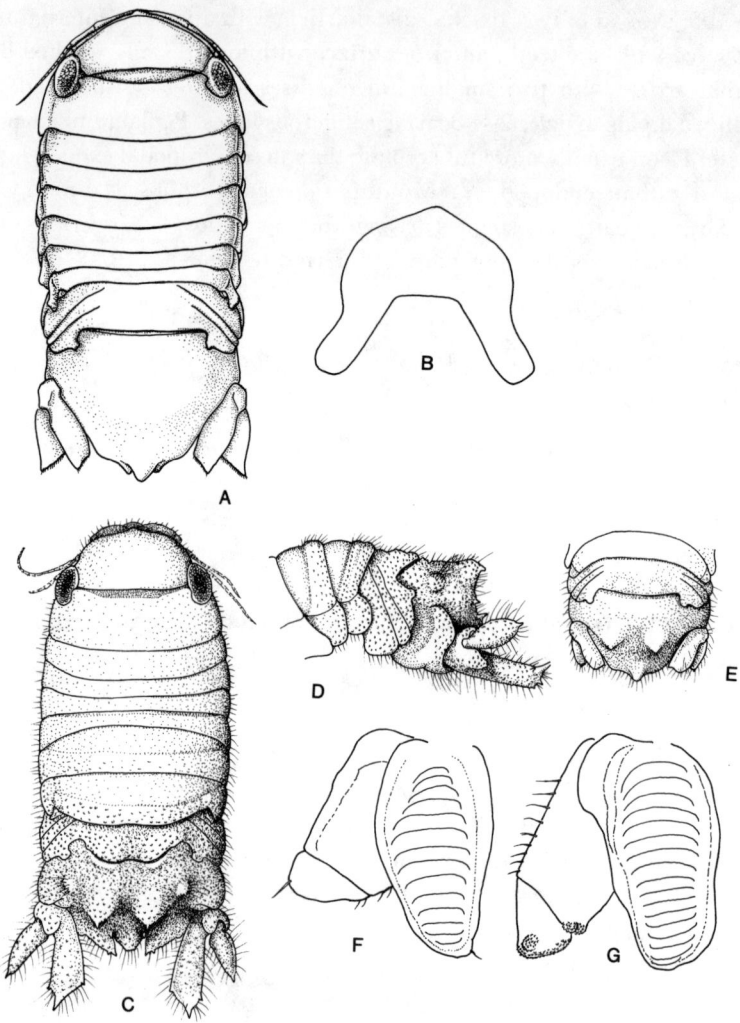

Figure 99. *"Cymodoce" barrerae:* A, ♀; B, frontal lamina. *Cymodoce ruetzleri:* C, ♂; D, pleon in lateral view, ♂; E, pleon ♀; F, pleopod 4; G, pleopod 5.

uropodal rami flattened; exopod with tiny apical tooth, endopod distally truncate-rounded, with small mediodistal tooth.

RECORDS Carrie Bow Cay, Belize, 0.5–13 m; in algal clumps, reef crest rubble, and seagrass flats.

Exosphaeroma Stebbing, 1900

DIAGNOSIS   Maxillipedal palp articles 2–4 produced medially into lobes. Pereonites 6 and 7 dorsally unarmed. Pleopod 3, exopod biarticulate. ♂: Penes short, separate. Copulatory stylet of pleopod 2 elongate, slender. Pleotelson lacking strong apical notch. ♀: Mouthparts not metamorphosed. Broodpouch of three pairs of oostegites on pereonites 2–4; oostegites short, not reaching midline. Brood held in four pairs of internal pouches.

Key to species of *Exosphaeroma*

1. Pleotelson with posterior margin entire, evenly convex 2
   Pleotelson with posterior margin faintly notched or trilobed 3

2. Frontal lamina with length less than 1.5 times greatest width   *diminuta*
   Frontal lamina with length almost two times greatest width
   .... *productatelson*

3. Pleotelson with posterior margin faintly trilobed, and with three low
     rounded tubercles anteriorly .... *yucatanum*
   Pleotelson with posterior margin faintly notched 4

4. Pleotelson posteriorly broadly notched; two rounded submedian
     tubercles on inflated midregion .... *antillense*
   Pleotelson with faint narrow notch posteriorly; lacking dorsal tubercles
   .... *alba*

Exosphaeroma alba Menzies and Glynn, 1968
  Figure 100A–C

DIAGNOSIS   ♂ 2.0 mm, ♀ 2.3 mm. Frontal lamina anteriorly broadly rounded, basally slightly wider than midlength. Pleotelson similar in ♂ and ♀; anterodorsally inflated and unornamented, posteriorly tapering to slight median notch, seen in dorsal view. Uropodal rami distally shallowly serrate, exopod 2.5 times longer than wide.

RECORDS   Puerto Rico, intertidal to 0.5 m; in algae on rocks, and under *Chiton tuberculatus* and *C. marmoratus*.

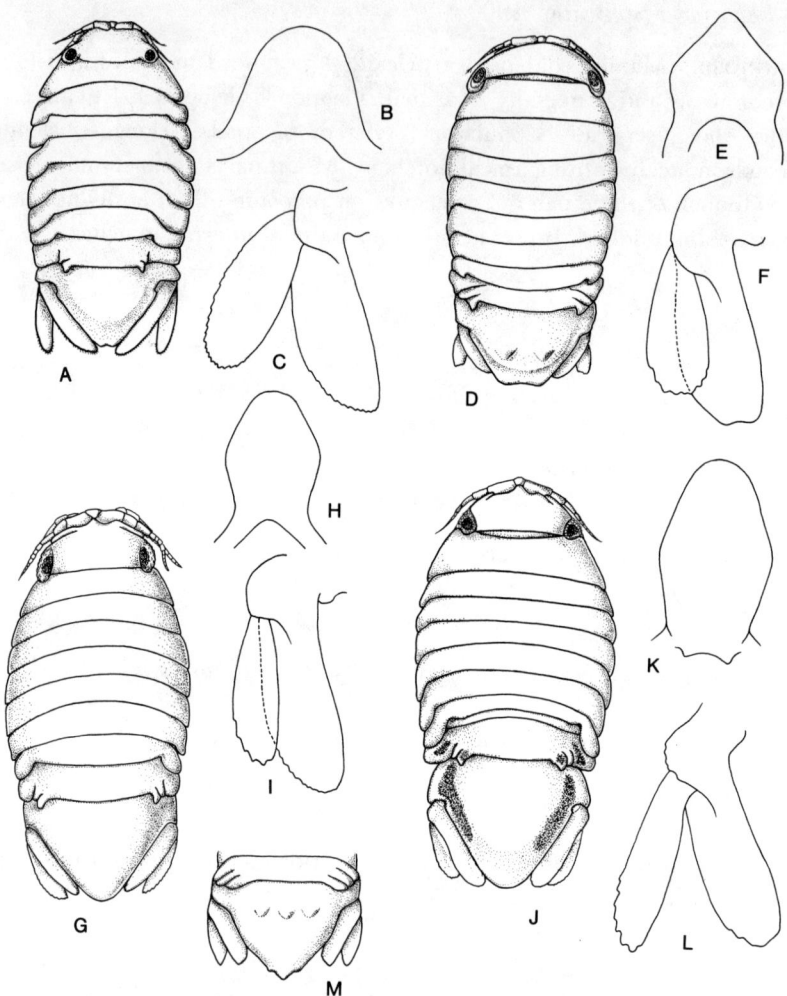

Figure 100. *Exosphaeroma alba: A; B*, frontal lamina; *C*, uropod. *Exosphaeroma antillense: D; E*, frontal lamina; *F*, uropod. *Exosphaeroma diminuta: G; H*, frontal lamina; *I*, uropod. *Exosphaeroma productatelson: J; K*, frontal lamina; *L*, uropod. *Exosphaeroma yucatanum: M*, pleon (from Richardson, 1905).

Exosphaeroma antillense Richardson, 1912d
Figure 100D,F

DIAGNOSIS ♀ 5.0 mm. Frontal lamina anteriorly tapering to subacute apex. Pleotelson with two broadly subconical submedian tubercles on inflated anterior area; posterior margin subtruncate to very faintly emarginate. Uropodal exopod distally crenulate, length slightly more than twice greatest width; endopod with faint distal notch.

RECORDS   Montego Bay, Jamaica.

REMARKS   The single ovigerous female holotype is the only known specimen of this species. The overlapping oostegites suggest that this may not be an *Exosphaeroma*.

Exosphaeroma diminuta Menzies and Frankenberg, 1966
Figure 100G–I

DIAGNOSIS   ♂ 2.2 mm. Frontal lamina widest at midlength, anteriorly truncate-rounded. Pleotelson with posterior margin broadly rounded. Uropodal rami not quite reaching pleotelsonic apex; exopod margin distally crenulate.

RECORDS   Chesapeake Bay to Florida; Venezuela; sand dwelling, intertidal and shallow subtidal.

Exosphaeroma productatelson Menzies and Glynn, 1968
Figure 100J–L

DIAGNOSIS   ♂ 2.5 mm, ♀ 1.5 mm. Sexes essentially similar. Frontal lamina widest at midlength, where slight shoulder apparent, anteriorly broadly rounded, 1.6 times longer than wide. Pleotelson unornamented, anteriorly inflated, posterior margin entire, evenly convex. Uropodal exopod distally shallowly serrate, almost four times longer than wide; endopod wider than exopod. Broad lateral patches of pigment on pleotelson in both sexes.

RECORDS   Puerto Rico, intertidal to 0.5 m, in algae on rocks; Texas, Gulf of Mexico.

Exosphaeroma yucatanum (Richardson, 1901)
Figure 100M

DIAGNOSIS   Frontal lamina anteriorly tapering from widest point to subacute apex, proximally narrower than at midlength. Pleotelson posteriorly obscurely trilobed, median lobe narrowly rounded, longest; three low rounded tubercles on pleotelson in anterior region.

RECORDS   Cape Catoche, Yucatan, Mexico, 48 m.

REMARKS   This species was described from a single specimen which has since been lost. The true generic placement of this species is thus undetermined and full description awaits the finding of more material.

Harrieta Kensley, 1987c

DIAGNOSIS   ♀ with mouthparts metamorphosed. Broodpouch of three pairs of oostegites on pereonites 2–4, overlapping in midline; brood held in five pairs of internal pouches. Uropodal rami subequal, lamellar in ♀, exopod twice length of endopod and oval in cross section in ♂. Pleopod 2 in ♂ with copulatory stylet articulating basally on endopod, curved, barely reaching apex of endopod. Penes basally fused, rami slender, elongate, tapering.

Harrieta faxoni (Richardson, 1905)
Figure 101A,B

DIAGNOSIS   ♂ 6.0 mm, ♀ 6.5 mm. ♂: Frontal lamina with broad slightly convex anterior margin. Two low rounded submedian tubercles on cephalon near posterior margin. Two rounded submedian tubercles on last pleonite. Pleotelson anteriorly inflated with two submedian tubercles; posterior margin trilobed. ♀: Essentially similar to ♂, but posterior margin of pleotelson less markedly trilobed, with median lobe longer, and uropodal rami subequal in length.

RECORDS   Florida to Texas, Gulf of Mexico, intertidal and subtidal in *Thalassia, Halodule,* and *Syringodium* seagrass beds, in salinities of 7‰ to 36‰.

Sphaeroma Bosc, 1802

DIAGNOSIS   Maxillipedal palp with three distal articles poorly developed, lacking lobes; fringe of robust plumose setae with swollen bases on internal margin of endite; distal margin of endite with simple setae. Pereopods 1–3 with plumose setae on ischium and merus. Posterior margin of pleotelson entire, similar in ♂ and ♀. Pleopod 3, exopod uniarticulate. Uropodal exopod with outer margin serrate. Able to conglobate. ♂: Penes short, rounded, separate. Pleopod 2, copulatory stylet articulating basally on endopod, slender, reaching well beyond rami. ♀: Mouthparts not meta-

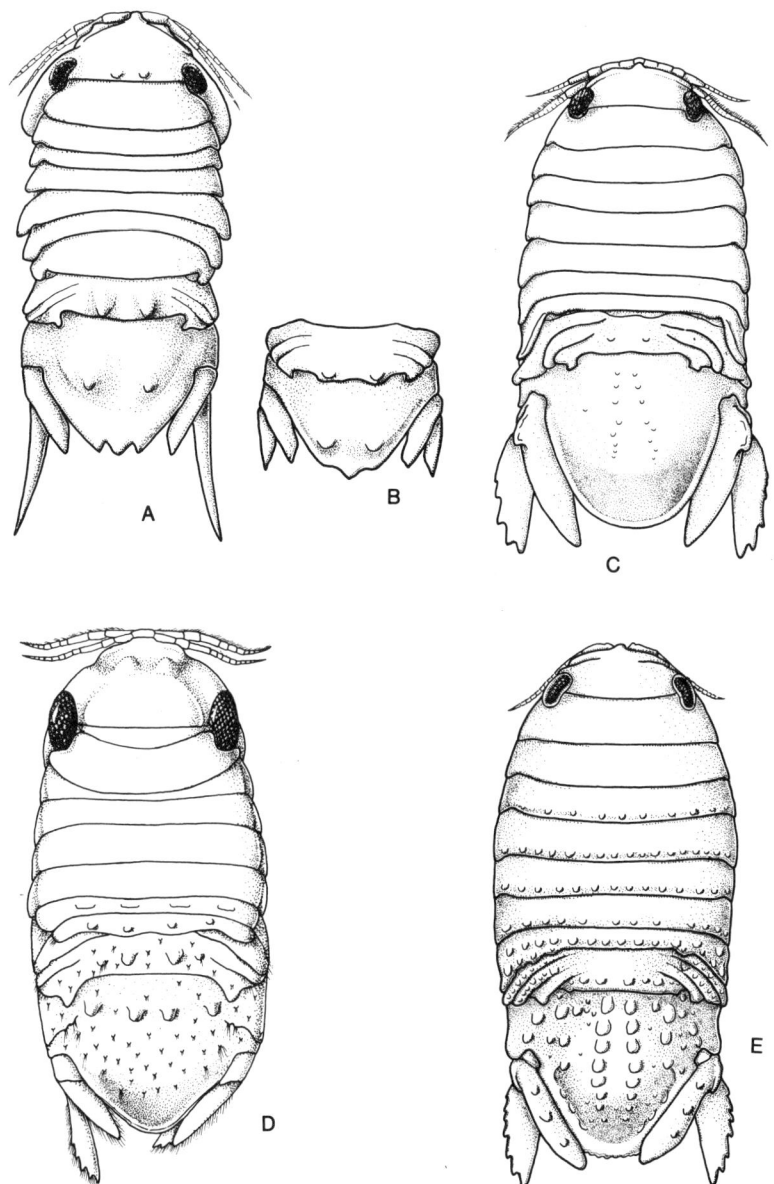

Figure 101. *Harrieta faxoni:* A, ♂; B, pleon ♀. *C, Sphaeroma quadridentata;* D, *Sphaeroma terebrans;* E, *Sphaeroma walkeri.*

morphosed. Three pairs of overlapping oostegites arising from pereonites 2–4 (but *S. terebrans* has anterior pair rudimentary).

REMARKS   The genus *Sphaeroma* is one of the few sphaeromatids in which the number of oostegites varies, from the diagnostic three pairs, through two normal pairs (as in *S. terebrans*), to having the oostegites completely absent (as in *S. annandalei*).

Jacobs (1987) has provided a useful reevaluation of the European, Mediterranean, and northwest African species of *Sphaeroma* and related genera.

Key to species of *Sphaeroma*

1. Pleotelson posteriorly bluntly triangular, with 4 strong anterior tubercles
   .... *terebrans*
   Pleotelson posteriorly broadly rounded 2

2. Pleotelson dorsally smooth or with few low tubercles ..... *quadridentata*
   Pleotelson dorsally with numerous strong tubercules .... *walkeri*

Sphaeroma quadridentata Say, 1818
   Figure 101C

DIAGNOSIS   ♂ 11.0 mm, ♀ 8.0 mm. Pleotelson anteriorly inflated, sometimes with few low rounded tubercles, posteriorly flattened to concave; posterior margin entire, broadly rounded.

RECORDS   New England to Florida; Gulf of Mexico, intertidal to 1 m, often in pilings and partially submerged dead tree trunks, and commonly associated with barnacles.

Sphaeroma terebrans Bate, 1866
   Figure 101D

DIAGNOSIS   ♂ 10.0 mm, ♀ 11.5 mm. Pereonite 7 with pair of submedian and pair of lateral tubercles. Dorsal pleon densely tuberculate. Posterior pleonite with pair of submedian acute tubercles. Pleotelson anteriorly with submedian pair and lateral pair of tubercles, posteriorly rounded-triangular.

RECORDS   Virginia to Florida; Belize; Cuba; Venezuela to Brazil; Gulf of Mexico.

Nigeria, east coast of southern Africa, India, Sri Lanka, Thailand, Indonesia, Philippines, Australia.

REMARKS   There is no agreement on whether this species is synonymous with *S. destructor* Richardson, 1897. This latter (if distinct) bores into wood pilings in estuarine waters, while *S. terebrans* is found in the prop roots of the red mangrove tree, *Rhizophora mangle*. In this habitat, the isopods are interpreted either as being destructive agents (e.g., Rehm and Humm, 1973) or as promoting increased root growth (Simberloff et al., 1978). It is unlikely that the bored wood itself is a source of food for the isopods; rather, as with the genus *Limnoria*, the food is probably detritus or fungi and bacteria growing on the wood fragments in the burrows or on the setae of the appendages.

Sphaeroma walkeri Stebbing, 1905
Figure 101E

DIAGNOSIS   ♂ 9.5 mm, ♀ 10.0 mm. Pereonites 3–7 with transverse row of large rounded tubercles. Last pleonite with row of prominent tubercles and smaller scattered tubercles laterally. Pleotelson anteriorly inflated, posteriorly concave and cuplike, with four irregular longitudinal rows of large tubercles plus many small scattered tubercles. Posterior margin rounded, entire to irregularly crenulate. Uropodal endopod with several rounded tubercles on dorsal surface; exopod with row of smaller tubercles on ventral surface.

RECORDS   Probably pan-tropical. Florida to Puerto Rico, intertidal.

Family Tridentellidae Bruce, 1984

DIAGNOSIS   Eyes well developed. Pereonites 2–7 with distinct coxae. Pleon consisting of five free pleonites plus pleotelson. Mandible with acute incisor; lacinia mobilis absent; molar present; palp of three articles. Maxilla 1, outer ramus styliform with three to five strong terminal spines, and several short recurved subapical spines. Maxilla 2 uniramous, biarticulate, bearing small sometimes tridentate spines or scales distally. Maxillipedal palp of five articles; endite slender, lamellar, usually lacking coupling hooks.

Tridentella Richardson, 1905

DIAGNOSIS   Body dorsally often bearing spines, tubercles, or carinae, more developed in ♂ than in ♀. Frontal lamina narrow, pentagonal. Antennular peduncle of three articles; antennal peduncle of five articles. Mandibular molar weakly sclerotized. Pereopods 1–3 weakly prehensile; pereopods 4–7 ambulatory. Copulatory stylet of pleopod 2 rodlike, arising proximally on mesial margin of endopod. Pleopod 5 endopod lacking marginal setae.

REMARKS   Delaney and Brusca (1985) provide useful taxonomic and distributional comments on the family Tridentellidae.

Tridentella virginiana (Richardson, 1900b)
Figure 102

DIAGNOSIS   ♂ 9.5 mm, ovigerous ♀ 9.5–11.0 mm. Cephalon and pereon dorsally smooth, pleon minutely granular. Uropodal rami with distal margins faintly dentate, apically narrowly rounded, endopod wider and slightly longer than exopod. Pleotelson basally wider than middorsal length; posterior margin broadly rounded to subtruncate.

RECORDS   Nova Scotia to Florida; off Georgia, 550 m; Gulf Stream off Key West, 220 m.

Suborder Gnathiidea Leach, 1814

DIAGNOSIS   Eyes usually well developed, rarely on short lateral processes, occasionally absent. Mandibles in ♂ greatly enlarged, projecting anteriorly from cephalon, not used in feeding. Mandibles lacking in ♀. Mouthparts of praniza larva styliform, with acute mandibles projecting anteriorly (see Figure 103D). Pereopod 1 modified, forming second pair of broad opercular maxillipeds covering mouthparts, referred to as pylopods. Pereopods 2–6 ambulatory. Pereonite 7 reduced, lacking pereopod. Pleonites separate, narrower than pereon. Uropods lateral, rami lamellar, forming tailfan with telson. Praniza larva with pereonites 4–6 enlarged, sometimes inflated. ♀ with pereonites 4–6 greatly inflated, forming broodpouch for internally brooded eggs (see Figure 103E).

REMARKS   The gnathiideans are entirely marine, most described species being from shallow waters. The males and females are frequently found in association with sponges and do not feed. The praniza larva is an efficient swimmer and has been recorded from shallow-water plankton, but is more

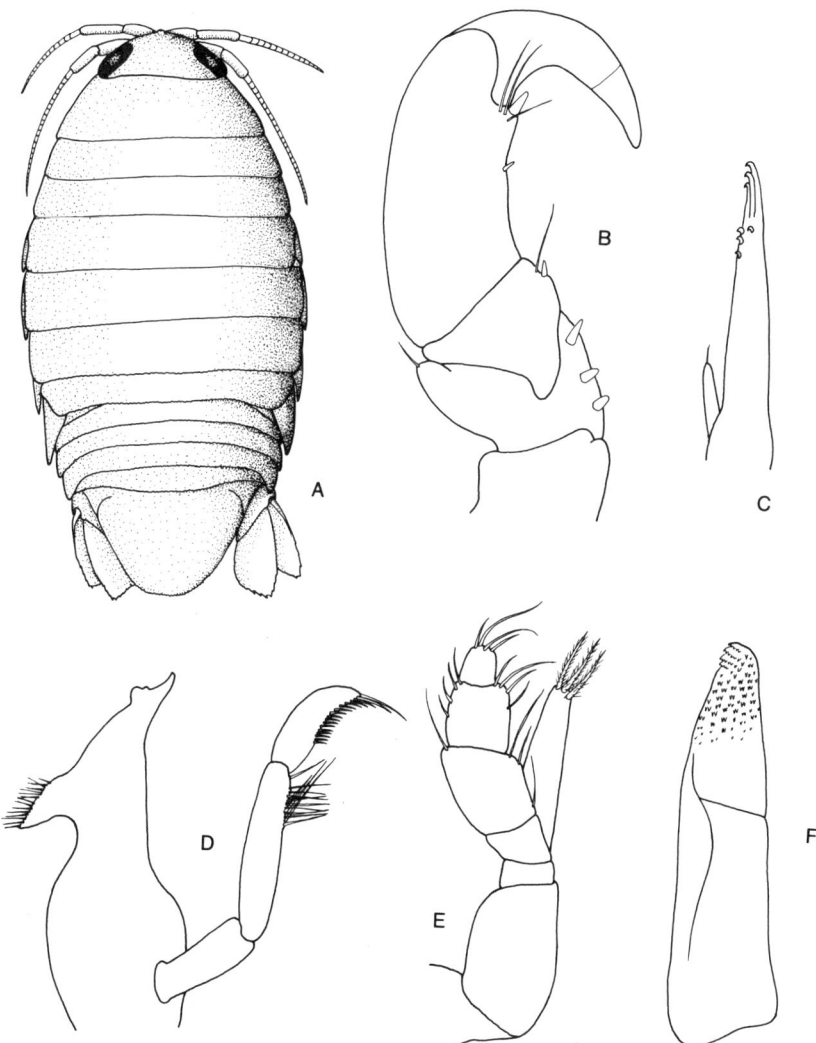

Figure 102. *Tridentella virginiana*: A, ♀; B, pereopod 1; C, maxilla 1; D, mandible; E, maxilliped; F, maxilla 2.

frequently encountered as a fish parasite, the favored site for sucking the host's blood being in the nares. Upton (1987a, 1987b) has shed light on the unusual life history of at least one gnathiid genus, *Paragnathia*.

The taxonomy of the Gnathiidae is based almost entirely on males, the praniza and females of most species being remarkably similar.

Family Gnathiidae Harger, 1879

DIAGNOSIS   As for the suborder Gnathiidea.

Gnathia Leach, 1814

DIAGNOSIS   In addition to features mentioned in diagnosis of suborder: Eyes present in most species. Pylopod with two small articles distal to broad opercular article 2, terminal article minute.

Key to species of *Gnathia* (♂ only)

1. Anterior margin of cephalon with medial process or slightly convex .. 2
   Anterior margin of cephalon concave or lacking medial process 9

2. Anterior margin of cephalon broadly triangular, projecting, with small lateral teeth .... *triospathiona*
   Anterior margin of cephalon not triangular and projecting 3

3. Cephalon and two free anterior pereonites dorsally granular 4
   Cephalon and two free anterior pereonites smooth 5

4. Lobe of outer margin of mandible notched; pereonite 5 twice wider than middorsal length .... *velosa*
   Lobe of outer margin of mandible rounded; pereonite 5 1.5 times wider than middorsal length 6

5. Anterior margin of cephalon with distinct medial process ..... *virginalis*
   Anterior margin of cephalon barely convex, lacking medial process
   .... *rathi*

6. Inner proximal lobe of mandible distinct 7
   Inner proximal lobe of mandible indistinct .... *samariensis*

7. Inner proximal lobe of mandible entire 8
   Inner proximal lobe of mandible with rounded toothlike marginal structures .... *johanna*

8. Pereonites 3–5 poorly defined .... *puertoricensis*
   Pereonites 3–5 clearly defined .... *magdalenensis*

9. Anterior margin of cephalon concave, lacking projections ..... *gonzalezi*
   Anterior margin of cephalon with four projections .... *beethoveni*

Gnathia beethoveni Paul and Menzies, 1971
Figure 103A

DIAGNOSIS ♂ 3.0 mm. Anterior margin of cephalon with two low tubercles flanking shallow medial notch plus slightly larger pair of lateral tubercles. Cephalon lacking dorsal tubercles. Pereonite 5 1.5 times wider than middorsal length. Uropodal endopods reaching beyond telsonic apex.

RECORDS Off Venezuela, 95 m. Colombia.

Gnathia gonzalezi Müller, 1988
Figure 103B

DIAGNOSIS ♂ 2.0 mm. Body smooth. Anterior margin of cephalon concave. Pereonites 3–5 distinct; pereonite 5 2.5 times wider than middorsal length. Cutting margin of mandible with four or five low rounded teeth.

RECORDS Colombia, 30 m.

Gnathia johanna Monod, 1926
Figure 103C

DIAGNOSIS ♂ 2.1 mm. Anterior margin of cephalon medially convex between pair of submedian tubercles. Pereonites 4 and 5 poorly separated. Proximomedial lobe of mandible having four or five rounded crenulations, with tiny seta between adjacent crenulations.

RECORDS U.S. Virgin Islands, 29–46 m; Colombia.

Gnathia magdalenensis Müller, 1988
Figure 103D

DIAGNOSIS ♂ 3.0 mm. Anterior margin of cephalon with three tubercles, median tubercle slightly shorter than submedian pair. Cephalon with few scattered low granulations dorsally. Pereonite 5 about 1.5 times wider than middorsal length. Proximomedial lobe of mandible entire.

RECORDS Carrie Bow Cay, Belize, intertidal; Colombia, 18 m.

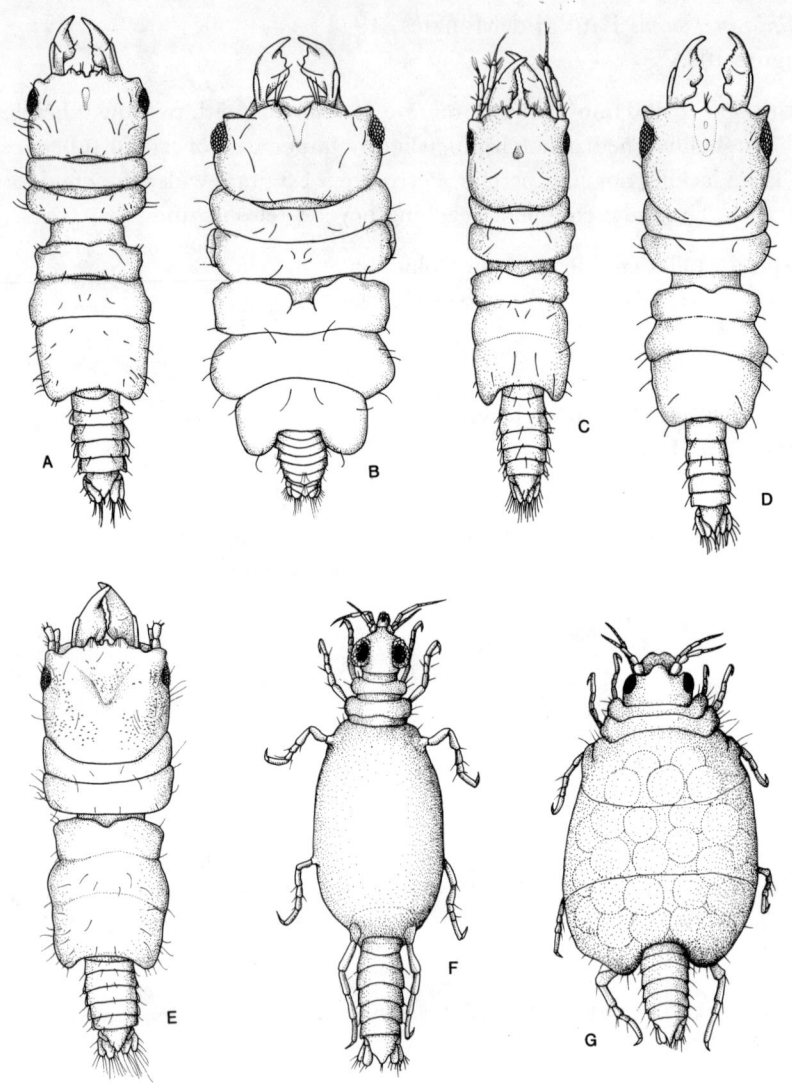

Figure 103. *Gnathia beethhoveni*: A, ♂. *Gnathia gonzalezi*: B, ♂ (after Müller, 1988). *Gnathia johanna*: C, ♂. *Gnathia magdalenensis*: D, ♂ (after Müller, 1988). *Gnathia puertoricensis*: E, ♂; F, praniza larva; G, ovigerous ♀.

Gnathia puertoricensis Menzies and Glynn, 1968
Figure 103E–G

DIAGNOSIS ♂ 3.0 mm, ovigerous ♀ 1.8 mm. Anterior margin of cephalon having three tubercles between mandibular bases, medial tubercle narrower than submedian pair. Dorsal integument finely granular, with coarser granules mediodorsal to eye. Pereonites 4 and 5 indistinctly separated. Mandible lacking proximomedial lobe.

RECORDS Carrie Bow Cay, Belize, intertidal to 2 m; Puerto Rico, intertidal; Cuba.

Gnathia rathi Kensley, 1984
Figure 104A

DIAGNOSIS ♂ 1.9 mm, ovigerous ♀ 2.2 mm. Frontal margin faintly convex to straight between single low lateral tubercle mesial to mandibular bases. Dorsal integument of cephalon and anterior two free pereonites coarsely granular. Lateral margins of telson faintly denticulate. Pereonites 4 and 5 poorly separated.

RECORDS Carrie Bow Cay, Belize, 1–36 m.

Gnathia samariensis Müller, 1988
Figure 104B

DIAGNOSIS ♂ 2.0 mm. Anterior margin of cephalon with three tubercles, median tubercle slightly shorter than submedian pair; dorsal integument smooth. Pereonites 4 and 5 well differentiated; pereonite 5 about 2.2 times wider than middorsal length. Mandible lacking proximomedial lobe.

RECORDS Colombia.

Gnathia triospathiona Boone, 1918
Figure 104C

DIAGNOSIS ♂ 8.8 mm. Anterior margin of cephalon with broad-based triangular projection bearing three low teeth; deep V-shaped depression posterior to anterior margin, with low flanking granulations.

RECORDS Off Key West, in Gulf Stream, 218 m.

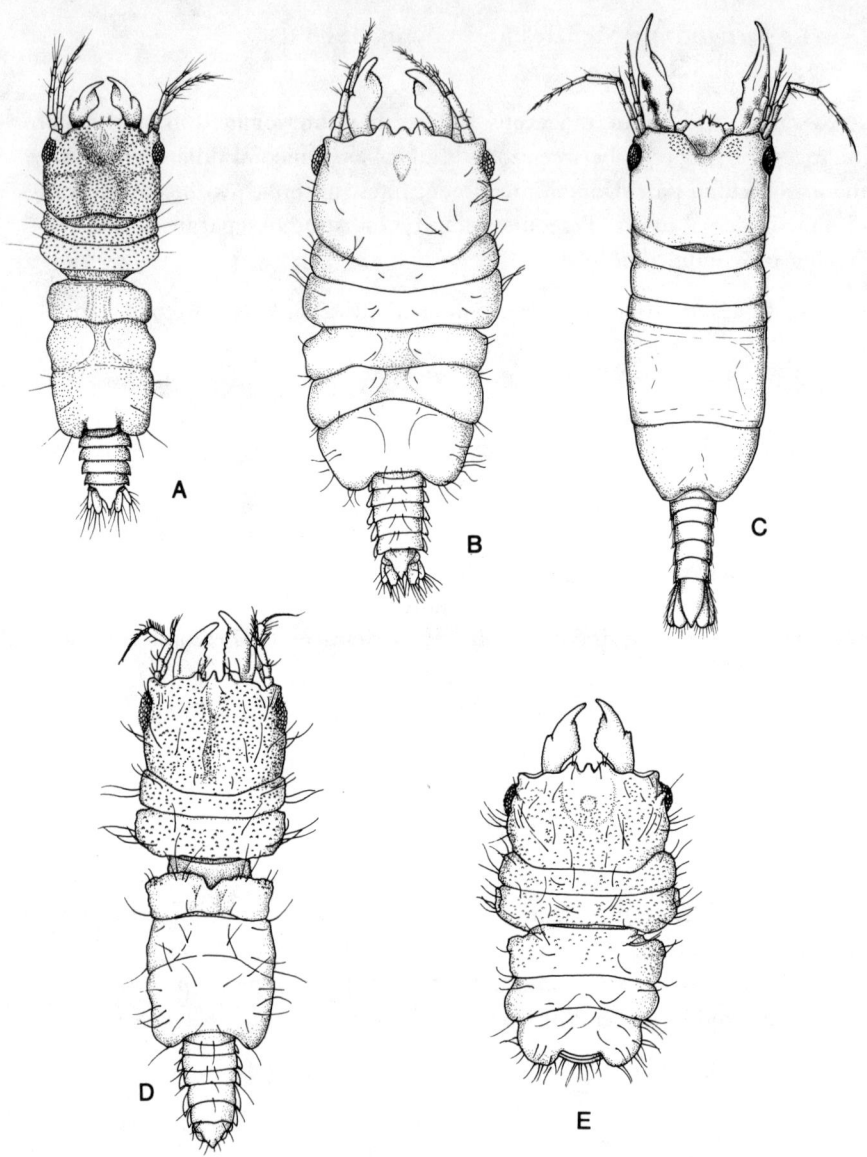

Figure 104. *Gnathia rathi:* A, ♂; *Gnathia samariensis:* B, ♂ (after Müller, 1988); *Gnathia triospathiona:* C, ♂; *Gnathia virginalis:* D, ♂; *Gnathia velosa:* E, ♂ (after Müller, 1988).

Gnathia velosa Müller, 1988
Figure 104E

DIAGNOSIS ♂ 1.5 mm. Anterior margin of cephalon with three tubercles, median tubercles slightly shorter and narrower than submedian pair. Dorsal integument of cephalon and anterior three pereonites granular. Pereonite 5 about 2.5 times wider than middorsal length. Lateral lobe of mandible notched.

RECORDS Colombia.

Gnathia virginalis Monod, 1926
Figure 104D

DIAGNOSIS ♂ 2.2 mm. Anterior margin of cephalon with three tubercles, median tubercles slightly longer than submedian pair. Dorsal integument of cephalon and anterior three pereonites granular. Pereonite 5 about 1.7 times wider than middorsal length. Lateral lobe of mandible rounded.

RECORDS U.S. Virgin Islands, 29 m; Colombia.

Suborder Microcerberidea Lang, 1961

DIAGNOSIS Cephalon free. Mandibles with reduced palp, or lacking palp. Maxillipedal palp of five articles. Pereon of seven free segments. Pereopod 1 subchelate; pereopods 2–7 ambulatory. Pleon of two free pleonites plus pleotelson. Pleopod 1 in ♂ usually absent. Pleopod 2 modified for copulation. Pleopod 3 uniramous, opercular. Pleopod 4 biramous. Pleopod 5 reduced. Uropods usually uniramous or biramous.

Family Microcerberidae Karaman, 1933b

DIAGNOSIS Eyes absent. Body elongate, slender. Antennular peduncle of three articles; antennal peduncle of six to eight articles. Mandibular palp of single article; molar reduced to single stout fringed spine. Maxilla 2 reduced to single ramus bearing two distal fringed lobes. Pereopods 2–7 ambulatory, dactyli biunguiculate.

REMARKS The species of the Microcerberidae are all very small (less than 2 mm total length) and are most often found in interstitial habitats. They have been recorded from marine, brackish, and freshwater environments.

The microcerbserideans were often classified with the Anthuridea, mainly because of the similarity in body shape. Wägele (1983) however, has convincingly demonstrated the asellotan affinities of the group.

Key to genera of Microcerberidae

1. Maxillipedal palp articles 2 and 3 enlarged; basis of pereopods lacking spinous process .... *Yvesia*
   Maxillipedal palp articles 2 and 3 not markedly enlarged; basis of pereopods with spinous process .... *Microcerberus*

Microcerberus Karaman, 1933b

DIAGNOSIS   Maxillipedal palp articles 2 and 3 not enlarged. Articles 2 and 3 of antennal peduncle with spinous process. Basis of pereopods with spinous process. Propodus of pereopod 2 with two denticulate proximal spines.

Microcerberus syrticus Kensley, 1984
   Figure 105A–E

DIAGNOSIS   ♂ 1.1 mm, ♀ 1.1 mm. Tergal lobes of pereonites 2–4 rounded. Apical lobe of ♂ pleopod 2 acute.

RECORDS   Carrie Bow Cay, Belize, interstitial in intertidal sand bar.

REMARKS   In addition to *M. syrticus*, six species of *Microcerberus* have been recorded from the Caribbean area: *M. littoralis* Chappuis and Delamare Deboutteville, 1956, from the Bahamas; *M. minutus* Coineau and Botosaneanu, 1973, from Cuba; *M. mirabilis* Chappuis and Delamare Deboutteville, 1956, from the Bahamas; *M. nunezi* Coineau and Botosaneanu, 1973, from Cuba; *M. renaudi* Chappuis and Delamare Deboutteville, 1956, from the Bahamas; *M. simplex* Coineau and Botosaneanu, 1973, from Cuba. The reader is referred to the original descriptions for separation of the species.

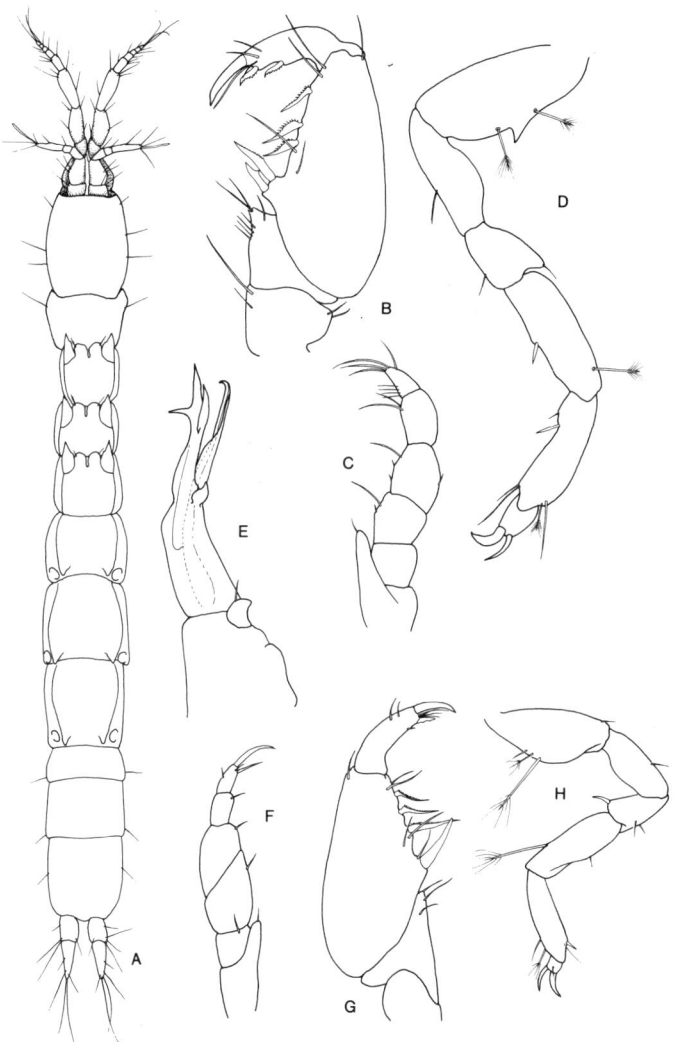

Figure 105. *Microcerberus syrticus:* A, ♂; B, pereopod 1; C, maxilliped; D, pereopod 2; E, pleopod 2 ♂. *Yvesia striata* (from Coineau and Botosaneanu, 1973): F, maxilliped; G, pereopod 1; H, pereopod 2.

Yvesia Coineau and Botosaneanu, 1973

Yvesia striata Coineau and Botosaneanu, 1973
Figure 105F–H

DIAGNOSIS ♀ 1.6 mm. Antennal peduncular articles 2 and 3 smooth, lacking spinous processes. Maxillipedal palp articles 2 and 3 enlarged. Bases of pereopods unarmed, lacking spinous processes. Propodus of pereopod 1 with single smooth proximal spine. Body having longitudinal ventrolateral striae.

RECORDS Oriente, Cuba, interstitial on beach.

Suborder Oniscidea Latreille, 1803

DIAGNOSIS Compound eyes usually present. Antennules usually very short. Antennae with 4- or 5-articulate peduncle; flagellum varying from few articles to multiarticulate. Mandibular palp present. Distal articles of maxillipedal palp often reduced. Coxae of pereonites 1–7 usually distinct, expanded. Pleopods respiratory, often with pseudotrachea; ♂ with pleopod 2, and sometimes pleopod 1 as well, modified for copulation. Uropods terminal or subterminal with terete rami, or ventral and opercular, with reduced rami.

REMARKS The Oniscidea includes all the isopods that have successfully invaded the terrestrial environment. While still in some degree reliant on external moisture, their morphological and behavioral adaptations have allowed them to live in almost all terrestrial habitats, from hot, dry deserts, through tropical rainforests and grasslands, to cold-temperate niches. Several forms have successfully inveigled themselves into termite or ant colonies, where with varying degrees of morphological adaptations they take advantage of the security of these habitats. A small number of species have evolved to live in more constantly wet habitats. Several species may be found in the marine intertidal, either living in and under piles of decomposing litter along the high-tide line, digging into beach sand, or sheltering in the damp cracks and crevices of rocky shores. A few may also be found in mangrove swamps.

A breakdown of families, genera, and species is not provided for this suborder, but those few species that are commonly encountered in intertidal habitats are dealt with individually. Schultz (1974, 1984) records several oniscidean isopods from the Caribbean area.

Key to genera and species of littoral Oniscidea

1. At least one uropodal ramus reaching well beyond outline of body ... 5
   Uropodal rami very short, not reaching beyond outline of body 2

2. Uropods ventral, not visible in dorsal view .... 3 *(Tylos)*
   Uropods visible in dorsal view .... *Armadilloniscus ninae*

3. Ventral extensions of pleonite 5 meeting in midline .... *Tylos niveus*
   Ventral extensions of pleonite 5 not meeting in midline 4

4. Ventral extensions of pleonite 5 very short, obsolete ..... *Tylos wegeneri*
   Ventral extensions of pleonite 5 well separated .... *Tylos marcuzzii*
   Ventral extensions of pleonite 5 just falling short of meeting in midline
   .... *Tylos latreillei*

5. Uropodal rami both elongate, subequal .... 6 *(Ligia)*
   Uropodal rami very unequal in length 8

6. Propodus of ♂ pereopod 1 with distal rounded lobe ...... *Ligia exotica*
   Propodus of ♂ pereopod 1 lacking rounded lobe 7

7. Apex of ♂ pleopod 2 club shaped .... *Ligia olfersii*
   Apex of ♂ pleopod 2 bifid .... *Ligia baudiniana*

8. Antennal flagellum of two articles .... *Rhyscotus texensis*
   Antennal flagellum of three articles .... 9 *(Vandeloscia)*

9. Endopod of ♂ pleopod 1 with large scalelike subapical process
   .... *Vandeloscia riedli*
   Endopod of ♂ pleopod 1 with small scalelike subapical process
   .... *Vandeloscia culebrae*

Armadilloniscus Ul'yanin, 1875

Armadilloniscus ninae Schultz, 1984
   Figure 106A

DIAGNOSIS   ♂ 3.2 mm, ♀ 4.1 mm. Uropodal sympod expanded to form part of body outline; rami set mesial to expanded base, with exopod half length of endopod.

RECORDS   Ambergris Cay, Belize; under damp objects along beach drift line.

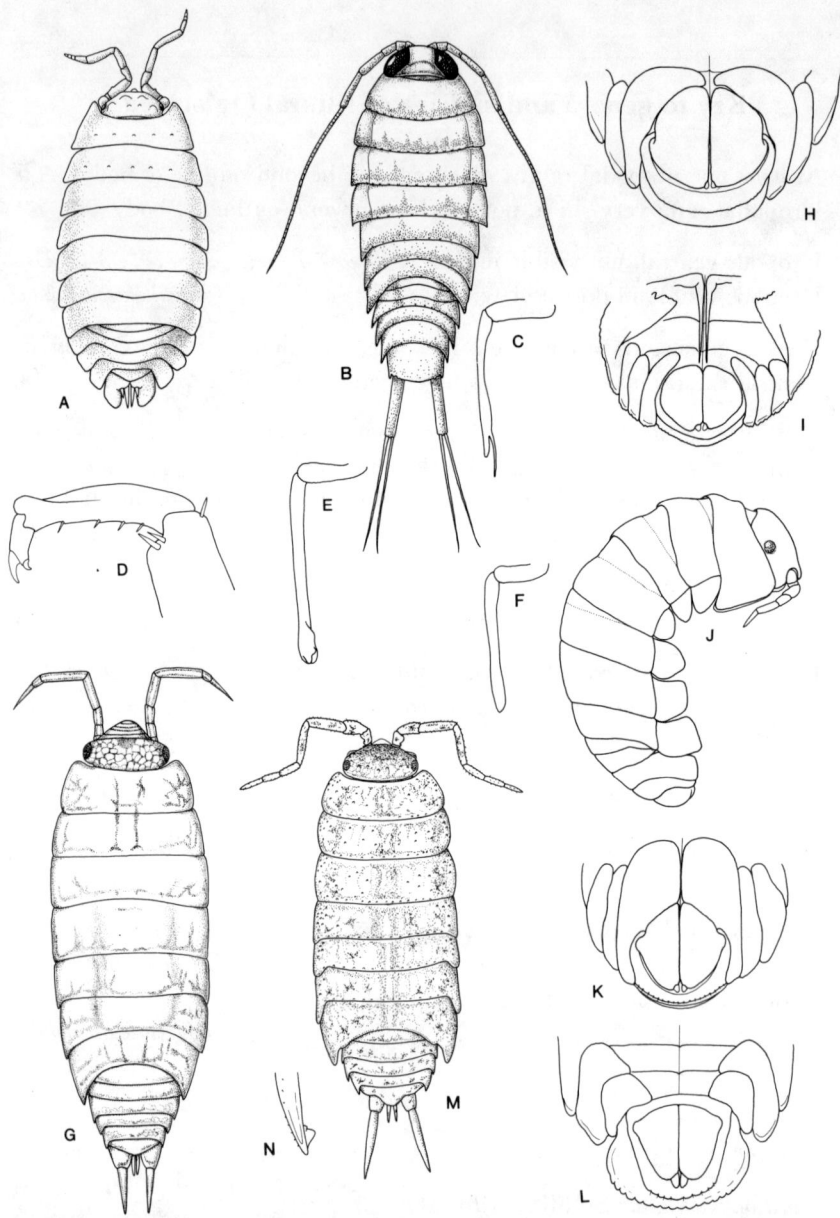

Figure 106. *A, Armadilloniscus ninae. Ligia baudiniana: B; C,* ♂ pleopod 2 endopod. *Ligia exotica: D,* dactylus and propodus of pereopod 1; *E,* ♂ pleopod 2 endopod. *Ligia olfersii: F,* ♂ pleopod 2 endopod. *G, Rhyscotus texensis. Tylos latreillei: H,* ventral pleon. *Tylos marcuzzi: I,* ventral pleon (from Schultz, 1984). *Tylos niveus: J,* lateral view; *K,* ventral pleon. *Tylos wegeneri: L,* ventral pleon. *Vandeloscia culebrae: M,* apex of pleopod 1 endopod. *N, Vandeloscia riedli.*

Ligia Fabricius, 1798
Ligia baudiniana H. Milne Edwards, 1840
  Figure 106B,C

DIAGNOSIS ♂ and ♀ up to 22 mm. Antennal flagellum elongate, multiarticulate. Apex of ♂ pleopod 2 bifid, with lateral lobe longer and more slender than mesial lobe. Uropods inserted terminally on pleotelson; sympods elongate-cylindrical; rami slender, elongate, subequal.

RECORDS Bermuda; Bahamas; U.S. Virgin Islands; Antigua; Carrie Bow Cay, Belize; Bonaire; Aruba; Trinidad; Tobago; Gulf of Mexico.

REMARKS As is typical in the genus *Ligia*, this species may be seen on rocks and sea walls, as well as piles of drift debris at low tide. When disturbed, they run rapidly, to shelter in damp crevices and hollows.

Ligia exotica Roux, 1828
  Figure 106D,E

DIAGNOSIS ♂ 28.5 mm, ovigerous ♀ 32.0 mm. Propodus of ♂ pereopod 1 with rounded lobe on inner distal surface. Apex of ♂ pleopod 2 club shaped, convoluted.

RECORDS New Jersey to Uruguay; Indo-Pacific.

Ligia olfersii Brandt, 1833
  Figure 106F

DIAGNOSIS ♂ 20.0 mm, ovigerous ♀ 24.0 mm. Apex of ♂ pleopod 2 simple, club shaped.

RECORDS South Florida to Rio de Janeiro, Brazil; Texas, Gulf of Mexico.

Rhyscotus Budde-Lund, 1885
Rhyscotus texensis (Richardson, 1905)
  Figure 106G

DIAGNOSIS ♂ and ♀ 6.0 mm. Antennal flagellum of two unequal articles. Uropodal endopod at least twice length of exopod, inserted distally on base, exopod inserted distally on base. Pleotelson broadly triangular.

RECORDS Carrie Bow Cay, Belize; Texas, Gulf of Mexico.

Tylos Latreille, 1826
Tylos latreillei Audouin, 1826
   Figure 106H

DIAGNOSIS   ♂ 12.8 mm, ♀ 13.0 mm. Ventral extensions of pleonite 5 not meeting in midline.

RECORDS   Bermuda; Cuba; Puerto Rico; Honduras. Mediterranean.

Tylos marcuzzii Soika, 1954
   Figure 106I

DIAGNOSIS   ♂ 6.6 mm. Antennal flagellum of four articles. Ventral extensions of pleonite 5 well separated.

RECORDS   Florida Keys; Bahamas; Leeward Islands; Ambergris Cay, Belize; under debris on sand beach drift line.

Tylos niveus Budde-Lund, 1885
   Figure 106J,K

DIAGNOSIS   ♂ 11.0 mm., ♀ 12.0 mm. Antennal flagellum of four articles. Ventral extensions of pleonite 5 expanded, medially contiguous.

RECORDS   Bahamas; Florida Keys; Cuba; Dominica; Lesser Antilles; Bonaire; Curaçao, under piles of decaying mangrove leaves at beach drift line; Carrie Bow Cay, Ambergris Cay, Belize, under deep piles of dead plant material on beach drift line; Tobago; Panama.
   Rio de Janeiro, Brazil.

Tylos wegeneri Vandel, 1952
   Figure 106L

DIAGNOSIS   ♂ 10.5 mm, ♀ 15 mm. Antennal flagellum of three articles. Ventral extensions of pleonites short or nearly absent. Pleonite 5 lacking free lateral margins.

RECORDS   Tobago; Venezuela, under decaying beach debris on drift line; Trinidad.

Vandeloscia Roman, 1977
Vandeloscia culebrae (Moore, 1901)
    Figure 106M

DIAGNOSIS ♂ 5.0 mm, ♀ 6.1 mm. Tiny lateral tubercles present on pereonites. Endopod of pleopod 1 in ♂ with small scalelike subapical process on laterally folded tip.

RECORDS Florida Keys; U.S. Virgin Islands; Puerto Rico; under decaying plant material, especially *Thalassia testudinea* accumulated along beach drift line.

Vandeloscia riedli (Strouhal, 1966)
    Figure 106N

DIAGNOSIS ♂ 5.9 mm, ♀ 6.0 mm. Tiny obsolete tubercles present on all pereonites. Endopod of ♂ pleopod 1 with large scalelike subapical process on laterally folded tip.

RECORDS Yucatan Peninsula, Mexico; Ambergris Cay, Belize; Barbuda; Venezuela; Brazil.

  Gulf of Aqaba; Red Sea; northeastern coast of Africa; Madagascar; Bay of Bengal; St. Helena Is.

Suborder Valvifera Sars, 1882

DIAGNOSIS Pereopodal coxae, in addition to usual dorsal coxal plates, expanded ventrally to form plates. Penes situated ventrally on articulation between pereon and pleon, or on pleonite 1. Pleonites and pleotelson variously fused. Uropods forming operculum covering over pleopods.

Key to families of Valvifera

1. Body often geniculate, flexed between pereonites 4 and 5; anterior pereopods setose for feeding, posterior pereopods ambulatory
    .... Arcturidae
   Body never geniculate; all pereopods ambulatory .... Idoteidae

REMARKS   Of the six families in the suborder, only two have been recorded in the Caribbean area, the Idoteidae and the Arcturidae.

Family Arcturidae Sars, 1897

DIAGNOSIS   Pereonite 1 either distinct, or completely or imcompletely fused with cephalon. Anterior four pairs of pereopods directed anteriorly, usually strongly setose; posterior three pairs of pereopods ambulatory, used for clinging to substrate. Body often bent between pereonites 4 and 5. Uropods usually biramous, with minute endopod concealed by larger exopod. Pleonites variously fused with pleotelson. Sexual dimorphism often marked.

REMARKS   Menzies and Kruczynski (1983) described three species of arcturids from the west coast of Florida, in depths of 55–73 m: *Arcturella spinata, Arcturella bispinata,* and *Edwinjoycea horologium.* These species are not covered here.

Key to genera of Arcturidae

1. Pereonite 1 not fused with cephalon; at least one free pleonite
   .... *Thermarcturus*
   Pereonite 1 fused with cephalon; pleonites fused with pleotelson
   .... *Astacilla*

Astacilla Cordiner, 1793

DIAGNOSIS   Antennae at least half length of body. Pereopod 1 with strong terminal claw on dactylus. Pereopods 2–4 lacking dactyli. Endopod of

Key to species of *Astacilla*

1. Body integument lacking ornamentation .... *cymodocea*
   Body integument with spines or tubercles 2
2. Pereonite 4 in ♂ and ♀ with strong middorsal tubercle; pairs of spines lacking on pereonites .... *regina*
   Pereonite 4 lacking strong middorsal tubercle; pairs of spines on all pereonites .... *lasallae*

pleopod 1 ♂ with median notch and three specialized setae; pleopod 2 copulatory stylet apically trifid. Pereonite 4 considerably longer than preceeding or following pereonite.

Astacilla cymodocea Menzies and Glynn, 1968
Figure 107A,B

DIAGNOSIS ♂ 6.4 mm, ovigerous ♀ 9.0 mm. Body cylindrical, ovigerous ♀ with pereonite 4 somewhat bulged, ♂ with pereonite 4 elongate-cylindrical. Shallow groove marking fusion between cephalon and pereonite 1. Pleonites fused with pleotelson, with two incomplete shallow dorsal grooves marking lines of fusion anteriorly. Pleotelson lacking any shoulders, posteriorly tapered to narrowly rounded apex.

RECORDS Florida Keys; Puerto Rico, 1.5 m, on *Cymodocea* sp. seagrass; Carrie Bow Cay, Belize, 1–2 m, on *Syringodium filiforme* seagrass.

REMARKS In life, *A. cymodocea* is bright green, blending in with its preferred substrate of seagrasses.

Astacilla lasallae Paul and Menzies, 1971
Figure 107C

DIAGNOSIS ♀ 3.5 mm. Cephalon with large rounded area bearing pair of spines; all pereonites and two anterior fused pleonites bearing pair of short submedian spines. Pleotelson with strong anterior shoulder, posteriorly triangular, tapering sharply to narrowly rounded apex.

RECORD Off Venezuela, 95 m.

REMARKS This species is known only from the small female holotype, and until a mature male and ovigerous female are found, it cannot be confidently diagnosed.

Astacilla regina Kensley, 1984
Figure 107D–G

DIAGNOSIS ♂ 6.5 mm, ovigerous ♀ 7.1 mm. Body strongly tuberculate, many tubercles acute. Cephalon with two submedian pairs of acute tubercles; fused pereonite 1 and pereonites 2 and 3 each with single middorsal acute tubercle. Pereonite 4 with strong middorsal tubercle situated in ante-

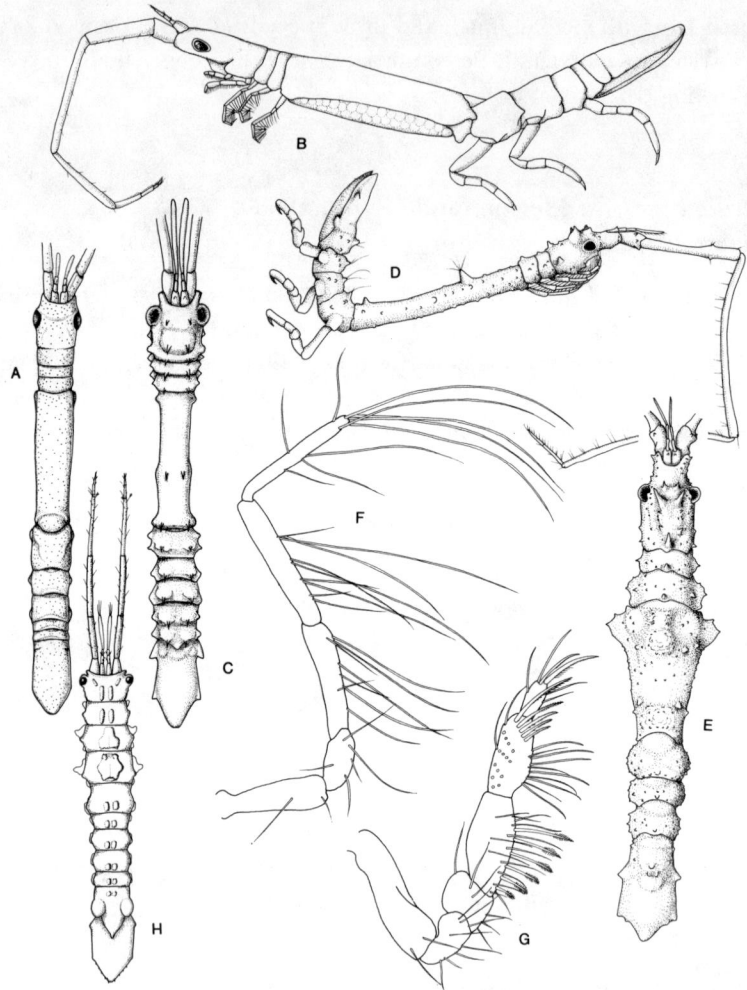

Figure 107. *Astacilla cymodocea:* A, ♂; B, ♀. *Astacilla lasallae:* C, ♂. *Astacilla regina:* D, ♂; E, ♀; F, pereopod 4; G, pereopod 1. *Thermarcturus venezuelensis:* H, ♀ (from Paul and Menzies, 1971).

rior half. Pleotelson with strong lateral shoulder in anterior half, second shoulder in posterior half, then tapering to rounded apex.

RECORDS   Carrie Bow Cay, Belize, on forereef slope, 27–36 m; Barbados, 100–400 m; St. Lucia, 2–3 m, associated with crinoids.

Thermarcturus Paul and Menzies, 1971

DIAGNOSIS   Pereonite 1 not fused with cephalon. Pereonite 4 subequal in length to pereonite 3, not markedly elongate. Pereopods 2–4 having dactyli but lacking elongate setae. Body cylindrical, flexed between pereonites 4 and 5. Pleon consisting of two free pleonites plus pleotelson.

Thermarcturus venezuelensis Paul and Menzies, 1971
   Figure 107H

DIAGNOSIS   ♀ 4.5 mm. Cephalon, all pereonites, and anterior two pleonites each with submedian pair of dorsal tubercles, those on pereonites 2 and 3 broad and expanded. Pleonite 2 with pair of bulbous lateral swellings, posterior margin triangular. Pleotelson with lateral shoulder anteriorly, posteriorly triangular.

RECORDS   Off Venezuela, 95 m.

REMARKS   Only the holotype (which seems to be lost) is known of this species. Considerable uncertainty exists regarding some of the features.

Family Idoteidae Fabricius, 1798
Subfamily Idoteinae Dana, 1852

DIAGNOSIS   Flagellum of antenna either multiarticulate; clavate, i.e., with large basal articles and with or without one to four reduced distal articles; or

Key to genera of Idoteinae

1. Antennal flagellum multiarticulate .... *Idotea*
   Antennal flagellum clavate 2

2. Pereopod 4 reduced, considerably smaller than pereopods 3 or 5 3
   Pereopod 4 not reduced, of similar size to pereopods 3 and 5
      .... *Erichsonella*

3. Pleon consisting of three complete and one incomplete pleonites plus
      pleotelson .... *Cleantioides*
   Pleon consisting of two complete and two incomplete pleonites plus
      pleotelson .... *Miratidotea*

vestigial. Maxillipedal palp consisting of five or fewer articles. Uropods uniramous or biramous, rami usually much smaller than sympod. Pleonites variously fused with pleotelson; number of fused pleonites often indicated by lateral sutures or furrows.

REMARKS   Brusca (1984) has reviewed the phylogeny, evolution, and biogeography of the subfamily Idoteinae, the only one of the five subfamilies recorded from the Caribbean.

Cleantioides Kensley and Kaufman, 1978

DIAGNOSIS   Antennal flagellum a single clavate article. Maxillipedal palp of four or five articles. Pereopod 4 somewhat reduced. Uropod uniramous. Pleon consisting of three complete and one incomplete pleonites plus pleotelson.

Cleantioides planicauda (Benedict, 1899)
Figure 108A

DIAGNOSIS   Ovigerous ♀ 5.5 mm. Body parallel sided. Maxillipedal palp of five articles. Pleotelson posteriorly broadly rounded, with obliquely truncate subcircular dorsal area in posterior half.

RECORDS   Maryland to Florida; Puerto Rico; Panama; Louisiana, Gulf of Mexico, intertidal to 44 m; often in hollow stems and roots of seagrasses, and tubes of the polychaete *Diopatra cuprea*.
   Oaxaca, Pacific Mexico.

REMARKS   *Cleantioides planicauda* has been recorded only once in the eastern Pacific, where it occurs with the more common *C. occidentalis* (Richardson, 1899).

Key to species of *Erichsonella*

1. Pereonites with dorsal spines 2
   Pereonites lacking dorsal spines .... *attenuata*
2. Pereonites 1–4 with middorsal and lateral spines .... *floridana*
   Pereonites 1–4 with middorsal spines only .... *filiformis*

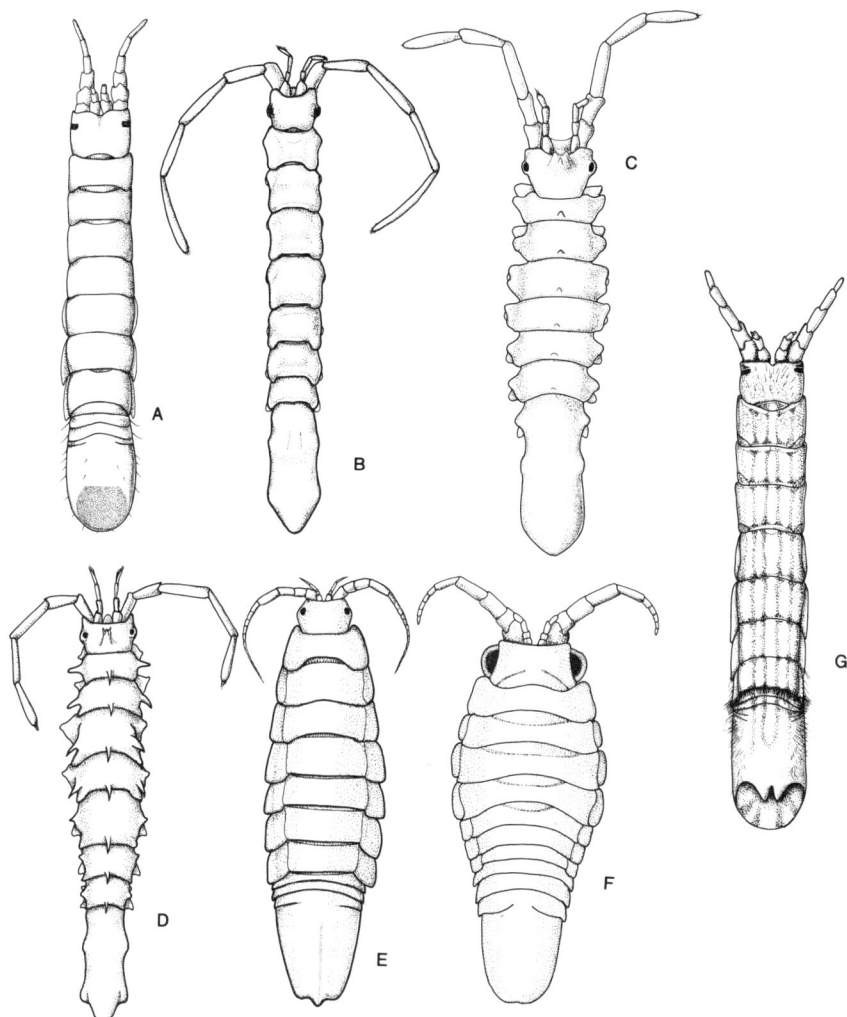

Figure 108. *A, Cleantioides planicauda; B, Erichsonella attenuata* ♂; *C, Erichsonella filiformis* ♂; *D, Erichsonella floridana* ♀; *E, Idotea balthica* ♂; *F, Idotea metallica* ♀; *G, Miratidotea bruscai* ♀.

Erichsonella Richardson, 1901

DIAGNOSIS   Antennal flagellum clavate. Maxillipedal palp of four articles. Uropod uniramous. Pleonites completely fused with pleotelson.

REMARKS   Pires (1984) reviewed the genus *Erichsonella* and did not recognize the subspecies *E. filiformis tropicalis* Menzies and Glynn, 1968.

Erichsonella attenuata (Harger, 1873)
Figure 108B

DIAGNOSIS ♂ 11.4 mm, ovigerous ♀ 12.0 mm. Body dorsally smooth. Cephalon lacking middorsal elevation. Antennule reaching only slightly beyond antennal peduncular article 2. Pleotelson with slight marginal bulge in anterior half, indicating ventrolateral articulation of uropod.

RECORDS Connecticut to Miami; Florida, Mississippi, Texas, Gulf of Mexico; intertidal to 2 m, usually associated with submerged seagrass and algal beds.

REMARKS While not recorded in the Florida Keys, this species does reach Miami, and continues into the Gulf of Mexico.

Erichsonella filiformis (Say, 1818)
Figure 108C

DIAGNOSIS ♂ 10.5 mm, ovigerous ♀ 8.2 mm. Body dorsally with bifid tubercle on cephalon, and low rounded middorsal tubercle on pereonites. Antennule reaching midlength of antennal peduncular article 3. Basis of pereopods 2–7 with larges tubercles. Pleotelson with distinct lateral shoulder in anterior half.

RECORDS Connecticut to Florida, shallow infratidal to 55 m; Bahamas; Turks and Caicos Islands; Puerto Rico; Quintana Roo, Yucatan Peninsula, Mexico, 60–109 m; Florida and Texas, Gulf of Mexico.
Brazil.

Erichsonella floridana Richardson, 1901
Figure 108D

DIAGNOSIS Ovigerous ♀ 10.0 mm. Antennule reaching distal end of antennal peduncular article 3. Cephalon with strong trifid tubercle. Pereonites 1–7 each with posteriorly directed spine near posterior margin; pereonites 1–4 each with lateral spine. Basis of pereopods 2–7 smooth.

RECORDS Florida Keys, intertidal to 2 m; Florida, Gulf of Mexico, intertidal mud flats.

Idotea Fabricius, 1798

DIAGNOSIS Antennal flagellum multiarticulate. Maxillipedal palp of four or five articles. Uropod uniramous. Pleon consisting of two complete and one incomplete pleonites plus pleotelson.

Key to species of *Idotea*

1. Posterior margin of pleotelson truncate .... *metallica*
   Posterior margin of pleotelson with distinct median lobe ...... *balthica*

Idotea balthica (Pallas, 1772)
  Figure 108E

DIAGNOSIS ♂ 24.5 mm, ovigerous ♀ 13.2–23.5 mm. Anterior margin of cephalon concave. Cephalon dorsally smooth. Pereonites evenly convex, smooth. Posterior margin of pleotelson with rounded median lobe.

RECORDS Worldwide in tropical to cold-temperate waters, often on floating seaweed, from surface to 357 m.

Idotea metallica Bosc, 1802
  Figure 108F

DIAGNOSIS ♂ 30.0 mm, ovigerous ♀ 22.2 mm. Cephalon with sinuous furrow in posterior half. Pereonites 2–4 laterally with rounded convex area close to coxae. Posterior margin of pleotelson truncate.

RECORDS Worldwide in tropical to cold-temperate waters, often on floating seaweed, from surface to 200 m.

Miratidotea Kensley, 1987a

DIAGNOSIS Antennal flagellum of single clavate article. Maxillipedal palp of four articles. Uropod uniramous. Pleon consisting of two complete and two incomplete pleonites plus pleotelson.

Miratidotea bruscai Kensley, 1987a
Figure 108G

DIAGNOSIS   Ovigerous ♀ 13.0 mm. Body parallel sided. Maxillipedal palp of four articles, terminal article very short. Pereopods 1–3 increasing in length posteriorly, pereopod 4 reduced, shorter than pereopod 5, and with dactylus spinelike, pereopods 5–7 increasing in length. Pleotelson consisting of two complete and two incomplete pleonites plus pleotelson; latter with broadly rounded posterior margin, and with bifid median process situated dorsal to posterior oblique-concave area.

RECORDS   Carrie Bow Cay, Belize, 1.5 m, in hollow root-internodes of seagrass *Syringodium filiforme*.

Zoogeography

FAUNAL PROVINCES

The area under discussion has been divided into several faunal regions or provinces, of which the Caribbean, West Indian, and Brazilian are the major ones (Briggs, 1974). The extent and boundaries of the provinces have been variously defined depending on the group of organisms under discussion. Inevitably, zones of overlap exist, but for the purposes of this discussion, the following rough limits have been used.

Brazilian Province: This province stretches from Cape Frio near Rio de Janeiro in Brazil to the mouth of the Orinoco River in Venezuela. The outflow of freshwater from the major rivers of this region has probably contributed to the isolation of the Brazilian coral reefs and their associated fauna from those of the Caribbean. This isolation is demonstrated by the considerable endemism of the Brazilian reef fauna and that of the Caribbean reef fauna, with very few species being common to both.

Caribbean Province: This province has two components, a northern part in peninsular Florida, that stretches from around Cape Kennedy on the east coast to Tampa or Sanibel Island on the west coast, and a southern component that runs from the mouth of the Orinoco River to around Cabo Rojo or Tampico on the gulf coast of Mexico. The northern Gulf of Mexico is excluded from this province and is characterized as being warm-temperate, rather than subtropical (Briggs, 1974:66).

West Indian Province: This includes all the islands of the West Indian chain, the Bahamas, and the isolated outrider, Bermuda. The West Indian Province closely approaches the Caribbean Province in the Yucatan Peninsula to the north, and between Grenada and Trinidad in the south. There is also some indication of the isolating effect on the Bahamas of the Florida Current through the Straits of Florida.

It has been suggested, on the basis of the molluscan fauna, that a relict of the Neogene Gatunian Province exists around northern Venezuela and Colombia (Petuch, 1982). While several isopod species have been recorded only from this area, these are all described in a single paper that covers a very small part of this region (Paul and Menzies, 1971). There is as yet too little evidence to explore the idea of this relict fauna further.

ANALYSIS OF THE ISOPOD FAUNA

In the following discussion, the West Indian and Caribbean provinces are treated as one, the isopod faunas offering little evidence to warrant a separate treatment of each.

It is a truism that for any discussion of the zoogeography of an area to have meaning, the true extent of the fauna must be known. With the area under review, collecting effort has been uneven, and the true faunal composition of many regions is still incompletely known. Obviously, any conclusions based on such incomplete data are approximate and subject to revision. Nevertheless, certain general patterns or trends emerge when the present isopod fauna is broken down into its components.

The deepwater isopod fauna of the Caribbean (i.e., from deeper than 200 m) has barely been explored, and little is to be gained from discussing the relatively few species known. A list of these deeper dwelling species is included (Table 4).

Although about 280 shallow-water species are covered by this work, certain categories of species must be excluded, for various reasons, before analysis can be attempted. Such excluded groups include the species of Oniscidea (being essentially terrestrial forms and not part of the marine regime); the cymothoid species and the species of Aegidae (being fish parasites for at least part of their life history, and whose distribution is complicated by the distribution and mobility of the hosts); the limnoriids (being wood-borers whose distribution is more a function of the distribution of floating wood); and the true cave species (which have a history more reflective of the geological history of the area than of the marine regime). The epicarideans have a distribution somewhat complicated by the distribution of their crustacean hosts and their pelagic epicaridean and cryptoniscan larvae. Nevertheless, the decapod hosts of the great majority of species covered here are Caribbean endemics, and inclusion of the epicarideans changes very little the overall patterns of distribution, as demonstrated by the two figures provided (Figure 109). After making these exclusions there remain about 166 species (218 with the epicarideans) that can be broken down into the following components (figures in brackets include epicarideans):

1. True Caribbean/Bahamian species—124 species, 74.8% [147, 67.5%]. These are the species recorded only from the Caribbean and the Bahamas. The term endemic is avoided, as too little is known of the actual distribution of many species. Of these species, 86 [87] have been recorded from a single locality.

2. Species occurring south of the discussion area, and extending into Brazil—5 species, 3.0% [9, 4.1%]. These low numbers indicate that the

TABLE 4. CARIBBEAN ISOPODS RECORDED FROM DEPTHS GREATER THAN 200 M

SUBORDER ANTHURIDEA
Family Paranthuridae
  Neoanthura coeca Menzies, 1956b. South of Jamaica, 1244 m
SUBORDER ASELLOTA
Family Dendrotiidae
  Dendrotion hanseni Menzies, 1956b. South of Jamaica, 1244 m
Family Desmosomatidae
  Desmosoma magnispina Menzies, 1962a. Bay of Panama, 1906 m
Family Echinothambematidae
  Echinothambema ophiuroides Menzies, 1956a. North of Puerto Rico Trench, 5104–5122 m
Family Eurycopidae
  Acanthocope spinosissima Menzies, 1956b. South of Jamaica, 1224 m
  Storthyngura pulchra caribbea (Benedict, 1901). Off Windward Islands, 1256 m
  Storthyngura snanoi Menzies, 1962a. Colombia abyssal plain, 4071 m
Family Haploniscidae
  Antennuloniscus dimeroceras (Barnard, 1920). North of Puerto Rico Trench, 5440–5410 m; South Atlantic off South and West Africa, 1400–3921 m; off Argentina, 5843 m
  Haploniscus unicornis Menzies, 1956a. North of Puerto Rico Trench, 5104–5122 m
  Hydroniscus quadrifrons Menzies, 1962a. North of Puerto Rico Trench, 5271–5684 m
Family Ischnomesidae
  Haplomesus tropicalis Menzies, 1962a. Colombia abyssal plain, 4071 m; off South Africa, 2526 m; Mediterranean
  Heteromesus bifurcatus Menzies, 1962a. Colombia abyssal plain, 4071 m
  Ischnomesus armatus Hansen, 1916. North of Puerto Rico Trench, 5494–5477 m; Davis Straits, 2702 m
  Ischnomesus caribbicus Menzies, 1962a. Off Panama, 1714 m
  Ischnomesus multispinis Menzies, 1962a. Off Panama, 975 m
Family Janiridae
  Abyssianira dentifrons Menzies, 1956a. North of Puerto Rico Trench, 5104–5122 m; off Argentina, 5024–5293 m; off southwest Africa, 4588 m
  Ianirella caribbica Menzies, 1956b. South of Jamaica, 1244 m
  Ianirella vemae Menzies, 1956a. Near Puerto Rico Trench, 5104–5122 m
  Spinianirella serrata Kensley and Heard, 1985. Off Puerto Rico, 350 m
Family Macrostylidae
  Macrostylis caribbicus Menzies, 1962a. Off Colombia, 2875–2941 m
  Macrostylis minutus Menzies, 1962a. North of Puerto Rico Trench, 5163–5494

(continued)

TABLE 4. (*Continued*)

Macrostylis setifer Menzies, 1962a. North of Puerto Rico Trench, 5477–5494 m
Macrostylis vemae Menzies, 1962a. North of Puerto Rico Trench, 5410–5684 m
Family Mesosignidae
Mesosignum kohleri Menzies, 1962a. Colombia abyssal plain, 2868–4076 m
Family Nannoniscidae
Nannoniscus camayae Menzies, 1962a. Off Panama, 1714 m
SUBORDER GNATHIIDEA
Family Gnathiidae
Akidognathia poteriophora Monod, 1926. Off U. S. Virgin Islands, 914 m.
SUBORDER VALVIFERA
Family Arcturidae
Antarcturus annaoides Menzies, 1956b. South of Jamaica, 1244 m
Arcturus caribbaeus Richardson, 1901. Off Aves Island, 1360 m
Arcturus purpureus Beddard, 1886. Off Leeward Islands, 900 m

Note: Records from deep water around Bermuda are not included.

great area of mixed-salinity waters resulting from the outflow of the Orinoco, Amazon, Tocantins, and Parnaiba rivers form an effective barrier to the movement of shallow-water isopod species.

3. Species having an amphi-Panamic distribution—7 species, 4.2% [8, 3.7%] (Table 5). In spite of the history of immergence and emergence of the Isthmus of Panama, this very small amphi-Panamic component in the Caribbean isopod fauna suggests that most of this fauna has evolved since the last emergence of the late Pliocene. Given the limited mobility of most isopod species, the Panama Canal seems to have played a minimal role in contributing to this component.

4. Species occurring outside of the western Atlantic (but excluding the amphi-Panamic species)—3 species, 1.8% [7, 3.2%].

5. The role of the Gulf of Mexico isopod fauna (see Clark and Robertson, 1982) in the composition of the Caribbean/Bahamian is complex and difficult to analyze. One hundred and thirteen species of shallow-water isopods have been recorded from the Gulf of Mexico (Table 6). This number would indicate that many species remain to be recorded in this region. Of these 113 species, 61 (54%) have also been reported from the Caribbean region. It is therefore possible that there exists a true Gulf of Mexico fauna, whose evolution was perhaps spurred by the relative isolation and reduction of the Gulf

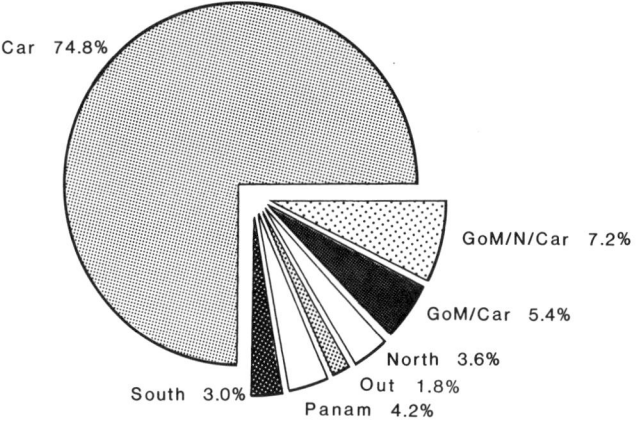

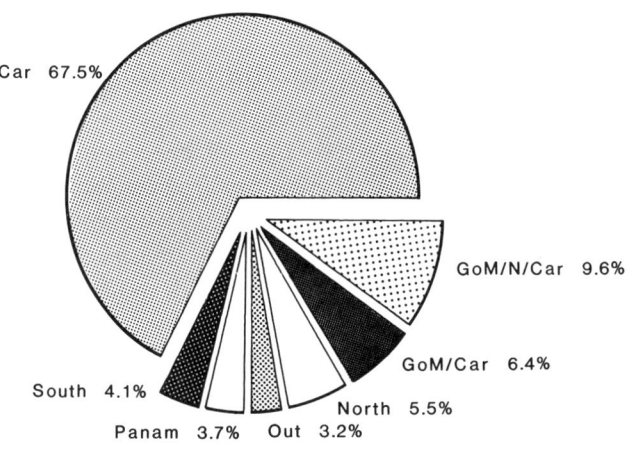

Figure 109. Relative proportions of the zoogeographic components of the Caribbean isopod fauna, with and without the parasitic Epicaridea. *Car*, Caribbean; *Out*, extra-western Atlantic; *GoM/Car*, Gulf of Mexico–Caribbean; *GoM/N/Car*, Gulf of Mexico-Northern-Caribbean; *North*, northern; *Panam*, amphi-Panamic; *South*, southern.

during a low-water stand (100 m below present sea level) during the Pleistocene. A significant proportion (about 27 species, 28%) of the Gulf of Mexico isopods are known from the eastern coast of the United States north

TABLE 5. SPECIES OF ISOPODS OCCURRING ON BOTH SIDES OF THE ISTHMUS OF PANAMA

Aega deshaysiana (H. Milne Edwards, 1840)	*Paradella dianae* (Menzies, 1962b)
Anopsilana browni (Van Name, 1936)	*Paraleptosphaeroma glynni* Buss and Iverson, 1981
Cleantioides planicauda (Benedict, 1899)	*Probopyrus pandalicola* (Packard, 1879)
Excirolana braziliensis Richardson, 1912	**Rocinela oculata* Harger, 1883
Excorallana tricornis (Hansen, 1890)	**Rocinela signata* Schioedte and Meinert, 1879
**Nerocila acuminata* Schioedte and Meinert, 1881	*Uromunna reynoldsi* Frankenberg and Menzies, 1966

* fish parasite or fish predator

of Cape Kennedy, which would indicate a significant cooler-water component. What proportion of originally Gulf species have spread into the Caribbean, and what proportion of Caribbean and temperate east coast species have entered the Gulf, cannot yet be assessed, given our incomplete knowledge of the Gulf fauna. Because of this unresolved situation, three categories of species have been separated: species ranging from north of Cape Kennedy into the Caribbean—6, 3.6% [12, 5.5%]; species occurring in the Gulf of Mexico and the Caribbean—9, 5.4% [14, 6.4%]; species occurring north of Cape Kennedy, in the Gulf, and in the Caribbean—12, 7.2% [21, 9.6%] The conclusion that the fauna of the Gulf of Mexico contains an endemic component, a Caribbean component, and a warm-temperate component was also reached by Topp and Hoff (1972), in an analysis of the pleuronectiform fishes of the Gulf.

THE BAHAMAS

The Florida Current flowing through the Straits of Florida has been suggested as a factor in reducing the movement of shallow-water fauna between peninsular Florida and the Florida Keys on the west and the Bahamas on the east (Briggs, 1974). Comparison of the number of isopod species on either side of the Straits of Florida (13 from the Bahamas, 50 from southern peninsular Florida and the Florida Keys) supports this view. Of the 13 species from the Bahamas, only four are "endemic," three of these being interstitial microcerberideans.

TABLE 6. ISOPOD SPECIES OCCURRING IN THE GULF OF MEXICO

SUBORDER ANTHURIDEA
Accalathura crenulata (Richardson, 1901)
Amakusanthura magnifica (Menzies and Frankenberg, 1966)
Cyathura polita (Stimpson, 1855)
Horoloanthura irpex Menzies and Frankenberg, 1966
Kupellonura formosa (Menzies and Frankenberg, 1966)
**Mesanthura floridensis* Menzies and Kruczynski, 1983
**Mesanthura hopkinsi* Hooker, 1985
**Mesanthura pulchra* Barnard, 1925
Paranthura floridensis Menzies and Kruczynski, 1983
Ptilanthura tricarina Menzies and Frankenberg, 1966
Skuphonura lindae Menzies and Kruczynski, 1983
**Xenanthura brevitelson* Barnard, 1925

SUBORDER ASELLOTA
Carpias floridensis Menzies and Kruczynski, 1983
Gnathostenetrioides pugio Hooker, 1985
**Joeropsis coralicola* Schultz and McCloskey, 1967
**Joeropsis rathbunae* Richardson, 1902
Mexicope kensleyi Hooker, 1985
Munnogonium wilsoni Hooker, 1985
**Pleurocope floridensis* Hooker, 1985
**Santia milleri* (Menzies and Glynn, 1968)
**Stenetrium stebbingi* Richardson, 1902

Uromunna hayesi Robertson, 1978
**Uromunna reynoldsi* Frankenberg and Menzies, 1966

SUBORDER EPICARIDEA
Allodiplophryxus floridanus Markham, 1985
**Aporobopyrina anomala* Markham, 1973
**Azygopleon schmitti* (Pearse, 1932)
**Bopyrina abbreviata* Richardson, 1904
**Bopyrione synalphei* Bourdon and Markham, 1980
**Cancricepon choprae* (Nierstrasz and Brender à Brandis, 1925)
Dactylokepon sulcipes Adkison, 1982
Eophryxus subcaudalis (Hay, 1917)
**Gigantione mortenseni* Adkison, 1984b
Gigantione uberlackerae Adkison, 1984b
**Hemiarthrus synalphei* (Pearse, 1950)
Hyperphrixus castrensis Markham, 1985
**Munidion longipedis* Markham, 1975a
Ovobopyrus alphezemiotes Markham, 1985
Parabopyrella mortenseni (Nierstrasz and Brender à Brandis, 1929)
**Parabopyrella richardsonae* (Nierstrasz and Brender à Brandis, 1929)
Parabopyriscus stellatus Markham, 1985
**Probopyria alphei* (Richardson, 1900b)
Probopyrinella heardi Adkison, 1984a

(*continued*)

TABLE 6. (*Continued*)

Probopyrinella latreuticola (Gissler, 1882)
Prodajus cf. *bigelowiensis* Schultz and Allen, 1982
Pseudione cognata Markham, 1985
Pseudione upogebiae Hay, 1917
**Schizobopyrina urocaridis* (Richardson, 1904)
**Stegophryxus hyptius* Thompson, 1902
**Synsynella choprae* (Pearse, 1932)
**Synsynella deformans* Hay, 1917
Synsynella integra Bourdon, 1981
**Urobopyrus processae* Richardson, 1904

SUBORDER FLABELLIFERA

*Aega deshaysiana (H. Milne Edwards, 1840)
*Aega ecarinata Richardson, 1898
Aega incisa Schioedte and Meinert, 1879
Alcirona krebsii Hansen, 1890
Ancinus depressus (Say, 1818)
Anilocra acuta Richardson, 1910
Anilocra laticauda H. Milne Edwards, 1840
**Bathynomus giganteus* A. Milne Edwards, 1879
**Cassidinidea ovalis* (Say, 1818)
Ceratothoa transversa (Richardson, 1900b)
Cirolana borealis Lilljeborg, 1851
**Cirolana obtruncata* Richardson, 1901
*Cirolana parva Hansen, 1890
Conilera cylindracea (Montagu, 1804)
**Cymothoa caraibica* Bovallius, 1885
*Cymothoa excisa Perty, 1833
*Cymothoa oestrum (Linnaeus, 1793)
*Cerceis carinata Glynn, 1970
*Eurydice convexa Richardson,1900b
Eurydice littoralis (Moore, 1901)
**Eurydice piperata* Menzies and Frankenberg, 1966

**Excirolana braziliensis* Richardson, 1912a
**Excirolana mayana* (Ives, 1891)
**Excorallana antillensis* (Hansen, 1890)
Excorallana mexicana Richardson, 1905a
**Excorallana tricornis* (Hansen, 1890)
*Harrieta faxoni (Richardson, 1905a)
**Limnoria tuberculata* Sowinsky, 1884
Lironeca ovalis (Say, 1818)
**Lironeca redmanni* Leach, 1818
Lironeca texana Pearse, 1952
Lironeca tropicalis Menzies and Kruczynski, 1983
*Nalicora rapax Moore, 1901
**Nerocila acuminata* Schioedte and Meinert, 1881
Olencira praegustator (Latrobe, 1802)
**Paracerceis caudata* (Say, 1818)
*Paradella dianae (Menzies, 1962b)
Paradynamene benjamensis Richardson, 1905
**Politolana polita* (Stimpson, 1853)
**Rocinela insularis* Schioedte and Meinert, 1879
*Rocinela oculata Harger, 1883
*Rocinela signata Schioedte and Meinert, 1879
*Serolis mgrayi Menzies and Frankenberg, 1966
**Sphaeroma quadridentata* Say, 1818
**Sphaeroma terebrans* Bate, 1866

SUBORDER GNATHIIDEA

Gnathia floridensis Menzies and Kruczynski, 1983

SUBORDER MICROCERBERIDEA

Microcerberus mexicanus Pennak, 1958

SUBORDER VALVIFERA
  Antarcturus floridanus (Richardson, 1900b)
  Arcturella bispinata Menzies and Kruczynski, 1983
  Arcturella spinata Menzies and Kruczynski, 1983
  Astacilla lauffi Menzies and Frankenberg, 1966
  Chiridotea excavata Harper, 1974
  **Cleantioides planicauda* (Benedict, 1899)
  Edotea lyonsi (Menzies and Kruczynski, 1983)
  Edotea montosa (Stimpson, 1853)
  Edwinjoycea horologium Menzies and Kruczynski, 1983
  **Erichsonella attenuata* (Harger, 1873)
  **Erichsonella filiformis* (Say, 1818)
  **Erichsonella floridana* Benedict, 1901
  Erichsonella isabelensis Menzies, 1951b
  *Idotea metallica Bosc, 1802

* species also occurring in the Caribbean
Note: Records for the Gulf of Mexico have been assembled from published literature; in most cases, actual material has not been examined.

BERMUDA

Twenty-nine species of isopods have been recorded from Bermuda (Table 7). Of these, nine are endemics (three being cave forms). The remaining 20 species have all been recorded from the Caribbean region, indicating a strong subtropical connection, in spite of the relatively high latitude (32°15′N). Although Bermuda is of Eocene or Oligocene age, the tropical fauna was probably decimated by the low temperatures of the last Pleistocene glaciation (Briggs, 1974:76).

CAVE ISOPODS

With the expanding efforts of cave divers, more and more true stygobiont forms are being found. Concurrently, discussion of the origin of cave fauna has spurred several theories, all invoking the geological history of the Caribbean area.

Among the isopods, cave forms have been found in four suborders, the Asellota, Anthuridea, Flabellifera, and Microcerberidea. Two valuable discussions on the origin of cave crustaceans may be found in Stock (1986) and Wägele (1985).

The only true cave asellote, *Atlantasellus cavernicolus* Sket, was collected from Bermuda.

TABLE 7. ISOPOD SPECIES OCCURRING AT BERMUDA

Alcirona krebsi Hansen, 1890	*Excorallana quadricornis* (Hansen, 1890)
**Anthomuda stenotelson* Schultz, 1979	*Joeropsis rathbunae* Richardson, 1902
**Apanthura harringtoniensis* Wägele, 1981	*Leidya bimini* Pearse, 1951
	Paracerceis caudata (Say, 1818)
**Arubolana aruboides* (Bowman and Iliffe, 1983)	*Paranthura infundibulata* Richardson, 1902
**Atlantasellus cavernicolus* Sket, 1979	*Pendanthura tanaiformis* Menzies and Glynn, 1968
Bopyrissa wolffi Markham, 1978	
Cancricepon choprae (Nierstrasz and Brender à Brandis, 1925)	*Parathelges piriformis* Markham, 1972b
	Parathelges tumidipes Markham, 1972b
Carpias bermudensis Richardson, 1902	*Probopyrinella latreuticola* (Gissler, 1882)
*Carpias minutus (Richardson, 1902)	
**Colanthura tenuis* Richardson, 1902	*Pseudione affinis* (Sars, 1882)
Colopisthus parvus Richardson, 1902	**Stegias clibanarii* Richardson, 1904
**Curassanthura bermudensis* Wägele, 1985	*Stenetrium stebbingi* Richardson, 1902
	Stenobermuda acutirostrata Schultz, 1979
Dynamenella perforata (Moore, 1901)	*Synsynella choprae* (Pearse, 1932)
Eurydice personata Kensley, 1987b	*Synsynella deformans* Hay, 1917

* recorded only from Bermuda

The anthuridean cave representatives are found in two families: the genus *Curassanthura* Kensley in the Paranthuridae, and the genus *Cyathura* subgenus *Stygocyathura* Botosaneanu and Stock in the Anthuridae (see Figure 110).

Three species of *Curassanthura* are known, one each from Curaçao, Bermuda, and Lanzarote in the Canary Islands. *Curassanthura halma* Kensley, from Curaçao, is an interstitial form found in hypersaline waters. *Curassanthura bermudensis* Wägele was found in water of about 26‰ salinity. The Lanzarote species, *C. canariensis* Wägele, came from seawater in a lava cave. Wägele (1985) suggests that this amphi-Atlantic distribution of *Curassanthura* is the result of plate tectonics separating an ancestral hypogean progenitor that had a Tethyan distribution.

The genus *Cyathura* has representatives in the sea, in estuarine-brackish habitats, and in freshwater caves, and is found in the Atlantic, Indian, and Pacific oceans. This widespread distribution suggests a very long history for the genus. Using the morphology of the male copulatory stylet, Wägele (1985) suggests that marine ancestors, having a Tethyan distribution, entered freshwater interstitial habitats. The series of regressions of sea level during the Pleistocene probably served further to isolate these freshwater forms.

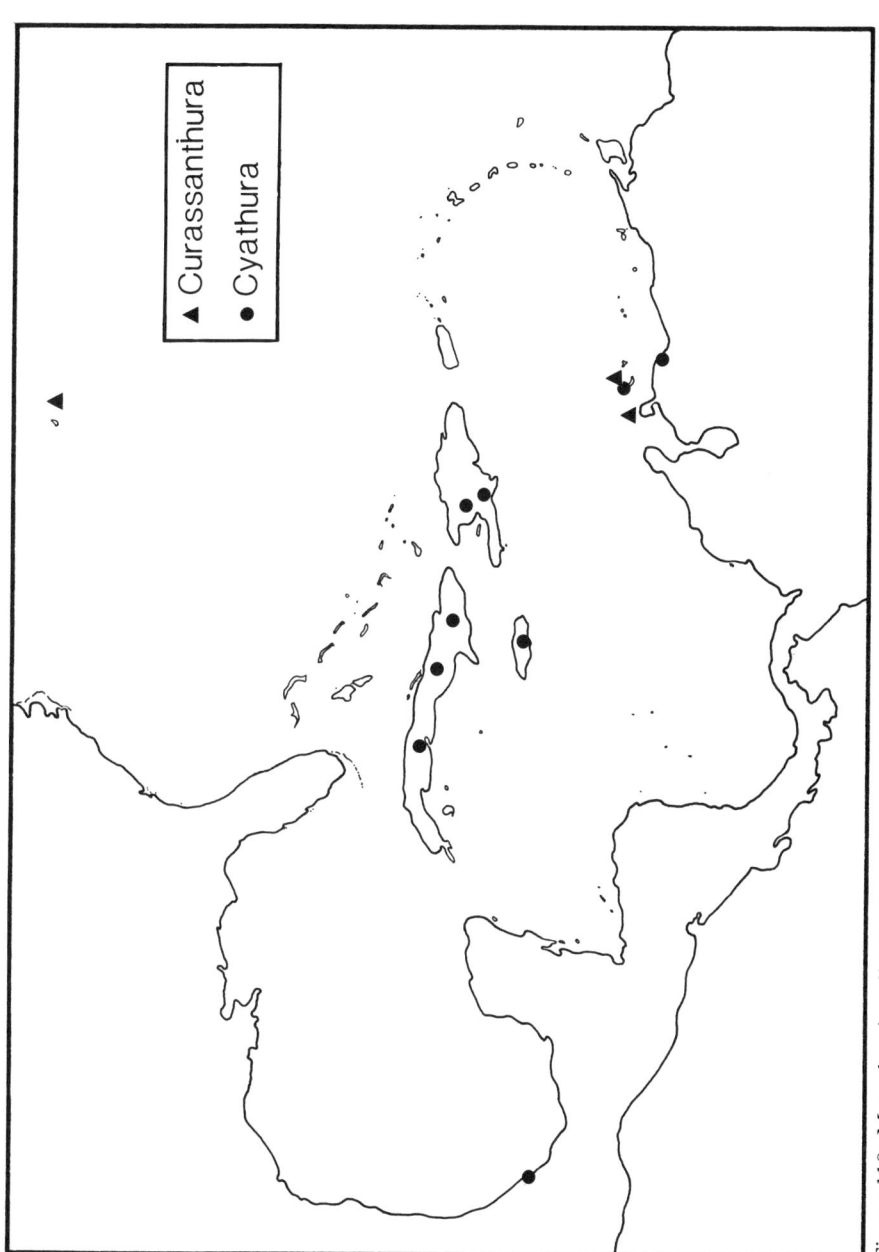
Figure 110. Map showing distribution of cave anthurideans.

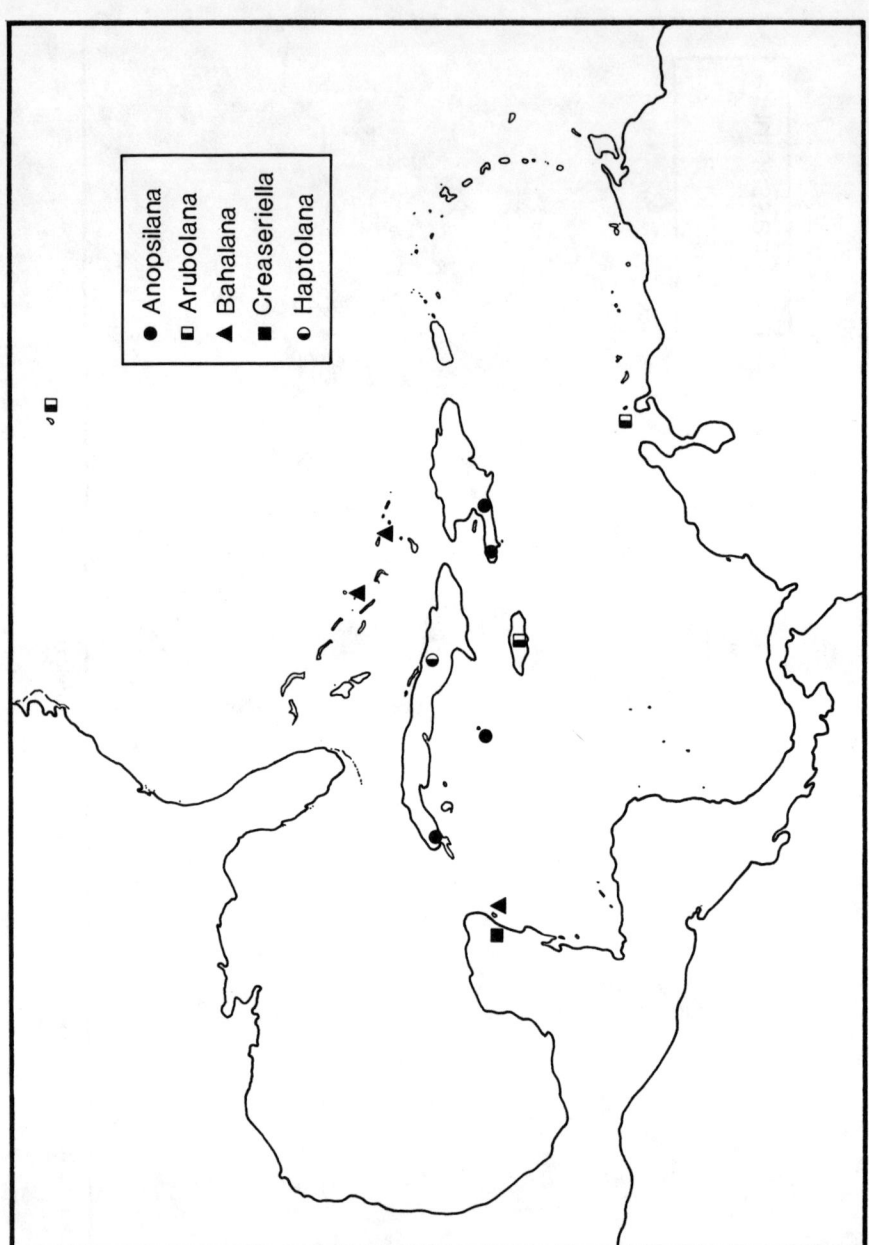

Figure 111. Map showing distribution of cave cirolanids.

The flabelliferan family Cirolanidae contains five stygobiont genera in the Caribbean: *Anopsilana, Arubolana, Bahalana, Creaseriella,* and *Haptolana* (Figure 111). Six other genera are known from the North American continent: *Antrolana, Cirolanides, Mexilana, Speocirolana, Sphaerolana,* and *Troglocirolana,* all of which, except *Antrolana* from the Appalachian Valley of Virginia, occur in Mexico and Texas (see Notenboom, 1981). A few of these forms occur in brackish water, but most are found in freshwater of caves. Cave cirolanids are also known from Palau, North and East Africa, Madagascar, Bulgaria, Greece, Jugoslavia, Israel, France, and Spain. This widespread distribution again suggests a Tethyan marine origin, with dispersal and isolation due to sea regressions.

The suborder Microcerberidea and the asellotan family Microparasellidae contain almost entirely interstitial forms, although few occur in caves. At least two genera, *Microcerberus* and *Angliera,* have very widespread distributions and are known from marine, brackish-water, and freshwater habitats, and may well have a history similar to that of *Cyathura.*

Appendix

Since the manuscript of this work was completed and sent to press, a few papers have appeared either describing new species, mentioning new records for the Caribbean and associated areas, or instituting a major new taxon. It was thought useful to include these, if only in an appendix, to make the work as current as possible. The relevant taxa are listed alphabetically, with the full citation given below.

Antheluridae Poore and Lew Ton, 1988
   Poore, G. C. B., and H. M. Lew Ton. 1988. Antheluridae, a new family of Crustacea (Isopoda: Anthuridea) with new species from Australia. *Journal of Natural History* 22:489–506.
   Within the geographical area covered by this work, only *Anthomuda* belongs to this new family.

Aporobopyrus collardi Adkison, 1988
   Adkison, D. L. 1988. *Pseudione parviramus* and *Aporobopyrus collardi*, two new species of Bopyridae (Isopoda: Epicaridea) from the Gulf of Mexico. *Proceedings of the Biological Society of Washington* 101(3):576–584.

Booralana tricarinata Camp and Heard, 1988
   Camp, D. K., and R. W. Heard. 1988. *Booralana tricarinata*, a new species of isopod from the western Atlantic Ocean (Crustacea: Isopoda: Cirolanidae). *Proceedings of the Biological Society of Washington* 101 (3):603–613.
   Originally recorded from the outer shelf and upper slope off the Little Bahama Bank and the Antilles Islands in 110–610 m, this species has since been recorded off Haiti in 620 m.

Bythognathia yucatanensis Camp, 1988
   Camp, D. K. 1988. *Bythognathia yucatanensis*, new genus, new species, from abyssal depths in the Caribbean Sea, with a list of gnathiid

species described since 1926 (Isopoda: Gnathiidae). *Journal of Crustacean Biology* 8(4):668–678.

Edotea samariensis Müller, 1988
Müller, H. G. 1988. Idoteidae aus N-Kolumbien mit Beschreibung von *Edotea samariensis* n. sp. (Crustacea: Isopoda: Valvifera). *Senckenbergiana biologica* 68(4/6):407–412.

Gnathia johanna Monod, 1926
Müller, H. G. 1988. Redescription of *Gnathia johanna*, 1926 (Isopoda) from St. John, Virgin Islands. *Bulletin Zoologisch Museum, Universiteit van Amsterdam* 11(15):129–133.

Phycolimnoria bacescui Ortiz and Lalana, 1988
Ortiz, M., and R. Lalana. 1988. Una nueva especie del genero *Phycolimnoria* (Isopoda, Limnoriidae) de aguas cubanas. *Revista de Investigaciones Marinas, La Habana* 9(2):37–42.

Pseudione parviramus Adkison, 1988
Adkison, D. L. 1988. *Pseudione parviramus* and *Aporobopyrus collardi*, two new species of Bopyridae (Isopoda: Epicaridea) from the Gulf of Mexico. *Proceedings of the Biological Society of Washington* 101(3):576–584.

Literature Cited

Adkison, D. L. 1982. Description of *Dactylokepon sulcipes* n. sp. (Crustacea: Isopoda: Bopyridae) and notes on *D. caribaeus*. *Proceedings of the Biological Society of Washington* 95(4):702–708.

———. 1984a. *Probopyrinella heardi* n. sp. (Isopoda: Bopyridae) a branchial parasite of the hippolytid shrimp *Latreutes parvulus* (Decapoda: Caridea). *Proceedings of the Biological Society of Washington* 97(3):550–554.

———. 1984b. Two new species of *Gigantione* Kossmann (Isopoda: Epicaridea: Bopyridae) from the western North Atlantic. *Proceedings of the Biological Society of Washington* 97(4):761–772.

Adkison, D. L., and Heard, R. W. 1978. Description of a new genus and species of Pseudioninae (Isopoda: Bopyridae) parasite of the hermit crab *Pagurus annulipes* (Stimpson) from North Carolina. *Proceedings of the Biological Society of Washington* 91(2):408–417.

Amar, R. 1957. *Gnathostenetroides laodicense* nov. gen. nov. sp. Type nouveau d'Asellota et classification des isopodes asellotes. *Bulletin de l'Institut Océanographique* 1100:1–10.

Argano, R. 1971. *Cyathura sbordonii*, nuova specie cavernicola del Messico sudorientale. Diagnosi preliminare (Crustacea, Isopoda, Anthuridae). *Fragmenta Entomologica* 7(4):303–304.

Audouin, V. 1826. Explication Sommaire des Planches de Crustacés de l'Egypte et de la Syrie. In J.-C. Savigny, *Description de l'Egypte ou Recueil des Observations et des Recherches qui ont été faites en Egypte pendant l'Expédition de sa Majesté l'Empereur Napoléon le Grand. Histoire Naturelle*, vol. 1, pp. 77–98. Paris: L'Imprimerie Impériale.

Barnard, K. H. 1914. Contributions to the crustacean fauna of South Africa. 3. Additions to the marine Isopoda, with notes on some previously incompletely known species. *Annals of the South African Museum* 10(11):325a–442.

———. 1920. Contributions to the crustacean fauna of South Africa. 6. Further additions to the list of marine Isopoda. *Annals of the South African Museum* 17(5):319–438.

———. 1925. A revision of the family Anthuridae (Crustacea Isopoda), with remarks on certain morphological peculiarities. *Journal of the Linnaean Society of London, Zoology* 36:109–160.

Bate, C. S. 1866. Carcinological gleanings, 2. *Annals and Magazine of Natural History* (3)17:24–31.

Bate, C. S., and J. O. Westwood. 1868. *A History of the British Sessile-eyed Crustacea*. lvi + 536 pp. London: John van Voorst.

Becker, G., and W.-D. Kampf. 1958. Funde der holzzerstörenden Isopodengattung *Limnoria* an der Festland Indiens und Neubgeschreibung von *Limnoria indica*. *Zeitschrift für Angewandte Zoologie* 45:1–9.

Beddard, F. E. 1886. Report on the Isopoda collected by H. M. S. Challenger during the years 1873–76. Part 2. *Reports of the Voyage of the Challenger* 17:1–178.

Benedict, J. E. 1899. [*Cleantis planicauda* Benedict, new species.] *In* Richardson, H. 1899. Key to the isopods of the Pacific coast of North America, with descriptions of twenty-two new species. *Proceedings of the United States National Museum* 21:815–869.

———. 1901. *In* Richardson, H. 1901. Key to the isopods of the Atlantic coast of North America with descriptions of new and little known species. *Proceedings of the United States National Museum* 23:493–579.

Bliss, D. E., editor-in-chief. 1982–1985. *The Biology of the Crustacea*, vols. 1–10. New York: Academic Press.

Bonnier, J. 1900. Contribution à l'étude des Epicarides. Les Bopyridae. *Travaux de la Station Zoologique de Wimereux* 8:1–396.

Boone, P. L. 1918. Description of ten new isopods. *Proceedings of the United States National Museum* 54:591–604.

———. 1921. Report on the Tanidacea and Isopoda, collected by the Barbados-Antigua Expedition from the University of Iowa in 1918. *University of Iowa Studies in Natural History* 9(5):91–98.

Boone, L. 1927. Crustacea from tropical east American seas. *Bulletin of the Bingham Oceanographic Collection* 1(2):1–147.

———. 1930. New decapod and isopod crustaceans from Gonave Bay, Haiti. *Zoologica, New York* 12(4):41–53.

Bosc, L. A. G. 1802. Histoire Naturelle des Crustacés. *In* G. L. L. de Buffon, *Histoire Naturelle de Buffon classée...d'aprés le systêm de Linné...par R. R. Castel...nouvelle edition.* Paris.

Botosaneanu, L. 1983. First record of an anthurid isopod, *Cyathura univam* sp. n., on the South American continent. *Bijdragen tot de Dierkunde* 53(2):247–254.

Botosaneanu, L., N. L. Bruce, and J. Notenboom. 1986. Isopoda: Cirolanidae. *In* L. Botosaneanu, ed., *Stygofauna Mundi*, pp. 412–422. Leiden: E. J. Brill/Dr. W. Backhuys.

Botosaneanu, L., and J. H. Stock. 1979. *Arubolana imula* n. gen., n. sp., the first hypogean cirolanid isopod crustacean found in the Lesser Antilles. *Bijdragen tot de Dierkunde* 49(2):227–233.

———. 1982. Les *Cyathura* stygobies (Isopoda, Anthuridea) et surtout celles des Grandes et des Petites Antilles. *Bijdragen tot de Dierkunde* 52(1):13–42.

Bourdon, R. 1972. Sur quelques Bopyridae (Crustacea, Isopoda) parasites de Galatheides. *Bulletin du Muséum national d'Histoire naturelle* (3)66, zool. 52:817–838.

———. 1981. Remarques sur le genre *Synsynella* Hay, avec description de *S. integra* n. sp. (Crustacea, Epicaridea, Bopyridae). *Bulletin du Muséum national d'Histoire naturelle, Paris* (4)3(A4):1143–1162.

Bourdon, R., and J. C. Markham. 1980. A new genus and species of bopyrid isopod infesting alpheid shrimps of the genus *Synalpheus* in the western Atlantic Ocean. *Zoologische Mededelingen* 55(19):221–230.

Bovallius, C. 1885. New or imperfectly known Isopoda. *Bihang till Kongliga Svenska Vetenskaps-Akademiens Handlingar* 10(11):1–32.

Bowman, T. E. 1956. Una especie nueva de *Bopyrella* (Crustacea: Isopoda) de Los Roques,

Venezuela. *Novedades Cientificas, Contribuciones Ocasionales del Museo de Historia Natural La Salle, Caracas, Venezuela (Zoologica)* 19:1–4.

———. 1965. *Cyathura specus*, a new cave isopod from Cuba. *Studies on the Fauna of Curaçao and other Caribbean Islands* 85:88–97.

———. 1966. *Haptolana trichostoma*, a new genus and species of troglobitic cirolanid isopod from Cuba. *International Journal of Speleology* 2:105–108.

———. 1981. *Thermosphaeroma milleri* and *T. smithi*, new sphaeromatid isopod crustaceans from hot springs in Chihuahua, Mexico, with a review of the genus. *Journal of Crustacean Biology* 1(1):105–122.

———. 1987. *Bahalana mayana*, a new troglobitic cirolanid isopod from Cozumel Island and the Yucatan Peninsula, Mexico. *Proceedings of the Biological Society of Washington* 100(3):659–663.

Bowman, T. E., and R. Franz. 1982. *Anopsilana crenata*, a new troglobitic cirolanid isopod from Grand Cayman Island, Caribbean Sea. *Proceedings of the Biological Society of Washington* 95(3):522–529.

Bowman, T. E., and T. M. Iliffe. 1983. *Bermudalana aruboides*, a new genus and species of troglobitic Isopoda (Cirolanidae) from marine caves on Bermuda. *Proceedings of the Biological Society of Washington* 96(2):291–300.

Bowman, T. E., and B. F. Morris. 1979. *Carpias* Richardson, 1902, a senior synonym of *Bagatus* Nobili, 1906, and the validity of *Carpias minutus* (Richardson, 1902) (Isopoda: Asellota: Janiridae). *Proceedings of the Biological Society of Washington* 92(3):650–657.

Boyle, P., and R. Mitchell. 1978. Absence of microorganisms in crustacean digestive tracts. *Science* 200(4346):1157–1159.

Brandt, J. F. 1833. Conspectus monographiae Crustaceorum Oniscodorum Latreillii. *Bulletin de la Société Impériale des Naturalistes de Moscou* 6:171–193.

Briggs, J. C. 1974. *Marine Zoogeography*. 475 pp. New York: McGraw-Hill Book Company.

Bruce, N. L. 1981. Cirolanidae (Crustacea: Isopoda) of Australia: Diagnoses of *Cirolana* Leach, *Metacirolana* Nierstrasz, *Neocirolana* Hale, *Anopsilana* Paulian & Debouttevilie, and three new genera—*Natatolana, Politolana,* and *Cartetolana*. *Australian Journal of Marine and Freshwater Research* 32:945–966.

———. 1984. A new family for the isopod crustacean genus *Tridentella* Richardson, 1905, with description of a new species from Fiji. *Zoological Journal of the Linnean Society* 80:447–455.

———. 1985. *Calyptolana hancocki*, a new genus and species of marine isopod (Cirolanidae) from Aruba, Netherlands Antilles, with a synopsis of Cirolanidae known from the Caribbean and Gulf of Mexico. *Journal of Crustacean Biology* 5(4):707–716.

———. 1986a. Cirolanidae (Crustacea: Isopoda) of Australia. *Records of the Australian Museum*, supplement 6:1–239.

———. 1986b. Revision of the isopod crustacean genus *Mothocya* Costa, in Hope, 1851 (Cymothoidae: Flabellifera), parasitic on marine fishes. *Journal of Natural History* 20(5):1089–1192.

Brusca, R. C. 1981. A monograph on the Isopoda Cymothoidae (Crustacea) of the eastern Pacific. *Zoological Journal of the Linnean Society* 73(2):117–199.

———. 1983. A monograph on the isopod family Aegidae in the tropical Eastern Pacific I. The genus *Aega*. *Allan Hancock Monographs in Marine Biology* 12:1–39.

———. 1984. Phylogeny, evolution and biogeography of the marine isopod Subfamily Idoteinae (Crustacea: Isopoda: Idoteidae). *Transactions of the San Diego Society of Natural History* 20(7):99–134.

Budde-Lund, G. 1885. *Crustacea Isopoda Terrestria per Familias et Genera et Species*. 319 pp. Copenhagen.

Buss, L. W., and E. W. Iverson. 1981. A new genus and species of Sphaeromatidae (Crustacea: Isopoda) with experiments and observations on its reproductive biology, interspecific interactions and color polymorphisms. *Postilla* 184:1–23.

Calman, W. T. 1904. On the classification of the Crustacea Malacostraca. *Annals and Magazine of Natural History* (7)13:144–158.

———. 1910. On two new species of wood-boring Crustacea from Christmas Island. *Annals and Magazine of Natural History* (8)5:181–186.

Carpenter, J. H. 1981. *Bahalana geracei* n. gen., n. sp., a troglobitic marine cirolanid isopod from Lighthouse Cave, San Salvador Island, Bahamas. *Bijdragen tot de Dierkunde* 51(2):259–267.

Carpenter, J. H., and G. J. Magniez. 1982. Deux asellotes stygobies des Indes Occidentales: *Neostenetroides stocki* n. gen., n. sp., et *Stenetrium* sp. *Bijdragen tot de Dierkunde* 52(2):200–206.

Carvacho, A. 1977. Isopodes de la Mangrove de la Guadeloupe, Antilles Françaises. *Studies on the Fauna of Curaçao and other Caribbean Islands* 54(174):1–24.

Chappuis, P.-A., and C. Delamare Deboutteville. 1955. Recherches sur les Crustacés souterrains. 7. Les isopodes psammiques de la Mediterranée. *Archives de Zoologie Expérimentale et Générale* 91(1):103–138.

———. 1956. Etudes sur la faune interstitielle des îles Bahamas recoltée par Madame Renaud-Debyser. 1. Copepodes et Isopodes. *Vie et Milieu* 7(3):373–396.

Clark, S. T., and P. B. Robertson. 1982. Shallow water marine isopods of Texas. *Contributions to Marine Science* 25(1):45–59.

Coineau, N., and L. Botosaneanu. 1973. Isopodes interstitiels de Cuba. *Résultats des expéditions biospéologiques cubano-roumaines à Cuba* 1:191–222.

Cordiner, C. 1793. *Remarkable ruins and romantic prospects, of North Britain. With ancient monuments and singular subjects of natural history*. 96 plates with letterpress. London: Mazell.

Costa, A. 1851. Caratteeri di alcuni de'generi e specie nouve segnete nel presente catalogo. In F. W. Hope, *Catalogo dei crostacei Italiani e di molti altri de Mediterranea*, pp. 41–48. Naples.

Coutière, H. 1908. Sur le *Synalpheion Giardi*, n. gen., n. sp., entonisciens parasite d'une Synalphée. *Comptes Rendus Hebdomadaires des Séances de l'Académie des Sciences, Paris* 146:1333–1335.

Creaser, E. P. 1936. Crustaceans from Yucatan. *Carnegie Institution of Washington Publication* 457:117–132.

Dana, J. D. 1852. On the classification of the Crustacea Choristopoda or Tetradecapoda. *American Journal of Science* (2)14:297–316.

Delaney, P.M. 1984. Isopods of the genus *Excorallana* Stebbing, 1904 from the Gulf of California, Mexico (Crustacea, Isopoda, Corallanidae). *Bulletin of Marine Science* 34(1):1–20.

Delaney, P. M., and R. C. Brusca. 1985. Two new species of *Tridentella* Richardson, 1905 (Isopoda: Flabellifera: Tridentellidae) from California, with a rediagnosis and comments on the family, and a key to the genera of Tridentellidae and Corallanidae. *Journal of Crustacean Biology* 5(4):728–742.

Fabricius, J. C. 1793. Volume 2 in *Entomologia systematica emendata et aucta...adjectis synonymis, locis, observationibus, descriptionibus,* 4 vols., 1792–1794. Copenhagen.

———. 1798. *Entomologia systematica emendata et aucta...adjectis synonymis, locis, obervationibus, descriptionibus. Supplementum. 1798.* Copenhagen.

Frankenberg, D., and R. J. Menzies. 1966. A new species of asellote marine isopod, *Munna (Uromunna) reynoldsi* (Crustacea: Isopoda). *Bulletin of Marine Science* 16(2):200–208.

Fresi, E., and U. Schieke. 1972. *Pleurocope dasyura* Walker, 1901, and the Pleurocopidae new family (Isopoda, Asellota). *Crustaceana, Supplement* 3:207–213.

George, R. Y., and J.-O. Strömberg. 1968. Some new species and new records of marine isopods from San Juan Archipelago, Washington, U.S.A. *Crustaceana* 14(3):225–254.

Gissler, C. F. 1882. *Bopyroides latreuticola*, a new species of isopod crustacean parasitic on a gulfweed shrimp. *American Naturalist* 16:591–594.

Glynn, P. W. 1970. A systematic study of the Sphaeromatidae (Crustacea: Isopoda) of Isla Margarita, Venezuela, with descriptions of three new species. *Memoria de la Sociedad de Ciencias Naturales La Salle* 30:1–48.

Glynn P. W., D. M. Dexter, and T. E. Bowman. 1975. *Excirolana braziliensis*, a Pan-American sand beach isopod: taxonomic status, zonation and distribution. *Journal of Zoology, London* 175:509–521.

Glynn, P. W., and C. S. Glynn. 1974. On the systematics on *Ancinus* (Isopoda, Sphaeromatidae), with the description of a new species from the tropical eastern Pacific. *Pacific Science* 28(4):401–422.

Grobben, C. 1892. Zur Kenntniss der Staumbaumer und der Systems der Crustaceen. *Sitzungsberichte der kaiserlichen Akademie der Wissenschaften. Mathematisch-Naturwissenschaftliche Classe, Wien* 101(1&2)1:237–274.

Haller, G. 1880. Ueber einige neue Cymothoinen. *Archiv für Naturgeschichte, Berlin* 46(1)375–395.

Hansen, H. J. 1890. Cirolanidae et familiae nonnullae propinquae Musei Hauniensis. Et Bidrag til Kundskaben om nogle Familier af isopode Krebsdyr. *Kongelige Danske Videnskabernes Selskabs Skrifter, 6te Raekke, Naturvidenskabelig og mathematisk Afdeling* 3:239–426.

———. 1904. On the morphology and classification of the Asellota-group of crustaceans, with descriptions of the genus *Stenetrium* Hasw. and its species. *Proceedings of the Zoological Society of London, 1904,* 2:302–331.

———. 1905a. On the morphology and classification of the Asellota-group of Crustacea, with descriptions of the genus *Stenetrium* Haswell, and its species. *Proceedings of the Zoological Society of London for 1904* 2(2):302–331.

———. 1905b. On the propagation, structure, and classification of the family Sphaeromidae. *Quarterly Journal of Microscopical Science* 49(1):69–135.

———. 1916. Crustacea Malacostraca 3. *Danish Ingolf Expedition* 3(5):1–262.

Harger, O. 1873. *[Erichsonia attenuata.] In* A. E. Verrill, S. I. Smith, and O. Harger,

Catalogue of the Marine invertebrate animals of the southern coast of New England, and adjacent waters, p. 276. *In* A. E. Verrill and S. I. Smith, Report upon the invertebrate animals of Vineyard Sound and adjacent waters, with an account of the physical features of the region. *In* S. F. Baird, *Report of Professor S. F. Baird, Commissioner of Fish and Fisheries, on the condition of the sea-fisheries of the south coast of New England in 1871 and 1872.* 478 pp. Washington, D.C.: Government Printing Office.

———. 1879. Notes on New England Isopoda. *Proceedings of the United States National Museum* No.75:157–165.

———. 1883. Reports on the results of dredging, under the supervision of Alexander Agassiz, on the east coast of the United States, during the summer of 1880, by the U.S. Coast Survey Steamer "Blake," Commander J. R. Bartlett, U.S.N., commanding. *Bulletin of the Museum of Comparative Zoology at Harvard College* 11(4):91–104.

Harper, D. E., Jr. 1974. *Chiridotea excavata* n. sp. (Crustacea, Isopoda) from marine waters of Texas. *Contributions to Marine Sciences, Texas A & M University* 18:229–239.

Harrison, K. 1984. The morphology of the sphaeromatid brood pouch (Crustacea: Isopoda: Sphaeromatidae). *Zoological Journal of the Linnean Society* 82:363–407.

Harrison, K., and D. M. Holdich. 1982. New eubranchiate sphaeromatid isopods from Queensland waters. *Memoirs of the Queensland Museum* 20(3):421–446.

Hartnoll, R. G. 1966. A new entoniscid from Jamaica (Isopoda, Epicaridea). *Crustaceana* 11(1):45–52.

Haswell, W. A. 1881. On some new Australian marine Isopoda. Part 1. *Proceedings of the Linnean Society of New South Wales* 5:476–478.

———. 1884. On a new crustacean found inhabiting the tubes of *Vermilia* (Serpulidae). *Proceedings of the Linnean Society of New South Wales* 9(3):676–680.

Hay, W. P. 1903. On a small collection of crustaceans from the island of Cuba. *Proceedings of the United States National Museum* 26:429–435.

———. 1917. A new genus and three new species of parasitic isopod crustaceans. *Proceedings of the United States National Museum* 51:569–574.

Hooker, A. 1985. New species of Isopoda from the Florida Middlegrounds (Crustacea: Peracarida). *Proceedings of the Biological Society of Washington* 98(1):255–280.

Hurley, D. E., and K. P. Jansen. 1977. The marine fauna of New Zealand: Family Sphaeromatidae (Crustacea Isopoda: Flabellifera). *New Zealand Oceanographic Institute Memoir* 63:1095.

Iverson, E. W. 1982. Revision of the isopod family Sphaeromatidae (Crustacea: Isopoda: Flabellifera) I. Subfamily names with diagnoses and key. *Journal of Crustacean Biology* 2(2):248–254.

Ives, J. E. 1891. Crustacea from the northern coast of Yucatan, the harbor of Vera Cruz, the west coast of Florida and the Bermuda Islands. *Proceedings of the Academy of Natural Sciences of Philadelphia*, 1891:176–207.

Jacobs, B. J. M. 1987. A taxonomic revision of the European, Mediterranean and NW. African species generally placed in *Sphaeroma* Bosc, 1802 (Isopoda: Flabellifera: Sphaeromatidae). *Zoologische Verhandelingen* 238:1–71.

Kaestner, A. 1967. *Invertebrate Zoology*, vol. 3, 523 pp. New York: Interscience Publisher. [Translated and adapted from the second German edition by H. W. Levi and L. R. Levi.]

Karaman, S. 1933a. Neue Isopoden aus unterirdischen Gewassern Jugoslawiens. *Zoologischer Anzeiger* 102(1/2):16–22.

———. 1933b. *Microcerberus stygius*, der dritte Isopod aus dem Grundwasser von Skopje, Jugoslawien. *Zoologischer Anzeiger* 102(5/6):165–169.

———. 1934. Beiträge zur Kenntnis des Isopoden-Familie Microparasellidae. *Mitteilungen über Hohlen- und Karstforschung* 1934:42–44.

Kensley, B. 1978. Five new genera of anthurid isopod crustaceans. *Proceedings of the Biological Society of Washington* 91(3):775–792.

———. 1980. Records of anthurids from Florida, Central America, and South America (Crustacea: Isopoda: Anthuridae). *Proceedings of the Biological Society of Washington* 93(3):725–742.

———. 1981. Amsterdam Expeditions to the West Indian Islands, Report 10. *Curassanthura halma*, a new genus and species of interstitial isopod from Curaçao, West Indies (Crustacea: Isopoda: Paranthuridae). *Bijdragen tot de Dierkunde* 51(1):131–134.

———. 1982. Anthuridea (Crustacea: Isopoda) of Carrie Bow Cay, Belize. *In* K. Rutzler and I. G. Macintyre, eds., The Atlantic Barrier Reef Ecosystem at Carrie Bow Cay, Belize, 1: Structure and Communities, pp. 321–352. *Smithsonian Contributions to Marine Sciences* 12, 539 pp.

———. 1983. The role of isopod crustaceans in the reef crest community at Carrie Bow Cay, Belize. *Marine Ecology* 5(1):29–44.

———. 1984. The Atlantic Barrier Reef Ecosystem at Carrie Bow Cay, Belize, III: New marine Isopoda. *Smithsonian Contributions to Marine Sciences* 24, iv + 81 pp.

———. 1987a. Further records of marine isopods from the Caribbean. *Proceedings of the Biological Society of Washington* 100(3):559–577.

———. 1987b. A re-evaluation of the systematics of K. H. Barnard's Review of anthuridean isopods. *Steenstrupia* 13(3):101–139.

———. 1987c. *Harrieta*, a new genus for *Cymodoce faxoni* (Richardson) (Crustacea: Isopoda: Sphaeromatidae). *Proceedings of the Biological Society of Washington* 100(4):

Kensley, B., and R. Heard. 1985. A new species of the genus *Spinianirella* Menzies (Crustacea: Isopoda: Janiridae) from the western Atlantic. *Proceedings of the Biological Society of Washington* 98(3):682–686.

Kensley, B., and H. W. Kaufman. 1978. *Cleantioides*, a new idoteid isopod genus from Baja California and Panama. *Proceedings of the Biological Society of Washington* 91(3):658–665.

Kensley, B., and M. Schotte. 1987. New records of isopod Crustacea from the Caribbean, the Florida Keys, and the Bahamas. *Proceedings of the Biological Society of Washington* 100(1):216–247.

Kensley, B., and P. Snelgrove. 1987. Records of marine isopod crustaceans associated with the coral *Madracis mirabilis* from Barbados. *Proceedings of the Biological Society of Washington* 100(1):186–197.

Koehler, R. 1885. Description d'un Isopode nouveau, le *Joeropsis brevicornis*. *Annales des Sciences Naturelles (Paris), Zoologie* (6)19:1–7.

Krøyer, H. 1839. Munna, en ny kraebsdyrslaegt. *Naturhistorisk Tidsskrift, Kjøbenhavn* 2:612–616.

Kussakin, O. G. 1967. Fauna of Isopoda and Tanaidacea in the coastal zones of the

Antarctic and Subantarctic waters. *Biological Reports of the Soviet Antarctic Expedition (1955–1958)* 3:220–389. [English translation by the Israel Program for Scientific Translations, Jerusalem, 1968.]

Lang, K. 1961. Contributions to the knowledge of the genus *Microcerberus* Karaman (Crustacea Isopoda) with a description of a new species from the central Californian coast. *Arkiv för Zoologi* (2)13(22):493–510.

Latreille, P. A. 1802. Histoire Naturelle des Crustacés et des Insectes. *In* Volume 3 of G. L. L. de Buffon, 1802–1805, *Histoire Naturelle, nouvelle edition, accompagnée des notes. Ouvrage rédigé par C. S. Sonnini*, 14 vols. Paris.

———. 1803. Histoire Naturelle des Crustacés et des Insectes. *In* Volume 5 of G. L. L. de Buffon, 1802–1805. *Histoire Naturelle, nouvelle edition, accompagnée des notes. Ouvrage rédigé par C. S. Sonnini*. 14 vols. Paris

———. 1806. *Genera Crustaceorum et Insectorum secundum ordinum naturalem in Familias disposita, iconibus exemplisque plurimis explicata*, vol. 1, 280 pp. Paris: Amand Koenig.

———. 1817. Les Crustacés, les Arachnides, et les Insectes. *In* G. L. C. F. D. Cuvier, *Le Regne Animal, distribué d'après son organisation, pour servir de base à l'histoire naturelle des animaux et d'introduction à l'anatomie comparée*, vol. 3. Paris.

———. 1826. *Explication sommaire des planches (Mollusques, Annelides, Crustacés, Arachnides, Insectes, Echinodermes, Zoophytes, Ascidies, Polypes, Hydrophytes, Oiseaux) dont les dessins ont été fournis par M. J. C. Savigny. Description de l'Egypte, ou recueil des observations et des recherches qui ont été faites en Egypte pendant l'expédition de l'armée Française (1798–1801)*. Paris.

———. 1831. *Cours d'Entomologique, ou de l'histoire naturelle des Crustacés, des Arachnides, des Myriapodes, et des Insectes*. 568 pp. Paris.

Latrobe, B. H. 1802. A drawing and description of the *Clupea tyrannus* and *Oniscus praegustator*. *Transactions of the American Philosophical Society* 5:77–81.

Leach, W. E. 1814. Crustaceology. In *Brewster's Edinburgh Encyclopedia*, vol. 7, pp. 383–439.

———. 1815. A tabular view of the external characters of four classes of animals, which Linné arranged under Insecta; with the description of the genera comprising three of these classes into order, etc., and descriptions of several new genera and species. *Transactions of the Linnean Society of London* 2:306–400.

———. 1818. *[Rocinela, Livoneca redmanni.] In* Volume 5 of F. Cuvier, ed., 1816–1830, *Dictionnaire des Sciences naturelles*. Paris & Strasbourg.

Leidy, J. 1855. Contributions towards a knowledge of the marine invertebrate fauna of the coasts of Rhode Island and New Jersey. *Journal of the Academy of Natural Sciences of Philadelphia* 3:135–152.

Lemos de Castro, A. 1959. Descriçao de uma nova especie do genero "Ancinus" Milne Edwards (Isopoda, Sphaeromidae). *Revista Brasileira de Biologia* 19(2):215–218.

Lilljeborg, W. 1851. Norger Crustacear. *Ofversigt af Kongliga Vetenskapsakademiens Forhandlingar, Stockholm* 8:19–25.

Linnaeus, C. 1793. *In* J. C. Fabricius, 1792–1794. *Entomologia systematica emendata et aucta...adjectis synonymis, locis, observationibus, descriptionibus*, 4 vols. Copenhagen.

Loyola e Silva, J. de. 1960. Sphaeromatidae do Litoral Brasileiro (Isopoda—Crustacea). *Boletim da Universidade do Paraná, Zoologia* 4:1–182.

Markham, J. C. 1972a. Two new genera of western Atlantic abdominally parasitizing

Bopyridae (Isopoda, Epicaridea), with a proposed new name for their subfamily. *Crustaceana, Supplement* 3:39–56.

———. 1972b. Four new species of *Parathelges* Bonnier, 1900 (Isopoda, Bopyridae), the first record of the genus from the western Atlantic. *Crustaceana, Supplement* 3:57–78.

———. 1973. Six new species of bopyrid isopods parasitic on galatheid crabs of the genus *Munida* in the Western Atlantic. *Bulletin of Marine Science* 23(3):613–648.

———. 1974. A new species of *Pleurocrypta* (Isopoda, Bopyridae), the first known from the western Atlantic. *Crustaceana* 26(3):267–272.

———. 1975a. A review of the bopyrid isopod genus *Munidion* Hansen, 1897, parasitic on galatheid crabs in the Atlantic and Pacific oceans. *Bulletin of Marine Science* 25(3):422–441.

———. 1975b. Bopyrid isopods infesting porcellanid crabs in the northwestern Atlantic. *Crustaceana* 28(3):257–270.

———. 1975c. New records of two species of parasitic isopods of the bopyrid subfamily Ioninae in the western Atlantic. *Crustaceana* 29(1):55–67.

———. 1975d. Two new species of *Asymmetrione* (Isopoda, Bopyridae) from the Western Atlantic. *Crustaceana* 29(3):255–265.

———. 1977. Description of new western Atlantic species of *Argeia* Dana with a proposed new subfamily for this and related genera (Crustacea Isopoda, Bopyridae). *Zoologische Mededelingen* 52(9):107–123.

———. 1978. Bopyrid isopods parasitizing hermit crabs in the northwestern Atlantic Ocean. *Bulletin of Marine Science* 28(1):102–117.

———. 1985. A review of the bopyrid isopods infesting caridean shrimps in the northwestern Atlantic Ocean, with special reference to those collected during the Hourglass cruises in the Gulf of Mexico. *Memoirs of the Hourglass Cruises* 7(3):1–156.

Menzies, R. J. 1951a. A new species of *Limnoria* (Crustacea: Isopoda) from Southern California. *Bulletin of the Southern California Academy of Sciences* 50(2):86–88.

———. 1951b. A new subspecies of marine isopod from Texas. *Proceedings of the United States National Museum* 101:575–579.

———. 1956a. New abyssal tropical Atlantic isopods, with observations on their biology. *American Museum Novitates* 1798:1–16.

———. 1956b. New bathyal Isopoda from the Caribbean with observations on their nutrition. *Breviora* 63:1–10.

———. 1957. The marine borer family Limnoriidae (Crustacea, Isopoda). *Bulletin of Marine Science of the Gulf and Caribbean* 7(2):101–200.

———. 1962a. The isopods of abyssal depths in the Atlantic Ocean. *Vema Research Series* 1:79–206.

———. 1962b. The marine isopod fauna of Bahia de San Quintin, Baja California, Mexico. *Pacific Naturalist* 3(11):337–348.

———. 1962c. The zoogeography, ecology, and systematics of the Chilean marine isopods. *Lunds Universitets Årsskrift*, N.F. Avd. 2, 57(11):1–162.

Menzies, R. J., and D. Frankenberg. 1966. *Handbook on the Common Marine Isopod Crustacea of Georgia.* University of Georgia Press, Athens, Georgia. 93 pp.

Menzies, R. J., and P. W. Glynn. 1968. The common marine isopod Crustacea of Puerto

Rico: A handbook for marine biologists. *Studies on the Fauna of Curaçao and other Caribbean Islands* 27(104):1–133.

Menzies, R. J., and W. L. Kruczynski. 1983. Isopod Crustacea (Exclusive of Epicaridea). *Memoirs of the Hourglass Cruises* 6(1):1–126.

Miers, E. J. 1880. On a collection of Crustacea from the Malaysian region. Part 4. Penaeidea, Stomatopoda, Isopoda, Suctoria, and Xiphosura. *Annals and Magazine of Natural History* (5)5:457.

Miller, M. A. 1941. The isopod Crustacea of the Hawaiian Islands, II. Asellota. *Occasional Papers of the Bernice P. Bishop Museum, Honolulu, Hawaii* 16(13):305–320.

Milne Edwards, A. 1879. Sur un isopode gigantesque, des grandes profondeurs de la mer. *Comptes Rendus Hebdomadaires des Séances de l'Académie des Sciences* 88:21–23.

Milne Edwards, H. 1840. *Histoire Naturelle des Crustacés, comprenant l'anatomie, la physiologie et la classification de ces animaux*, vol. 3. Paris.

Monod, T. 1926. Les Gnathiidae. Essai monographique (morphologie, biologie, systématique). *Mémoires de la Société des Sciences Naturelles du Maroc* 13:1–667.

Montagu, G. 1804. Description of several marine animals (Cancer rhomboidalis, C. maxillaris, C. phasma, C. palmatus, Oniscus hirsutus, etc) found on the south coast of Devonshire. *Transactions of the Linnean Society, London* 7:61–85.

Moore, H. F. 1901. Report on Porto Rican Isopoda. *U.S. Fish Commission Bulletin for 1900* 2:161–176.

Moreira, P. S. 1972. Species of *Eurydice* (Isopoda, Flabellifera) from southern Brazil. *Boletim do Instituto Paulista de Oceanografico, São Paula* 21:69–91.

Müller, H.-G. 1988. The genus *Gnathia* Leach (Isopoda) from the Santa-Marta area, northern Colombia, with a review of Gnathiidae from the Caribbean Sea and Gulf of Mexico. *Bijdragen tot de Dierkunde* 58(1):88–104.

Negoescu, I. 1979. *Cyathura cubana* sp. n. (Isopoda, Anthuridea) from the Caribbean Sea (Cuban waters). *Travaux du Museum d'Histoire naturelle Grigore Antipa* 20:157–164.

Negoescu Vlădescu, I. 1983. A study of genus *Cyathura* from the Cuban freshwaters with the description of a new cave species: *C. orghidani* (Isopoda, Anthuridae). *Resultats des expéditions biospéologiques cubano-roumaines à Cuba* 4:39–45.

Nierstrasz, H. F. 1931. Die Isopoden der Siboga-Expedition. III. Isopoda Genuina II. Flabellifera. *Siboga Expedition Monographie* 32c:123–232.

Nierstrasz, H. F., and G. A. Brender à Brandis, 1925. Bijdrage tot de kennis der fauna van Curaçao. Epicaridea. *Bijdragen tot de Dierkunde* 24:1–8.

———. 1929. Papers from Dr. Th. Mortensen's Pacific Expedition 1914–16. 48. Epicaridea 1. *Videnskabelige Meddelelsers fra Dansk Naturhistorisk Forening i Kjøbenhavn* 87:1–44.

———. 1931. Papers from Dr. Th. Mortensen's Pacific Expedition 1914–16. 57. Epicaridea 2. *Videnskabelige Meddelelsers fra Dansk Naturhistorisk Forening i Kjøbenhavn* 91:147–225.

Nordenstam, A. 1933. Marine Isopoda of the families Serolidae, Idotheidae, Pseudidotheidae, Arcturidae, Parasellidae and Stenetriidae mainly from the South Atlantic. *Further Zoological Results of the Swedish Antarctic Expedition 1901–1903* 3(1):1–284.

Norman, A. M., and T. R. R. Stebbing. 1886. Crustacea Isopoda of the 'Lightning', 'Porcupine', and 'Valorous' Expeditions, Part 1. *Transactions of the Zoological Society of London* 12(4):119–133.

Notenboom, J. 1981. Amsterdam Expeditions to the West Indies Islands, report 12. Some new hypogean cirolanid isopod crustaceans from Haiti and Mayaguana (Bahamas). *Bijdragen tot de Dierkunde* 51(2):313–331.

———. 1984. *Arubolana parvioculata* n. sp. (Isopoda, Cirolanidae) from the interstitial of an intermittent river in Jamaica, with notes on *A. imula* Botosaneanu & Stock and *A. aruboides* (Bowman & Iliffe). *Bijdragen tot de Dierkunde* 54(1):51–65.

Nunomura, N. 1977. Marine Isopoda from Amakusa, Kyushu (I). *Publications from the Amakusa Marine Biological Laboratory* 4:71–90.

Ortiz, M., and R. Lalana. 1980. Una nueva especie de isópodo (Crustacea, Isopoda), de los manglares de la costa sur de Cuba. *Revista Investigaciones Marinas* 1(2–3):160–174.

Ortiz, M., R. Lalana, and O. Gómez. 1987. Lista de especies y bibliografía de los isópodos (Crustacea, Peracarida) de Cuba. *Revista de Investigaciones Marinas* 8(3):29–37.

Packard, A. S. 1879. *Zoology for Students and General Readers.* 719 pp. New York: Henry Holt & Co.

Pallas, P. S. 1772. In fasc. 9, *Spicilegia Zoologica (quibus novae...et obscurae animalium species...illustrantur).* 2 vols. (1767-) 1774–1780. Berlin.

Paul, A. Z., and R. J. Menzies. 1971. Sub-tidal isopods of the Fosa de Cariaco, Venezuela, with descriptions of two new genera and twelve new species. *Boletin de Instituto Universidade Oriente* 10(1):29–48.

Paulian, R., and C. Delamare Deboutteville. 1956. Un cirolanide cavernicole à Madagascar (Isopode). *Mémoires de l'Institut Scientifique de Madagascar* (A)9:85–88.

Pearse, A. S. 1932. New bopyrid isopod crustaceans from Dry Tortugas, Florida. *Proceedings of the United States National Museum* 81(1):1–6.

———. 1950. Bopyrid isopods from the coast of North Carolina. *Journal of the Elisha Mitchell Scientific Society* 66:41–43.

———. 1951. Parasitic Crustacea from Bimini, Bahamas. *Proceedings of the United States National Museum* 101(3280):341–372.

———. 1952. Parasitic Crustacea from the Texas coast. *Publications of the Institute of Marine Sciences, University of Texas* 2(2):5–42.

Pearse, A. S., and H. A. Walker. 1939. Two new parasitic isopods from the eastern coast of North America. *Proceedings of the United States National Museum* 87:19–23.

Pennak, R. W. 1958. A new micro-isopod from a Mexican marine beach. *Transactions of the American Microscopical Society* 77:298–303.

Pennant, T. 1777. *British Zoology.* 4th Edition, vol. 4.

Perty, J. A. 1833. In *Delectus animalium articulatorum quae in itinere per Brasiliam annis 1817–20...collegerunt...J. B. de Spix...et...C. F. P. de Martius, digessit, descripsit, pingenda curavit M. Perty,* vol. 3. Monaco.

Petuch, E. J. 1982. Geographical heterochrony: contemporaneous coexistence of Neogene and Recent molluscan faunas in the Americas. *Palaeogeography, Palaeoclimatology, Palaeoecology* 37:277–312.

Pires, A. M. S. 1980. Revalidation and redescription of the genus *Carpias* Richardson, 1902 (Isopoda, Asellota). *Crustaceana* 39(1):95–103.

———. 1981. *Carpias harrietae* (Isopoda, Asellota), a new species from Florida. *Crustaceana* 40(2):206–212.

———. 1982. Taxonomic revision of *Bagatus* (Isopoda, Asellota) with a discussion of ontogenetic polymorphism in males. *Journal of Natural History* 16(2):227–259.

———. 1984. Taxonomic revision and phylogeny of the genus *Erichsonella* with a discussion on *Ronalea* (Isopoda, Valvifera). *Journal of Natural History* 18(5):665–683.

Poore, G. C. B. 1984. Redefinition of *Munna* and *Uromunna* (Crustacea: Isopoda: Munnidae), with descriptions of five species from coastal Victoria. *Proceedings of the Royal Society of Victoria* 96(2):61–81.

Poore, G. C. B., and Lew Ton, H. M. 1988. *Amakusanthura* and *Apanthura* (Crustacea: Isopoda: Anthuridae) with new species from tropical Australia. *Memoirs of the Museum of Victoria* 49:107–147.

Racovitza, E. G. 1908. *Ischyromene Lacazei* n.g., n. sp. Isopode mediterranéen de la famille des Spheromides (Note prèliminaire). *Archives de Zoologie Expérimentale et Générale (4)9, notes et revue* 3:LX–LXIV.

Rafinesque-Schmaltz, C. S. 1815. *Analyse de la Nature ou Tableau de l'Univers et des Corps Organisés*. 224 pp. Palerma.

Ray, D. L. 1959 (Editor). *Marine boring and fouling organisms*. 536 pp. Seattle: University of Washington Press.

Rehm, A., and H. J. Humm. 1973. *Sphaeroma terebrans:* A threat to the mangroves of southwestern Florida. *Science* 182:173–174.

Richardson, H. 1897. Description of a new species of *Sphaeroma*. *Proceedings of the Biological Society of Washington* 11:105–107.

———. 1898. Description of four new species of *Rocinela* with a synopsis of the genus. *Proceedings of the American Philosophical Society* 37(157):8–17.

———. 1899. Key to the isopods of the Pacific coast of North America, with descriptions of twenty-two new species. *Proceedings of the United States National Museum* 21:815–869.

———. 1900a. Results of the Branner-Agassiz Expedition to Brazil 2. The isopod Crustacea. *Proceedings of the Washington Academy of Sciences* 2:157–159.

———. 1900b. Synopses of North American invertebrates. 7. The Isopoda. *American Naturalist* 34:207–230.

———. 1901. Key to the isopods of the Atlantic coast of North America with descriptions of new and little known species. *Proceedings of the United States National Museum* 23:493–579.

———. 1902. The marine and terrestrial isopods of the Bermudas, with descriptions of new genera and species. *Transactions of the Connecticut Academy of Sciences* 11:277–310.

———. 1904. Contributions to the natural history of the Isopoda. *Proceedings of the United States National Museum* 27:1–89.

———. 1905. A monograph on the isopods of North America. *Bulletin of the United States National Museum* 54, liii + 727 pp.

———. 1910. Description of a new species of Anilocra from the Atlantic coast of North America. *Proceedings of the United States National Museum* 39:137–138.

———. 1912a. Descriptions of a new genus of isopod crustaceans, and of two new species from South America. *Proceedings of the United States National Museum* 43:201–204.

———. 1912b. Description of a new isopod crustacean belonging to the genus *Livoneca* from the Atlantic coast of Panama. *Proceedings of the United States National Museum* 42:173–174.

———. 1912c. Marine and terrestrial isopods from Jamaica. *Proceedings of the United States National Museum* 42:187–194.

Rioja, E. 1953. Estudios carcinológicos. XXX. Observaciones sobre los cirolánidos cavernícolas de México (Crustáceos, Isópodos). *Anales del Instituto de Biología, México* 24(1):147–170.

Robertson, P. B. 1978. A new species of asellote marine isopod *Munna (Uromunna) hayesi* (Crustacea: Isopoda) from Texas. *Contributions in Marine Science* 21(1):39–46.

Robins, C. R., R. M. Bailey, C. E. Bond, J. R. Brooker, E. A. Lachner, R. N. Lea, and W. B. Scott. 1980. A list of common and scientific names of fishes from the United States and Canada (4th edition). *American Fisheries Society, Special Publication* 12:1–174.

Roman, M.-L. 1977. Les oniscoides halophiles de Madagascar (Isopoda, Oniscoidea). *Beaufortia* 26:107–152.

Roux, P. 1828. Crustacés de la Méditerranée et de son littoral, décrits et lithographiés par lui-même. *Annales des Sciences Naturelles* 16: plate 13.

Sars, G. O. 1882. Oversigt af Norges Crustacea. *Forhandlinger i Videnskabsselskabet i Kristiania 1882* 18:1–124.

———. 1897. *An account of the Crustacea of Norway, vol. 2, pts. 3–8. Isopoda.* 103 pp. Bergen.

———. 1899. *An account of the Crustacea of Norway, vol. 2, pts. 13–14. Isopoda.* 270 pp. Bergen.

Say, T. 1818. An account of the Crustacea of the United States. *Journal of the Academy of Natural Sciences of Philadelphia* 1(2):393–401, 423–433.

Schioedte, J. C., and F. Meinert. 1879. Symbolae ad Monographiam Cymothoarum Crustaceorum Isopodum Familiae 1. Aegidae. *Naturhistorisk Tidsskrift* (3)12:321–414.

———. 1881. Symbolae ad Monographiam Cymothoarum Crustaceorum Isopodum Familiae 2. Anilocridae. *Naturhistorisk Tidsskrift* (3)13:1–166.

———. 1883. Symbolae ad Monographiam Cymothoarum Crustaceorum Isopodum Familiae 3. Saophridae. 4. Ceratothoinae. *Naturhistorisk Tidsskrift* (3)13:281–378.

———. 1884. Symbolae ad Monographiam Cymothoarum Crustaceorum Isopodum Familiae 4. Cymothoidae. Trib. II. Cymothoinae. Trib. III. Livonecinae. *Naturhistorisk Tidsskrift* (3)14:221–454.

Schram, F. R. 1986. *Crustacea.* xiv + 606 pp. New York, Oxford: Oxford University Press.

Schultz, G. A. 1969. *How to know the marine isopod crustaceans.* vii + 359 pp. Dubuque, Iowa: Wm. C. Brown Co.

———. 1974. Terrestrial isopod crustaceans (Oniscoidea) mainly from the West Indies and adjacent regions. I. *Tylos* and *Ligia*. *Studies on the Fauna of Curaçao and other Caribbean Islands* 45(149):162–173.

———. 1977. Anthurids from the west coast of North America, including a new species and three new genera (Crustacea, Isopoda). *Proceedings of the Biological Society of Washington* 90(4):839–848.

———. 1979. A new Asellota (Stenetriidae) and two, one new, Anthuridea (Anthuridae) from Bermuda (Crustacea, Isopoda). *Proceedings of the Biological Society of Washington* 91(4):904–911.

———. 1984. Three new and five other species of Oniscoidea from Belize, Central America (Crustacea: Isopoda). *Journal of Natural History* 18:3–14.

Schultz, G. A., and D. M. Allen. 1982. *Prodajus bigelowiensis*, new species (Isopoda: Epicaridea: Dajidae) parasite of *Mysidopsis bigelowi* (Mysidacea) from coastal New Jersey, with observations on infestation. *Journal of Crustacean Biology* 2(2):296–302.

Schultz, G. A., and L. R. McCloskey. 1967. Isopod crustaceans from the coral *Oculina arbuscula* Verrill. *Journal of the Elisha Mitchell Scientific Society* 83(2):103–113.

Simberloff, D., B. J. Brown, and S. Lowrie. 1978. Isopod and insect root borers may benefit Florida mangroves. *Science* 201:630–632.

Sivertsen, E., and L. B. Holthuis. 1980. The marine isopod Crustacea of the Tristan da Cunha Archipelago. *Gunneria* 35:1–128.

Sket, B. 1979. *Atlantasellus cavernicolus* n. gen., n. sp. (Isopoda Asellota, Atlantasellidae n. fam.) from Bermuda. *Bioloski Vestnik, Ljubljana* 7(2):175–183.

Soika, A. G. 1954. Studi di Ecologia e Biogeografia 12. Ecologia, Sistematica, Biogeografia ed Evoluzione del *Tylos latreillei* Auct. (Isop. Tylidae). *Bollettino del Museo Civico di Storia Naturale di Venezia* 7:63–83.

Sowinsky, V. K. 1884. Contribution to the crustacean fauna of the Black Sea. *Mémoires de la Société des Naturalistes de Kieff* 7:225–288. [In Russian.]

Stebbing, T. R. R. 1900. On Crustacea brought by Dr. Willey from the South Seas. *Willey's Zoological Results* 5:605–690.

———. 1904. Marine Crustaceans. 12. Isopoda, with description of a new genus. In J. S. Gardiner, *The Fauna and Geography of the Maldive and Laccadive Archipelagoes, being the account of the work carried on and of the collections made by an expedition during the years 1899 and 1900*, vol. 2, pp. 699–720. Cambridge: Cambridge University Press.

———. 1905. Report to the Government of Ceylon on the pearl oyster fisheries of the Gulf of Manaar. Report on the Isopoda collected by Professor Herdman, at Ceylon, in 1902. *Ceylon Pearl Oyster Fisheries, 1905, supplementary reports* 23:1–64.

Stimpson, W. 1853. Synopsis of the marine Invertebrata of Grand Manan, or the region about the Bay of Fundy, New Brunswick. *Smithsonian Contributions to Knowledge* 6:39–44.

———. 1855. Descriptions of some new marine Invertebrata. *Proceedings of the Academy of Natural Sciences of Philadelphia* 7:385–394.

Stock, J. H. 1977. Microparasellidae (Isopoda, Asellota) from Bonaire. *Studies on the fauna of Curaçao and other Caribbean islands* 51(168):69–91.

———. 1986. Two new amphipod crustaceans of the genus *Bahadzia* from 'blue holes' in the Bahamas and some remarks on the origin of the insular stygofaunas of the Atlantic. *Journal of Natural History* (4)20:921–933.

Stone, I., and R. W. Heard. 1989. *Excorallana delaneyi* n. sp. (Crustacea: Isopoda: Excorallanidae) from the northeastern Gulf of Mexico, with observations on adult characters and sexual dimorphism in related species of *Excorallana* Stebbing, 1904. *Gulf Research Reports*) 8(2):199–211.

Stork, H. A. 1940. A new fresh-water isopod from Curaçao. *Studies on the Fauna of Curaçao, Aruba, Bonaire and the Venezuelan Islands* 10:147–150.

Strouhal, H. 1966. Eine neue Halophile *Stenophiloscia* aus den Rotmeergebiete (Isop. terr.). *Annalen die Naturhistorischen Museums, Wien* 69:323–333.

Tattersall, W. M. 1905. The marine fauna of the coast of Ireland. Part 5. Isopoda. *Scientific Investigations for 1904, Fisheries Branch, Ireland* 2:1–90.

Thompson, M. T. 1902. A new isopod parasitic on the hermit crab. *U. S. Fish Commission Bulletin for 1901*, pp. 53–56.

Topp, R. W., and F. H. Hoff, Jr. 1972. Flatfishes (Pleuronectiformes). *Memoirs of the Hourglass Cruises* 4(2):1–135.

Ul'yanin, B. N. 1875. [Crustacea of Turkestan]. *Imperatorskoe Obshchestvo Lyubitelei Estestvoznaniya Antropologhii i Etnoghrafii*, vol. 2, pt. 6. Moscow. [In Russian.]

Upton, N. P. D. 1987a. Asynchronous male and female life cycles in the sexually dimorphic, harem-forming isopod *Paragnathia formica* (Crustacea: Isopoda). *Journal of Zoology, London* 212:677–690.

———. 1987b. Gregarious larval settlement within a restricted intertidal zone and sex differences in subsequent mortality in the polygynous saltmarsh isopod *Paragnathia formica* (Crustacea: Isopoda). *Journal of the Marine Biological Association of the United Kingdom* 67(3):663–678.

Vandel, A. 1952. Etude des isopodes terrestres récoltés au Vénézuela par le Dr. G. Marcuzzi, suivie de considerations sur le peuplement du Continent de Gondwana. *Memorie del Museo Civico di Storia Naturale di Verona* 3:59–203.

Vanhöffen, E. 1914. Die Isopoden der Deutschen Südpolar-Expedition 1901–1903. *Deutsche Südpolar-Expedition 1901–1903, 25 (Zoologie)* 7:447–598.

Van Name, W. G. 1936. The American land and fresh-water isopod Crustacea. *Bulletin of the American Museum of Natural History* 71:1–535.

Veillet, A. 1945. Recherches sur le parasitisme des crabes et des Galathées par les Rhizocephales et les Epicarides. *Annales de l'Institut océanographique, Paris*, new series 22(4):193–341.

Wägele, J.-W. 1979. Morphologische Studien an *Eisothistos* mit Beschreibung von drei neuen Arten (Crustacea, Isopoda, Anthuridea). *Mitteilungen aus dem Zoologischen Museum der Universität Kiel* 1(2):1–19.

———. 1981. Zur Phylogenie der Anthuridea (Crustacea, Isopoda). Mit Beiträgen zur Lebensweise, Morphologie, Anatomie und Taxonomie. *Zoologica* 132:1–127.

———. 1982. On *Apanthuretta lathridia* n. sp. (Crustacea, Isopoda, Anthuridea) from Cuba. *Bijdragen tot de Dierkunde* 52(1):43–48.

———. 1983. On the origin of the Microcerberidae (Crustacea: Isopoda). *Zeitschrift für zoologische Systematik und Evolutionsforschung* 21(4):249–262.

———. 1985. New west Atlantic localities for the stygobiont paranthurid *Curassanthura* (Crustacea, Isopoda, Anthuridea) with description of *C. bermudensis* n. sp. *Bijdragen tot de Dierkunde* 55(2):324–330.

Walker, A. O. 1901. Contributions to the Malacostracan fauna of the Mediterranean. *Journal of the Linnean Society of London, Zoology* 28:290–307.

Waterman, T. H., ed. 1960. *The Physiology of Crustacea*, vols. 1, 2, 670, 681 pp. New York & London: Academic Press.

Williams, E. H., and L. B. Williams. 1980. Four new species of *Renocila* (Isopoda: Cymothoidae), the first reported from the New World. *Proceedings of the Biological Society of Washington* 93(3):573–592.

———. 1982. *Mothocya bohlkeorum*, new species (Isopoda: Cymothoidae) from West Indian cardinalfishes (Apogonidae). *Journal of Crustacean Biology* 2(4):570–577.

———. 1985a. A new cymothoid isopod, *Glossobius hemiramphi*, from the mouth of the ballyhoo, *Hemiramphus brasiliensis* (Linnaeus) (Exocoetidae), in the Caribbean Sea. *Crustaceana* 48(2):147–152.

———. 1985b. *Cuna insularis* n. gen. and n. sp. (Isopoda: Cymothoidae) from the gill chamber of the sergeant major *Abudefduf saxatilis* (Linnaeus) (Osteichthyes) in the West Indies. *Journal of Parasitology* 71(2):209–214.

———. 1986. *Kuna Nomen Novum* for *Cuna* Williams and Williams, 1985, preoccupied by *Cuna* Hedley, 1902. *Journal of Parasitology* 72(6):879.

Williams, L. B., and E. H. Williams. 1981. Nine new species of *Anilocra* (Crustacea: Isopoda: Cymothoidae) external parasites of West Indian coral reef fishes. *Proceedings of the Biological Society of Washington* 94(4):1005–1047.

Wilson, G. D. F. 1987. The road to the Janiroidea: Comparative morphology and evolution of the asellote isopod crustaceans. *Zeitschrift für zoologische Systematik und Evolutionsforschung* 25(4):257–280.

Index

abbreviata, Bopyrina, 110, 267
Abudefduf saxatilis, 171, 175, 186
abudefdufi, Anilocra, 171, 175
Abyssianira dentifrons, 263
Acanthocope spinosissima, 263
acanthophora, Domecia, 111
Acanthopleura granulata, 215
acanthura, Anopsilana, 124, 125
acanthuri, Anilocra, 171, 175, 176
Acanthurus
   bahianus, 171, 176
   chirurgus, 171, 176
acanthurus, Macrobrachium, 112
Accalathura, 64
   crenulata, 64, 65, 267
   setosa, 64, 65
Achelion occidentalis, 110
acuminata, Nerocila, 171, 172, 173, 190, 266, 268
   forma *acuminata,* 190
   forma *aster,* 190
acuta, Anilocra, 268
acutirostrata, Stenobermuda, 106, 270
acutirostris, Processa, 113
acutitelson, Dynamenella, 213, 214
Aega, 116, 117
   antillensis, 117
   incisa, 268
Aega (Aega), 116, 117
   deshaysiana, 117, 266, 268
   ecarinata, 117, 268
Aega (Rhamphion), 116, 117
   dentata, 117, 119
   tenuipes, 117, 119
Aegidae, 115, 116, 262
Aetobatus narinari, 166
affinis
   Pseudione, 113, 270
   Upogebia, 112
Agaricia, 88, 166
agaricicola, Metacirolana, 153, 154
Agarna, 170, 173
   cumulus, 173

Agelas, 88
Akidognathia poteriophora, 264
alba, Exosphaeroma, 229
albidoida, Cirolana, 132, 133
Alcirona, 157, 158
   insularis, 158
   krebsii, 158, 268, 270
algicola, Carpias, 82, 83, 87
Allodardanus bredini, 112
Allodiplophryxus floridanus, 267
alphei, Probopyria, 112, 267
Alpheus
   armillatus, 112
   formosus, 111, 112
   heterochaelis, 112
   normanni, 112
   viridari, 112
alphezemiotes, Ovobopyrus, 267
Alutera schoepfi, 171, 190
Amakusanthura, 17, 18
   geminsula, 18
   lathridia, 18, 20
   magnifica, 18, 20, 267
   signata, 18, 21
   significa, 18, 23
amazonicum, Macrobrachium, 112
Ambidexter symmetricus, 113
americana, Dasyatis, 122, 166
americanus, Periclimenes, 111, 113
Amphiroa, 219
amyle, Licranthura, 43
analis, Lutjanus, 122, 172, 183
Anchoa lamprotaenia, 171, 187
Ancininae, 204
Ancinus, 205
   belizensis, 205
   brasiliensis, 205, 206
   depressus, 268
andrewsi, Paralimnoria, 199
Angliera, 90, 91, 273
   dubitans, 91
   psamathus, 91
   racovitzai, 91

293

angulata
   Dynamene, 214
   Dynamenella, 214
angustifrons, Hexapanopeus, 111
Anilocra, 170, 174, 175
   abudefdufi, 171, 175
   acanthuri, 171, 175, 176
   acuta, 268
   chaetodontis, 171, 175, 177
   chromis, 172, 175, 177
   haemuli, 172, 173, 175, 177
   holacanthi, 172, 175, 179
   holocentri, 175, 179
   laticauda, 268
   myripristis, 172, 175, 180
   partiti, 173, 175, 180
annandalei, Sphaeroma, 234
annaoides, Antarcturus, 264
annulicornis, Pandalus, 113
annulipes, Pagurus, 113
anomala, Aporobopyrina, 110, 267
anops, Creaseriella, 137
Anopsilana, 124, 273
   acanthura, 124, 125
   browni, 124, 125, 266
   crenata, 124, 125
   cubensis, 124, 126
   jonesi, 124, 127
   radicicola, 124, 127
Antarcturus
   annaoides, 264
   floridanus, 269
Antennuloniscus dimeroceras, 263
Antheluridae, 275
Anthomuda, 17, 23, 291
   stenotelson, 23, 270
Anthuridae, 16, 270
Anthuridea, 2, 14, 15, 16, 244, 263, 269
antiguai, Plesionika, 113
antillense, Exosphaeroma, 229, 230
antillensis
   Aega, 117
   Dromidia, 111
   Excorallana, 161, 162, 268
   Paranthura, 69
Antrolana, 273
Apanthura, 17, 25
   cracenta, 25
   crucis, 25, 26
   harringtoniensis, 25, 26, 270
Apanthuroides, 17, 26
   millae, 27
Aphysina fistularis, 223

Apogon
   lachneri, 171, 188
   maculatus, 171, 193
   townsendi, 171, 193
Apogonidae, 187
Aporobopyrina, 110
   anomala, 110, 267
Aporobopyrus, 110
   collardi, 275
   curtatus, 110
arbuscula, Oculina, 88
Archosargus probatocephalus, 122
Arcturella, 252
   bispinata, 252, 269
   spinata, 252, 269
Arcturidae, 251, 252, 264
Arcturus
   caribbaeus, 264
   purpureus, 264
arenatus, Priacanthus, 173, 183
Argeia, 110
   atlantica, 110
Argeiinae, 107
Arius felis, 171, 190
Armadilloniscus, 247
   ninae, 247
armatus
   Ischnomesus, 263
   Petrolisthes, 110
armillatus, Alpheus, 112
arndti, Dies, 207
aruboides, Arubolana, 144, 145, 270
Arubolana, 144, 273
   aruboides, 144, 145, 270
   imula, 144, 145
   parvioculata, 144, 145
ascensionis, Holocentrus, 172, 179
Aselloidea, 75
Asellota, 15, 15, 73, 263, 267, 269
Astacilla, 252
   cymodocea, 252, 253
   lasallae, 252, 253
   lauffi, 269
   regina, 252, 253
Astalione, 110
   cruciaria, 110
aster, forma, *Nerocila acuminata*, 190
Astrapogon stellatus, 171, 188
Asymmetrione, 110
   clibanarii, 110
   desultor, 110
Athelginae, 107
Atherinidae, 187

Atlantasellidae, 75
Atlantasellus, 75
  cavernicolus, 75, 269, 270
atlantica
  Argeia, 110
  Megalops, 172, 183
attenuata, Erichsonella, 256, 258, 269
aurolineatum, Haemulon, 172, 178
Azygopleon, 110
  schmitti, 110, 267
bacescui, Phycolimnoria, 276
Bagatus, 83
Bahalana, 124, 127, 128, 273
  cardiopus, 128
  geracei, 128
  mayana, 128
bahianus, Acanthurus, 171, 176
bajonado, Calamus, 122
Balanopleon, 110
  tortuganus, 110
Balistes vetula, 122
balthica, Idotea, 259
barbadensis, Micropanope, 111
barbarae, Geogerceis, 215
barnardi
  Dies, 207
  Ischyromene, 218
  Paranthura, 69, 71
barracuda, Sphyraena, 122
barrerae, Cymodoce, 227
Bathynomus, 123, 129
  giganteus, 131, 268
Batrachoides surinamensis, 171, 190
baudiniana, Ligia, 247, 249
beethoveni, Gnathia, 238, 239
belizensis, Ancinus, 205
Belonidae, 187
benjamensis, Paradynamene, 268
berbicensis, Excorallana, 161, 162
bermudensis
  Carpias, 82, 83, 270
  Curassanthura, 67, 270
  Hemiramphus, 188
bicornutus, Microphrys, 110
bifasciatus, Joeropsis, 88
bifurcatus, Heteromesus, 263
bigelowiensis, cf., *Prodajus*, 268
bimini, Leidya, 111, 270
bispinata, Arcturella, 252, 269
bivittata, Mesanthura, 47
blackfordi, Lutjanus, 122
bohlkeorum, Mothocya, 171, 173, 187, 188
bonaci, Mycteroperca, 122

bonairensis, Pagurus, 110, 113
bonelli, Macrobrachium, 112
bonnieri, Pandalus, 113
Booralana tricarinata, 275
Bopyrella, 110
  harmopleon, 110
Bopyridae, 107, 114
Bopyrina, 110
  abbreviata, 110, 267
Bopyrinae, 107
Bopyrinella, 110
  thorii, 110
Bopyrione, 110
  synalphei, 110, 267
Bopyrissa, 110
  wolffi, 110, 270
Bopyroidea, 107
Bopyrophryxinae, 107
borealis, Cirolana, 268
Bothus lunatus, 122
bousfieldi, Synalpheus, 110
bowmani
  Renocila, 173, 191
  Stenetrium, 100
brachydactylus, Carpias, 82, 84
brasiliensis
  Ancinus, 205, 206
  Hemiramphus, 172, 184
braziliensis, Excirolana, 150, 266, 268
bredini, Allodardanus, 112
brevicarpus, Synalpheus, 110
brevidactylus, Pagurus, 112, 113
brevipes, var., *Paracerceis caudata*, 219
brevitelson, Xenanthura, 62, 267
brooksi, Synalpheus, 110, 111, 113
browni, Anopsilana, 124, 125, 266
bruscai, Miratidotea, 260
buccanella, Lutjanus, 122
Bythognathia yucatanensis, 275
Cabirops, 110
Calamus
  bajonado, 122
  calamus, 122
  penna, 122
calamus, Calamus, 122
Callispongia plicifera, 221
Calyptolana, 124, 131
  hancocki, 132
camayae, Nannoniscus, 264
canaliculata, Processa, 113
canariensis, Curassanthura, 270,
Cancricepon, 110
  choprae, 110, 267, 270

INDEX

Cancrion carolinus, 111
capistratus, Chaetodon, 171, 177
caraibica, Cymothoa, 182, 268
Caranx, 122, 171
  hippos, 171, 183
  latus, 171, 183
  ruber, 171, 183
carbonarium, Haemulon, 172, 178
carcinus, Macrobrachium, 112
cardiopus, Bahalana, 128
caribaeus, Dactylokepon, 111
caribbaeus, Arcturus, 264
caribbea, Storthyngura pulchra, 263
caribbica
  Ianirella, 263
  Malacanthura, 45
caribbicus
  Ischnomesus, 263
  Macrostylis, 263
caribea, Uromunna, 94
carinata, 'Cerceis,' 211, 268
carolii, Metaphrixus, 111
carolinense, Tozeuma, 112
carolinus, Cancrion, 111
Carpias, 82, 83
  algicola, 82, 83, 87
  bermudensis, 82, 83, 270
  brachydactylus, 82, 84
  floridensis, 267
  harrietae, 82, 84
  minutus, 82, 84, 270
  punctatus, 82, 85
  serricaudus, 82, 87
  triton, 82, 87
Cassidinidea, 207, 208
  mosaica, 208
  ovalis, 207, 208, 268
Cassidininae, 204, 207
castrensis, Hyperphrixus, 267
caudata, Paracerceis, 219, 268, 270
Caulerpa, 219
cavalla, Scomberomorus, 173, 187
cavernicolus, Atlantasellus, 75, 269, 270
cephalus, Mugil, 172, 190
Ceratothoa, 170, 180
  deplanata, 180
  transversa, 268
Cerceis, 210, 211
  carinata, 211, 268
Chaetodipterus faber, 171, 190
Chaetodon
  capistratus, 171, 177
  ocellatus, 171, 177

  sedentarius, 171, 177
  striatus, 171, 177
chaetodontis, Anilocra, 171, 175, 177
Chalixanthura, 17, 27
  lewisi, 27, 29
  scopulosa, 27, 29
Chilomycterus schoepfi, 172, 190
Chiridotea excavata, 269
chirurgus, Acanthurus, 171, 176
Chiton
  marmoratus, 229
  tuberculatus, 229
choprae
  Cancricepon, 110, 267, 270
  Synsynella, 113, 268, 270
Chromis
  cyaneus, 172, 177
  multilineatus, 172, 177
chromis, Anilocra, 172, 175, 177
chrysargyreum, Haemulon, 172, 178
chrysoptera, Orthopristis, 172, 183
chrysurus, Ocyurus, 172, 183
ciliatus, Monacanthus, 172, 190
circumsaltanus, Loki, 111
Cirolana, 123, 132
  albidoida, 132, 133
  borealis, 268
  crenulitelson, 132, 133
  minuta, 132, 135
  obtruncata, 132, 135, 268
  parva, 132, 135, 268
Cirolanidae, 1, 115, 122, 123, 139, 273
Cirolanides, 273
Cirolaninae, 123, 139
cirratum, Ginglymostoma, 122, 158
clarkae, Phycolimnoria, 201
Clastotoechus vanderhorsti, 110
Cleantioides, 255, 256
  occidentalis, 256
  planicauda, 256, 266, 269
clibanarii
  Asymmetrione, 110
  Stegias, 113, 270
Clibanarius
  tricolor, 110, 112, 113
  vittatus, 110, 112
coeca, Neoanthura, 263
cognata, Pseudione, 268
cohenae, Paracerceis, 219, 220
Colanthura, 64, 65
  tenuis, 65, 270
colini, Renocila, 171, 191
collardi, Aporobopyrus, 275

Colopisthus, 144, 146
  parvus, 147, 270
confixa, Cortezura, 31
Conilera cylindracea, 268
Conilerinae, 123, 139
conklini, Phaeoptyx, 173, 188
constricta, Munida, 112
convexa, Eurydice, 147, 148, 149, 268
coralicola, Joeropsis, 88, 267
Corallanidae, 115, 157
corallicola, Minyanthura, 53
corallinus, Pylopagurus, 112
Cortezura, 17, 29
  confixa, 31
  penascoensis, 31
cracenta, Apanthura, 25
crassa, Virganthura, 73
Creaseriella, 124, 137, 273
  anops, 137
crenata, Anopsilana, 124, 125
crenulata, Accalathura, 64, 65, 267
crenulitelson, Cirolana, 132, 133
cromis, Pogonias, 173, 190
cruciaria, Astalione, 110
crucis, Apanthura, 25, 26
cruentatus, Epinephelus, 172, 179
crumenophthalmus, Selar, 173, 183
Cryptoniscoidea, 107
Crytoniscidae, 107, 109
cubana, Cyathura (Cyathura), 33
cubense, Munidion, 111
cubensis
  Anopsilana, 124, 126
  Rocinela, 119, 120
cuborientalis, Cyathura (Stygocyathura), 34, 35
culebrae, Vandeloscia, 247, 251
cumanensis, Malacanthura, 45
cumulus, Agarna, 173
cuprea, Diopatra, 256
curacaoensis, Hippolyte, 110
Curassanthura, 64, 67, 270
  bermudensis, 67, 270
  canariensis, 270
  halma, 67, 68, 270
curassavica, Cyathura (Stygocyathura), 35
curri, Haliophasma, 41
curtatus, Aporobopyrus, 110
cuvieri, Galeocerdo, 122
cyaneus, Chromis, 172, 177
Cyathura, 17, 31, 270, 273
  polita, 267
  (Cyathura), 31
    cubana, 33

  (Stygocyathura), 31, 33, 35, 270
    cuborientalis, 34, 35
    curassavica, 35
    hummelincki, 35
    motasi, 35, 36
    orghidani, 35, 36
    parapotamica, 35, 36
    salpiscinalis, 35, 38
    sbordonii, 35, 38
    specus, 35, 38
    univam, 35, 38
Cyclograpsus interger, 111
cylindracea, Conilera, 268
Cymodoce, 226, 227
  barrerae, 227
  ruetzleri, 227
Cymodocea, 223, 253
cymododea, Astacilla, 252, 253
Cymothoa, 170, 172, 182
  caraibica, 182, 268
  excisa, 172, 173, 182, 268
  oestrum, 171, 172, 173, 182, 183, 268
Cymothoidae, 115, 169, 170
Cynoscion, 172, 183
  nebulosus, 172, 183
Dactylokepon
  caribaeus, 111
  sulcipes, 267
Dajidae, 107
Dardanus fucosus, 1122
Dasyatis americana, 122, 166
decorata, Mesanthura, 53
deformans, Synsynella, 110, 113, 268, 270
delaneyi, Excorallana, 160, 161
Dendrotiidae, 263
Dendrotion hanseni, 263
dentata, Aega, 117, 119
dentifrons, Abyssianira, 263
deplanata, Ceratothoa, 180
depressa, Panoplax, 111
depressus, Ancinus, 268
deshaysiana, Aega, 117, 266, 268
Desmosoma magnispina, 263
Desmosomatidae, 263
destructor, Sphaeroma, 235
desultor, Asymmetrione, 110
dianae, Paradella, 224, 266, 268
Dicropleon
  periclimenis, 111
Dictyota, 219
Dies, 207
  arndti, 207
  barnardi, 207

dimeroceras, Antennuloniscus, 263
diminuta, Exosphaeroma, 229, 231
diogenes, Petrochirus, 110
Diopatra cuprea, 256
Diplophryxus, 111
Discerceis, 210, 211
　linguicauda, 213
dispar, Paraliomera, 111
distorta, Leidya, 111
Domecia
　acanthophora, 111
　hispida, 111
Dromidia antillensis, 111
dubitans, Angliera, 91
Dynamene angulata, 214
Dynamenella, 210, 213, 214, 224
　acutitelson, 213, 214
　　var. *glabrothorax*, 214
　　var. *typica*, 214
　angulata, 214
　perforata, 213, 215, 270
　quadrilirata, 213, 215
Dynameninae, 204, 210, 211
ecarinata, Aega, 117, 268
Echinothambema ophiuroides, 263
Echinothambematidae, 263
edithae, Paracerceis, 218, 221
Edotea
　lyonsi, 269
　montosa, 269
　samariensis, 276
edulis, Processa, 113
edwardsi, Plesionika, 113
Edwinjoycea horologium, 252, 269
eglanteria, Raja, 122
Eisothistos, 16, 38, 39
　petrensis, 39
　teri, 39
ensis, Plesionika, 113
Entoniscidae, 107, 109
Entophilinae, 107
Eophrixus subcaudalis, 111, 267
Epicaridea, 4, 14, 107, 267
Epinephelus, 122, 172, 183
　cruentatus, 172, 179
　fulvus, 172, 179
　guttatus, 172, 179
　itajara, 122, 172, 190
　morio, 122
Erichsonella, 255, 256, 257
　attenuata, 256, 258, 269
　filiformis, 256, 258, 269
　　tropicalis, 257

　floridana, 256, 258, 269
　isabelensis, 269
Eriphia gonagra, 111
Eubranchiatae, 203
Eurycopidae, 263
Eurydice, 143, 147
　convexa, 147, 148, 149, 268
　littoralis, 148, 149, 268
　personata, 147, 149, 270
　piperata, 147, 149, 268
Eurydicinae, 123, 139, 143
excavata, Chiridotea, 269
Excirolana, 144, 149, 150
　braziliensis, 150, 266, 268
　mayana, 150, 153, 268
excisa, Cymothoa, 172, 173, 182, 268
Excorallana, 157, 159, 161
　antillensis, 161, 162, 268
　berbicensis, 161, 162
　delaneyi, 160, 161
　fissicauda, 161, 162
　mexicana, 160, 161, 268
　oculata, 161, 163
　quadricornis, 161, 165, 270
　sexticornis, 161, 165
　subtilis, 160
　tricornis, 266, 268
　　occidentalis, 167
　　tricornis, 161, 165
　warmingii, 161, 167
exilipes, Palaemonetes, 113
Exocoettus, 172, 184
Exosphaeroma, 226, 229, 231
　alba, 229
　antillense, 229, 230
　diminuta, 229, 231
　productatelson, 229, 231
　yucatanum, 229, 231
exotica, Ligia, 247, 249
faber, Chaetodipterus, 171, 190
fasciata, Mesanthura, 47, 49
faustinum, Macrobrachium, 112
faxoni, Harrieta, 232, 268
felis, Arius, 171, 190
filiforme, Syringodium, 253, 260
filiformis, Erichsonella, 256, 258, 269
fimbriata
　Pleurocryptella, 112
　Processa, 113
fissicauda, Excorallana, 161, 162
fistularis, Aphysina, 223
Flabellifera, 2, 14, 114, 115, 268, 269
flavolineatum, Haemulon, 122, 172, 178

flinti, Munida, 111
floridana
   Erichsonella, 256, 258, 269
   Pleurocrypta, 112
floridanus
   Allodiplophryxus, 267
   Antarcturus, 269
   Thor, 110, 111
floridensis
   Carpias, 267
   Gnathia, 268
   Mesanthura, 53, 267
   Paranthura, 69, 71, 267
   Pleurocope, 98, 267
foetens, Synodus, 173, 183
foliatus, Parathelges, 112
formosa, Kupellonura, 267
formosus, Alpheus, 111, 112
fritzmuelleri, Synalpheus, 110, 111
fucorum, Latreutes, 112
fucosus, Dardanus, 112
fulvus, Epinephelus, 172, 179
furcifer, Paranthias, 173, 179
Galathea rostrata, 112
galathinus, Petrolisthes, 110
Galeocerdo cuvieri, 122
geminsula, Amakusanthura, 18
Geocerceis, 210, 215
   barbarae, 215
geracei, Bahalana, 128
Gerres rhombeus, 172, 187
giardi, Synalpheion, 113
giganteus, Bathynomus, 131, 268
Gigantione
   mortenseni, 111, 267
   uberlackerae, 267
Ginglymostoma cirratum, 122, 158
glabrothorax, var., *Dynamenella acutitelson,* 214
Glossobius, 170, 172, 183, 184
   hemiramphi, 172, 184
   impressus, 172, 184
glynni
   Paracerceis, 218, 221
   Paraleptosphaeroma, 210, 266
Gnathia, 238
   beethoveni, 238, 239
   floridensis, 268
   gonzalezi, 238, 239
   johanna, 238, 239, 276
   magdalenensis, 238, 239
   puertoricensis, 238, 239
   rathi, 238, 241
   samariensis, 238, 241

   triospathiona, 238, 241
   velosa, 238, 243
   virginalis, 238, 243
Gnathiidae, 237, 238
Gnathiidea, 14, 236, 238, 268
Gnathostenetroides, 77
   laodicense, 78
   pugio, 77, 267
Gnathostenetroididae, 77
Gnathostenetroidoidea, 77
gonagra, Eriphia, 111
gonzalezi, Gnathia, 238, 239
goodei, Synalpheus, 110, 111
gracilis, Natatolana, 140
granulata, Acanthopleura, 215
guttatus, Epinephelus, 172, 179
haemuli, Anilocra, 172, 173, 175, 177
Haemulon
   aurolineatum, 172, 178
   carbonarium, 172, 178
   chrysargyreum, 172, 178
   flavolineatum, 122, 172, 178
   macrostomum, 172, 179
   plumieri, 172, 179
   sciurus, 172, 179
   steindachneri, 122
halia, Metacirolana, 153, 154
Halimeda, 88, 90, 219, 223
Haliophasma, 17, 41
   curri, 41
   irmae, 43
   valeriae, 41, 42
halma, Curassanthura, 67, 68, 270
Halodule, 232
hancocki, Calyptolana, 132
hanseni, Dendrotion, 263
Haplomesus tropicalis, 263
Haploniscidae, 263
Haploniscus unicornis, 263
Haptolana, 124, 137, 273
   somala, 138
   trichostoma, 138
harmopleon, Bopyrella, 110
Harrieta, 226, 232
   faxoni, 232, 268
harrietae, Carpias, 82, 84
harringtoniensis, Apanthura, 25, 26, 270
harrisii, Rithropanopeus, 111
hayesi, Uromunna, 267
heardi, Probopyrinella, 267
Hemiarthrinae, 107
Hemiarthrus synalphei, 111, 267
Hemibranchiatae, 203

hemiramphi, Glossobius, 172, 184
Hemiramphidae, 187
Hemiramphus
  bermudensis, 188
  brasiliensis, 172, 184
hemphilli, Synalpheus, 110, 111
hendleri, Pendanthura, 56
herbstii, Panopeus, 111
herrerai, Microcharon, 93
heterocarpus, Plesionika, 113
heterochaelis, Alpheus, 112
Heteromesus bifurcatus, 263
Hexapanopeus angustifrons, 111
Hippolyte
  curacaoensis, 110
  pleuracanthus, 110, 111
  zostericola, 110
hippos, Caranx, 171, 183
Hirundichthys speculifer, 172, 184
hispida, Domecia, 111
holacanthi, Anilocra, 172, 175, 179
Holacanthus tricolor, 172, 179
holocentri, Anilocra, 175, 179
Holocentrus ascensionis, 172, 179
hopkinsi, Mesanthura, 47, 51, 267
Horoloanthura irpex, 267
horologium, Edwinjoycea, 252, 269
hummelincki, Cyathura (Stygocyathura), 35
Hydroniscus quadrifrons, 263
Hyperphrixus castrensis, 267
Hypoconcha
  sabulosa, 111
  spinosissima, 111
Hyporhamphus unifasciatus, 172, 188
hyptius, Stegophryxus, 113, 268
Hyssuridae, 16, 58, 60
Ianirella
  caribbica, 263
  vemae, 263
Idotea, 255, 259
  balthica, 259
  metallica, 259, 269
Idoteidae, 251, 252, 255
Idoteinae, 255, 256
Iliacantha
  liodactyla, 111
  subglobosa, 111
imbricata, Parapagurion, 112
impressa, Politolana, 140
impressus, Glossobius, 172, 184
imswe, Kupellonura, 60
imula, Arubolana, 144, 145
incisa, Aega, 268

indica, Limnoria, 194, 195
infundibulata, Paranthura, 69, 71, 270
insulae, Limnoria, 195
insularis
  Alcirona, 158
  Kuna, 170
  Rocinela, 119, 120, 268
integra, Synsynella, 268
interger, Cyclograpsus, 111
intermedius, Palaemonetes, 113
Ioninae, 107
Iridopagurus, 112, 113
  iris, 113
iris, Iridopagurus, 113
irmae, Haliophasma, 43
irpex, Horoloanthura, 267
irrasa, Munida, 111
irritans, Munidion, 111
isabelensis, Erichsonella, 269
Ischnomesidae, 263
Ischnomesus
  armatus, 263
  caribbicus, 263
  multispinis, 263
Ischyromene, 210, 214, 217
  barnardi, 218
itajara, Epinephelus, 122, 172, 190
jacobus, Myripristis, 172, 180
jacqueti, Sclerocrangon, 110
Janiridae, 80, 81, 263
Janiroidea, 75, 79, 80, 81
Joeropsidae, 80, 87
Joeropsis, 87, 88
  bifasciatus, 88
  coralicola, 88, 267
  personatus, 88, 90
  rathbunae, 88, 90, 267, 270
johanna, Gnathia, 238, 239, 292
jonesi, Anopsilana, 124, 127
kadiakensis, Palaemonetes, 113
kensleyi, Mexicope, 81, 267
kohleri, Mesosignum, 264
krebsi, Alcirona, 158, 268, 270
Kuna, 170, 184
  insularis, 171, 186
Kupellonura, 60
  formosa, 267
  imswe, 60
lachneri, Apogon, 171, 188
Lachnolaimus maximus, 122
Laminaria, 201
lamprotaenia, Anchoa, 171, 187
laodicense, Gnathostenetroides, 78

lasallae, Astacilla, 252, 253
lata, Parabopyrella, 112
lathridia, Amakusanthura, 18, 20
laticauda, Anilocra, 268
laticeps, Skuphonura, 58
latreillei, Tylos, 247, 250
Latreutes fucorum, 112
latreuticola, Probopyrinella, 112, 268, 270
latus, Caranx, 171, 183
lauffi, Astacilla, 269
Laurencia, 219
Leidya
  bimini, 111, 270
  distorta, 111
Leiostomus xanthurus, 172, 183, 187, 190
Lepisosteus spatula, 172, 190
leptorhynchus, Pandalus, 113
lewisi, Chalixanthura, 27, 29
Licranthura, 16, 43
  amyle, 43
Ligia, 247, 249
  baudiniana, 247, 249
  exotica, 247, 249
  olfersii, 247, 249
Limnoria, 193, 194, 195, 235
  indica, 194, 195
  insulae, 195
  multipunctata, 195, 196
  pfefferi, 195, 198
  platycauda, 195, 198
  saseboensis, 195, 198
  simulata, 195, 198
  tripunctata, 199
  tuberculata, 194, 195, 199, 268
  unicornis, 195, 199
Limnoriidae, 114, 193
lindae, Skuphonura, 267
linguicauda, Discerceis, 213
liodactyla, Iliacantha, 111
Lironeca, 170, 172, 186
  ovalis, 268
  redmanni, 172, 173, 186, 268
  tenuistylis, 171, 186, 187
  texana, 268
  tropicalis, 268
littoralis
  Eurydice, 148, 149, 268
  Microcerberus, 244
Loki circumsaltanus, 111
longicarpus
  Pagurus, 110, 113
  Synalpheus, 110, 111, 113

longicaudatus, Periclimenes, 113
longipedis, Munidion, 112, 267
longipes, Munida, 112
looensis, Mesanthura, 47, 51
lunatus, Bothus, 122
Lutjanus
  analis, 122, 172, 183
  blackfordi, 122
  buccanella, 122
  mahogoni, 172, 183
  synagris, 172, 183
lyonsi, Edotea, 269
Lysmata
  rathbunae, 112
  wurdemanni, 112
Macrobrachium
  acanthurus, 112
  amazonicum, 112
  bonelli, 112
  carcinus, 112
  faustinum, 112
  ohione, 112
  olfersii, 112
  surinamicum, 112
Macrocystis, 201
macrostomum, Haemulon, 172, 179
Macrostylidae, 263
Macrostylis
  caribbicus, 263
  minutus, 263
  setifer, 264
  vemae, 264
maculatus
  Scomberomorus, 173, 187
  Sphoeroides, 173, 190
Madracis, 29, 41, 88, 90, 166
  mirabilis, 29, 41
magdalenensis, Gnathia, 238, 239
magnifica, Amakusanthura, 18, 20, 267
magnispina, Desmosoma, 263
mahogoni, Lutjanus, 172, 183
Malacanthura, 17, 43
  caribbica, 45
  cumanensis, 45
mangle, Rhizophora, 235
manningi, Thor, 111
marcuzzii, Tylos, 247, 250
margarita, Pontonia, 113
marginatus, Petrolisthes, 110
marmoratus, Chiton, 229
martia, Plesionika, 113
maximus, Lachnolaimus, 122

mayana
   Bahalana, 128
   Excirolana, 150, 153, 268
mcclendoni, Synalpheus, 110, 111
Megalops atlantica, 172, 183
menziesi, Metacirolana, 153, 154
Mesanthura, 17, 45
   bivittata, 47
   decorata, 53
   fasciata, 47, 49
   floridensis, 53, 267
   hopkinsi, 47, 51, 267
   looensis, 47, 51
   paucidens, 47, 51
   pulchra, 47, 52, 267
   punctillata, 47, 53
   reticulata, 47, 53
Mesosignidae, 264
Mesosignum kohleri, 264
Metacirolana, 144, 153
   agaricicola, 153, 154
   halia, 153, 154
   menziesi, 153, 154
   sphaeromiformis, 153, 154
metallica, Idotea, 259, 269
Metaphrixus carolii, 111
mexicana, Excorallana, 160, 161, 268
mexicanus, Microcerberus, 269
Mexicope, 80, 81
   kensleyi, 81, 267
Mexilana, 273
mgrayi, Serolis, 202, 268
Microcerberidae, 243, 244
Microcerberidea, 14, 243, 269, 273
Microcerberus, 244, 273
   littoralis, 244
   mexicanus, 269
   minutus, 244
   mirabilis, 244
   nunezi, 244
   renaudi, 244
   simplex, 244
   syrticus, 244
Microcharon, 90, 91
   herrerai, 93
   phreaticus, 93
   sabulum, 91
Micropanope barbadensis, 111
Microparasellidae, 80, 90, 273
Microphrys bicornutus, 110
miles, Munida, 112
millae, Apanthuroides, 27
milleri, Santia, 99, 267

minocule, Stenetrium, 100, 102
minus, Synalpheus, 110, 113
minuta, Cirolana, 132, 135
minutus
   Carpias, 82, 84, 270
   Macrostylis, 263
   Microcerberus, 244
Minyanthura, 16, 53
   corallicola, 53
mirabilis
   Madracis, 29, 41
   Microcerberus, 244
Miratidotea, 255, 259
   bruscai, 260
Monacanthus ciliatus, 172, 190
montagui, Pandalus, 113
montosa, Edotea, 269
morio, Epinephelus, 122
mortenseni
   Gigantione, 111, 267
   Parabopyrella, 112, 267
mosaica, Cassidinidea, 208
motasi, Cyathura (Stygocyathura), 35, 36
Mothocya, 170, 187
   bohlkeorum, 171, 173, 187, 188
   nana, 172, 187, 188
Mugil cephalus, 172, 190
multilineatus, Chromis, 172, 177
multipunctata, Limnoria, 195, 196
multispinis, Ischnomesus, 263
Munida
   constricta, 112
   flinti, 111
   irrasa, 111
   longipes, 112
   miles, 112
   schroederi, 112
   simplex, 110
   stimpsoni, 111
   valida, 110
Munidion
   cubense, 111
   irritans, 111
   longipedis, 112, 267
Munna, 93
   petronastes, 94
Munnidae, 80, 93
Munnogonium, 96
   wilsoni, 96
Mycteroperca
   bonaci, 122
   venenosa, 122
Myripristis jacobus, 172, 180

myripristis, Anilocra, 172, 175, 180
Nalicora, 157, 168
    rapax, 169, 268
nana, Mothocya, 172, 187, 188
Nannoniscidae, 264
Nannoniscus camayae, 264
narinari, Aetobatus, 166
Natatolana, 139
    gracilis, 140
nebulosus, Cynoscion, 172, 183
Nemanthura, 43
Neoanthura coeca, 263
Neopanope
    packardii, 111
    texana sayi, 111
Neostenetroides, 77, 78
    stocki, 78
Nerocila, 170, 172, 188
    acuminata, 171, 172, 173, 190, 266, 268
        forma *acuminata,* 190
        forma *aster,* 190
ninae, Armidilloniscus, 246, 247
niveus, Tylos, 247, 250
normanni, Alpheus, 112
northropi, Palaemon, 113
nunezi, Microcerberus, 244
nuttingi, Paracerceis, 218, 223
obtruncata, Cirolana, 132, 135, 268
occidentalis
    Achelion, 110
    Cleantioides, 256
    Excorallana tricornis, 167
    Parathelges, 112
ocellatus, Chaetodon, 171, 177
oculata
    Excorallana, 161, 163
    Rocinela, 119, 120, 266, 268
Oculina arbuscula, 88
Ocyurus chrysurus, 172, 183
oestrum, Cymothoa, 171, 172, 173, 182, 183, 268
ohione, Macrobrachium, 112
Olencira praegustator, 268
olfersii
    Ligia, 247, 249
    Macrobrachium, 112
Oncilorpheus, 124, 139
    stebbingi, 139
Oniscidea, 1, 4, 14, 15, 246, 247, 262
ophiuroides, Echinothambema, 263
Orbioninae, 107
orghidani, Cyathura (Stygocyathura), 35, 36

Orthopristis
    chrysoptera, 172, 183
    ruber, 122, 173, 179
ovalis
    Cassidinidea, 207, 208, 268
    Lironeca, 268
Ovobopyrus alphezemiotes, 267
oxyophthalmus, Paguristes, 112
Pachygrapsus transversus, 111
packardii, Neopanope, 111
Paguristes
    oxyophthalmus, 112
    tortugae, 112
Pagurus
    annulipes, 113
    bonairensis, 110, 113
    brevidactylus, 112, 113
    longicarpus, 110, 113
    provenzanoi, 110, 112, 113
Palaemon
    northropi, 113
    pandaliformis, 113
Palaemonetes
    exilipes, 113
    intermedius, 113
    kadiakensis, 113
    paludosus, 113
    pugio, 113
    vulgaris, 113
paludosus, Palaemonetes, 113
pandalicola, Probopyrus, 112, 266
pandaliformis, Palaemon, 113
Pandalus
    annulicornis, 113
    bonnieri, 113
    leptorhynchus, 113
    montagui, 113
pandionis, Synalpheus, 111, 113
Panopeus herbstii, 111
Panoplax depressa, 111
Parabopyrella
    lata, 112
    mortenseni, 112, 267
    richardsonae, 112, 267
    thomasi, 112
Parabopyriscus stellatus, 267
Paracerceis, 210, 218, 223
    caudata, 219, 268, 270
        var. *brevipes,* 219
    cohenae, 219, 220
    edithae, 218, 221
    glynni, 218, 221
    nuttingi, 218, 223

Paradella, 210, 214, 223
　dianae, 224, 266, 268
　plicatura, 223, 224
　quadripunctata, 223, 224
　tumidicauda, 223, 226
Paradynamene benjamensis, 268
Paragnathia, 237
Paraleptosphaeroma, 207, 208
　glynni, 210, 266
Paralimnoria, 193, 199
　andrewsi, 199
　　forma A, 200
　　forma B, 201
　　forma *typica*, 200
Paraliomera dispar, 111
Paramunnidae, 80, 96
Paranthias furcifer, 173, 179
Paranthura, 64, 69
　antillensis, 69
　barnardi, 69, 71
　floridensis, 69, 71, 267
　infundibulata, 69, 71, 270
Paranthuridae, 16, 64, 263, 270
Parapagurion imbricata, 112
Parapagurus, 112
parapotamica, Cyathura (Stygocyathura), 35, 36
Parathelges
　foliatus, 112
　occidentalis, 112
　piriformis, 112, 270
　tumidipes, 112, 270
partiti, Anilocra, 173, 175, 180
partitus, Pomacentrus, 173, 180
parva, Cirolana, 132, 135, 268
parvioculata, Arubolana, 144, 145
parviramus, Pseudione, 276
parvus, Colopisthus, 147, 270
patulipalma, Stenetrium, 100, 102
paucidens, Mesanthura, 47, 51
pectiniger, Synalpheus, 110, 111, 113
penascoensis, Cortezura, 31
Pendanthura, 17, 56
　hendleri, 56
　tanaiformis, 56, 270
penna, Calamus, 122
perforata, Dynamenella, 213, 215, 270
Periclimenes
　americanus, 111, 113
　longicaudatus, 113
periclimenis, Dicropleon, 111
personata, Eurydice, 147, 149, 270
personatus, Joeropsis, 88, 90
petrensis, Eisothistos, 39

Petrochirus diogenes, 110
Petrolisthes
　armatus, 110
　galathinus, 110
　marginatus, 110
petronastes, Munna, 94
pfefferi, Limnoria, 195, 198
Phaeoptyx
　conklini, 173, 188
　pigmentaria, 173, 188
phreaticus, Microcharon, 93
Phreatocoidea, 14
Phycolimnoria, 193, 201
　bacescui, 276
　clarkae, 201
Phyllodurinae, 107
pigmentaria, Phaeoptyx, 173, 188
piperata, Eurydice, 147, 149, 268
piriformis, Parathelges, 112, 270
planicauda, Cleantioides, 256, 266, 268
Platybranchiatae, 203
platycauda, Limnoria, 195, 198
Plesionika
　antiguai, 113
　edwardsi, 113
　ensis, 113
　heterocarpus, 113
　martia, 113
pleuracanthus, Hippolyte, 110, 111
Pleurocope, 97
　floridensis, 98, 267
Pleurocopidae, 80, 81, 96
Pleurocrypta floridana, 112
Pleurocryptella fimbriata, 112
plicatura, Paradella, 223, 224
plicifera, Callispongia, 221
plumieri, Haemulon, 172, 179
Pogonias cromis, 173, 190
polita
　Cyathura, 267
　Politolana, 140, 143, 268
Politolana, 139, 140
　impressa, 140
　polita, 140, 143, 268
Pomacentrus partitus, 173, 180
Pontonia margarita, 113
Porcellana sayana, 110
Porites, 88, 90
poteriophora, Akidognathia, 264
praegustator, Olencira, 268
Priacanthus arenatus, 173, 183
probatocephalus, Archosargus, 122
Probopyria alphei, 112, 267

Probopyrinella
  heardi, 267
  latreuticola, 112, 268, 270
Probopyrus pandalicola, 112, 266
Processa
  acutirostris, 113
  canaliculata, 113
  edulis, 113
  fimbriata, 113
  tenuipes, 113
processae, Urobopyrus, 113, 268
Prodajus cf. *bigelowiensis*, 268
productatelson, Exosphaeroma, 229, 231
provenzanoi, Pagurus, 110, 112, 113
psamathus, Angliera, 91
Pseudasymmetrione, 113
Pseudione
  affinis, 113, 270
  cognata, 268
  parviramus, 276
  upogebiae, 268
Pseudioninae, 107
Ptilanthura tricarina, 267
puertoricensis, Gnathia, 238, 241
pugilator, Uca, 111
pugio
  Gnathostenetroides, 77, 267
  Palaemonetes, 113
pulchra
  caribbea, Storthyngura, 263
  Mesanthura, 47, 52, 267
punctatus, Carpias, 82, 85
punctillata, Mesanthura, 47, 53
purpureus, Arcturus, 264
Pylopagurus, 110
  corallinus, 112
quadricornis, Excorallana, 161, 165, 270
quadridentata, Sphaeroma, 234, 268
quadrifrons, Hydroniscus, 263
quadrilirata, Dynamenella, 213, 215
quadripunctata, Paradella, 223, 223
racovitzai, Angliera, 91
radicicola
  Anopsilana, 124, 127
  Xylolana, 157
Raja eglanteria, 122
rapax, Nalicora, 169, 268
rathbunae
  Joeropsis, 88, 90, 267, 270
  Lysmata, 112
rathi, Gnathia, 238, 241
redmanni, Lironeca, 172, 173, 186, 268
regalis, Scomberomorus, 173, 187

regina, Astacilla, 252, 253
renaudi, Microcerberus, 244
Renocila, 170, 191
  bowmani, 173, 191
  colini, 171, 191
  waldneri, 173, 191, 193
reticulata, Mesanthura, 47, 53
reynoldsi, Uromunna, 94, 95, 266, 267
Rhamphion, see *Aega (Ramphion)*
Rhizophora mangle, 235
rhombeus, Gerres, 172, 187
Rhyscotus, 247, 249
  texensis, 247, 249
richardsonae, Parabopyrella, 112, 267
ricordi, Sesarma, 111
riedli, Vandeloscia, 247, 251
Rithropanopeus harrisii, 111
Rocinela, 116, 119
  cubensis, 119, 120
  insularis, 119, 120, 268
  oculata, 119, 120, 266, 268
  signata, 119, 120, 266, 268
rostrata, Galathea, 112
ruber
  Caranx, 171, 183
  Orthopristis, 122, 173, 179
ruetzleri, Cymodoce, 227
sabulosa, Hypoconcha, 111
sabulum, Microcharon, 91
salpiscinalis, Cyathura (Stygocyathura), 35, 38
samariensis
  Edotea, 276
  Gnathia, 238, 251
Santia, 98
  milleri, 99, 267
Santiidae, 80, 98
Sargassum, 84, 201
saseboensis, Limnoria, 195, 198
saxatilis, Abudefduf, 171, 175, 186
sayana, Porcellana, 110
sayi, Neopanope texana, 111
sbordonii, Cyathura (Stygocyathura), 35, 38
Schizobopyrina urocaridis, 113, 268
schmitti, Azygopleon, 110, 267
schoepfi
  Alutera, 171, 190
  Chilomycterus, 172, 190
schroederi, Munida, 112
sciurus, Haemulon, 172, 179
Sclerocrangon jacqueti, 110
Scomberomorus
  cavalla, 173, 187

Scomberomorus (cont.)
  maculatus, 173, 187
  regalis, 173, 187
scopulosa, Chalixanthura, 27, 29
sedentarius, Chaetodon, 171, 177
Selar crumenophthalmus, 173, 183
Serolidae, 114, 115, 201
Serolis, 202
  mgrayi, 202, 268
Serranus tigrinus, 173, 191, 193
serrata, Spinianirella, 263
serratum, Stenetrium, 100, 102
serricaudus, Carpias, 82, 87
Sesarma ricordi, 111
seticornis, Stenorhynchus, 110
setifer, Macrostylis, 264
setosa, Accalathura, 64, 65
sexticornis, Excorallana, 161, 165
signata
  Amakusanthura, 18, 21
  Rocinela, 119, 120, 266, 268
significa, Amakusanthura, 18, 23
simplex
  Microcerberus, 244
  Munida, 110
simulata, Limnoria, 195, 198
Skuphonura, 17, 58
  laticeps, 58
  lindae, 267
snanoi, Storthyngura, 263
somala, Haptolana, 138
Sparisoma viride, 122
spathulicarpus, Stenetrium, 100, 104
spatula, Lepisosteus, 172, 190
speculifer, Hirundichthys, 172, 184
specus, Cyathura (Stygocyathura), 35, 38
Speocirolana, 273
Sphaerolana, 273
Sphaeroma, 226, 232, 234
  annandalei, 234
  destructor, 235
  quadridentata, 234, 268
  terebrans, 234, 235, 268
  walkeri, 234, 235
Sphaeromatidae, 114, 115, 202, 204
Sphaeromatinae, 204, 226
sphaeromiformis, Metacirolana, 153, 154
Sphoeroides maculatus, 173, 190
Sphyraena barracuda, 122
spinata, Arcturella, 252, 269
Spinianirella serrata, 263

spinosissima
  Acanthocope, 263
  Hypoconcha, 111
stebbingi
  Oncilorpheus, 139
  Stenetrium, 100, 104, 267, 270
Stegias clibanarii, 113, 270
Stegophryxus hyptius, 113, 268
steindachneri, Haemulon, 122
stellatus
  Astrapogon, 171, 188
  Parabopyriscus, 267
Stenetriidae, 99
Stenetrioidea, 99
Stenetrium, 99, 100
  bowmani, 100
  minocule, 100, 102
  patulipalma, 100, 102
  serratum, 100, 102
  spathulicarpus, 100, 104
  stebbingi, 100, 104, 267, 270
Stenobermuda, 99, 106
  acutirostrata, 106, 270
Stenorhynchus seticornis, 110
stenotelson, Anthomuda, 23, 270
stimpsoni, Munida, 111
stocki, Neostenetroides, 78
Storthyngura
  pulchra caribbea, 263
  snanoi, 263
striata, Yvesia, 246
striatus, Chaetodon, 171, 177
Stygocyathura, see *Cyathura (Stygocyathura)*
subcaudalis, Eophrixus, 111, 267
subglobosa, Iliacantha, 111
subtilis, Excorallana, 160
sulcipes, Dactylokepon, 267
surinamensis, Batrachoides, 171, 190
surinamicum, Macrobrachium, 112
symmetricus, Ambidexter, 113
synagris, Lutjanus, 172, 183
synalphei
  Bopyrione, 110, 267
  Hemiarthrus, 111, 267
Synalpheion giardi, 113
Synalpheus
  bousfieldi, 110
  brevicarpus, 110
  brooksi, 110, 111, 113
  fritzmuelleri, 110, 111
  goodei, 110, 111
  hemphilli, 110, 111

longicarpus, 110, 111, 113
mcclendoni, 110, 111
minus, 110, 113
pandionis, 111, 113
pectiniger, 110, 111, 113
Synodus foetens, 173, 183
Synsynella, 110, 113
  choprae, 113, 268, 270
  deformans, 110, 113, 268, 270
  integra, 268
Syringodium, 166, 232
  filiforme, 253, 260
syrticus, Microcerberus, 244
tanaiformis, Pendanthura, 56, 270
Tecticeps, 204
Tecticipitinae, 204
tenuipes
  Aega, 117, 119
  Processa, 113
tenuis, Colanthura, 65, 270
tenuistylis, Lironeca, 171, 186, 187
terebrans, Sphaeroma, 234, 235, 268
teri, Eisothistos, 39
testudinea, Thalassia, 251
texana, Lironeca, 268
texensis, Rhyscotus, 247, 249
Thalassia, 166, 232
  testudinea, 251
Thermarcturus, 252, 255
  venezuelensis, 255
thomasi, Parabopyrella, 112
Thor
  floridanus, 110, 111
  manningi, 111
thorii, Bopyrinella, 110
tigrinus, Serranus, 173, 191, 193
tortugae, Paguristes, 112
tortuganus, Balanopleon, 110
Tozeuma carolinense, 112
transversa, Ceratothoa, 268
transversus, Pachygrapsus, 111
tricarina, Ptilanthura, 267
tricarinata, Booralana, 275
trichostoma, Haptolana, 138
tricolor
  Clibanarius, 110, 112, 113
  Holacanthus, 172, 179
tricornis, Excorallana, 161, 165, 266, 268
  occidentalis, 167
  tricornis, 161, 165
Tridentella, 236
  virginiana, 236

Tridentellidae, 115, 235, 236
triospathiona, Gnathia, 238, 241
tripunctata, Limnoria, 199
triton, Carpias, 82, 87
Troglocirolana, 273
tropicalis
  Erichsonella filiformis, 257
  Haplomesus, 263
  Lironeca, 268
tuberculata, Limnoria, 194, 195, 199, 268
tuberculatus, Chiton, 229
tumidicauda, Paradella, 223, 226
tumidipes, Parathelges, 112, 270
Turbinaria, 166, 219
Tylos, 247, 250
  latreillei, 247, 250
  marcuzzii, 247, 250
  niveus, 247, 250
  wegeneri, 247, 250
typica, var., *Dynamenella acutitelson,* 214
uberlackerae, Gigantione, 267
Uca, 111
  pugilator, 111
unicornis
  Haploniscus, 263
  Limnoria, 195, 199
unifasciatus, Hyporhamphus, 172, 188
univam, Cyathura (Stygocyathura), 35, 38
Upogebia affinis, 112
upogebiae, Pseudione, 268
Urobopyrus processae, 113, 268
urocaridis, Schizobopyrina, 113, 268
Uromunna, 93, 94
  caribea, 94
  hayesi, 267
  reynoldsi, 94, 95, 266, 267
valeriae, Haliophasma, 41, 42
valida, Munida, 110
Valvifera, 14, 251, 264, 269
Vandeloscia, 247, 251
  culebrae, 247, 251
  riedli, 247, 251
vanderhorsti, Clastotoechus, 110
velosa, Gnathia, 238, 243
vemae
  Ianirella, 263
  Macrostylis, 264
venenosa, Mycteroperca, 122
Venezanthura, 31
venezuelensis, Thermarcturus, 255
vetula, Balistes, 122

Virganthura, 64, 73
    crassa, 73
virginalis, Gnathia, 238, 243
virginiana, Tridentella, 236
viridari, Alpheus, 112
viride, Sparisoma, 122
vittatus, Clibanarius, 110, 112
vulgaris, Palaemonetes, 113
waldneri, Renocila, 173, 191, 193
walkeri, Sphaeroma, 234, 235
warmingii, Excorallana, 161, 167
wegeneri, Tylos, 247, 250
wilsoni, Munnogonium, 96
wolffi, Bopyrissa, 110
wurdemanni, Lysmata, 112
xanthurus, Leiostomus, 172, 183, 187, 190
Xenanthura, 60
    brevitelson, 62, 267
Xylolana, 123, 143, 144, 156
    radicicola, 157
yucatanensis, Bythognathia, 275
yucatanum, Exosphaeroma, 229, 231
Yvesia, 246
    striata, 246
zostericola, Hippolyte, 110